Sustainable Biogas Production

About the Series

The *Woodhead Series in Bioenergy* publishes titles on the topic of bioenergy as it relates to renewable sources of energy. The series notably covers biological resources, chemical and biological processes, and biomass products for energy production, along with the environmental, social, management, and economic aspects of bioenergy.

Series Editor:

Prof. Yves Gagnon, Université de Moncton, Moncton, Canada

Previous Volumes in the Series:

Kishor Kumar Sadasivuni and Mithra Geetha, Biofuel Cells and Energy Generation, 2025, 9780443216022

Krishna Kumar Jaiswal, Bhaskar Singh and Amit K. Jaiswal, Microalgal Biofuels: Sustainable Production, Conversion, and Life Cycle Assessment, 2025, 9780443241109

Jeyabalan Sangeetha and Devarajan Thangadurai, Microalgal Biomass for Bioenergy Applications, 2024, 9780443139277

Lina María Grajales Agudelo, Juan Carlos Valdés Serra and Erich Collicchio, Agroenergy, 2024, 9780443214301

Stefano Ferrari Interlenghi, José Luiz de Medeiros and Ofélia de Queiroz Fernandes Araújo, Sustainability of Methylic and Ethylic Biodiesel Production Routes, 2024, 9780443219429

Hwai Chyuan Ong, Islam Md Rizwanul Fattah and Indra Mahlia, Waste Valorization for Bioenergy and Bioproducts, 2024, 9780443191718

Woodhead Series in Bioenergy

Sustainable Biogas Production

Transforming Waste into Energy and Fertilisers

Katarzyna Chojnacka

Department of Advanced Material Technologies, Wroclaw University of Science and Technology, Wrocław, Poland

Dawid Skrzypczak

Department of Advanced Material Technologies, Wroclaw University of Science and Technology, Wrocław, Poland

Series Editor

Dr. Yves Gagnon

ELSEVIER

WP
WOODHEAD PUBLISHING
An imprint of Elsevier

Woodhead Publishing is an imprint of Elsevier
50 Hampshire Street, 5th Floor, Cambridge, MA 02139, United States
125 London Wall, London EC2Y 5AS, United Kingdom

ISBN: 978-0-443-40557-0 (print)
ISBN: 978-0-443-40558-7 (online)

For information on all Woodhead Publishing publications
visit our website at https://www.elsevier.com/books-and-journals

Publisher: Priscilla Braglia
Acquisitions Editor: Ruchi Bhardwaj
Editorial Project Manager: Akansha Rathi
Production Project Manager: Maria Bernard
Cover Designer: Bakyalakshmi S

Typeset by MPS Limited, Chennai, India

Contents

Foreword

The energy sector is undergoing a significant transition to meet growing energy demands, reduce greenhouse gas emissions, support climate change mitigation and protect the planet for future generations. A major component of this energy transition is the integration of large-scale renewable energy sources, including bioenergy, into the overall energy mix of countries.

Bioenergy refers to the conversion of biomass, derived from recently living organic materials, into electricity, heat and transportation fuels. Alongside wind and solar energy, bioenergy is poised to play a significant role in the global energy transition.

To support both the research community and the commercial bioenergy sector, the Woodhead Series in Bioenergy publishes research-driven and application-focused titles that explore bioenergy as a renewable source of energy. The series covers biological resources, chemical and biological processes and biomass-based products for energy production, all framed within the principles of sustainability, encompassing environmental, social, managerial and economic dimensions.

Anaerobic digestion is an important component of the energy transition and is a fundamental technology to transform waste into energy. Beyond the energy production, anaerobic digestion produces a digestate that can also be transformed into valuable bioproducts. When considering sustainable development approaches, the digestate should not be considered as a waste but rather as a resource for bioenergy and the production of bioproducts, notably as biofertilisers.

Sustainable Biogas Production: Transforming Waste into Energy and Fertilisers provides a broad scope description of bioenergy production through anaerobic digestion, from the production of biogas to the valorisation of the digestate. Along with case studies, this book also covers sustainability issues of biogas production, including policy and regulatory frameworks for biogas production.

This book is authored by a well-established research team, Prof. Dr. Hab. Eng. Katarzyna Chojnacka and Prof. Dr. Eng. Dawid Skrzypczak of the Wroclaw University of Science and Technology in Poland, who bring their substantial experiences in anaerobic digestion and digestate management in a comprehensive book anchored in environmental and economic sustainability.

Considering its broad scope treatment of anaerobic digestion and the strong focus on adding value to the digestate and sustainability principles, this book has the potential to be a reference in the fields of biogas and digestate management and should be of interest to all working with anaerobic digestion and the sustainable production of biogas and biofertilisers, in academia, research and industry.

Prof. Yves Gagnon PEng, DSc
Université de Moncton, Canada
Series Editor, Woodhead Series in Bioenergy

Introduction to biogas technology

1

1.1 Introduction to biogas technology

1.1.1 Overview of biogas production

Biogas technology is rooted in the biological conversion of organic wastes into an energy-rich gas and nutrient-rich residues. It has emerged as a cornerstone of renewable energy and sustainable waste management, aligning with decarbonisation and circular-economy goals (Kougias and Angelidaki, 2018; Gustafsson and Anderberg, 2021; Burg et al., 2023). The biogas process not only generates fuel (methane-rich gas) but also treats waste, reducing pollution and enabling nutrient recycling via digestate use and nutrient recovery (Trotta et al., 2023; Pranckietienė et al., 2023; Pilarski et al., 2025).

1.1.1.1 Anaerobic digestion process

AD is a multi-stage fermentation in which diverse microorganisms decompose organic matter in the absence of oxygen through hydrolysis, acidogenesis, acetogenesis and methanogenesis, producing biogas and a nutrient-rich digestate (Kougias and Angelidaki, 2018; Shah et al., 2024). Biogas consists primarily of methane (CH_4) and carbon dioxide (CO_2), with trace gases such as hydrogen sulphide and ammonia; typical compositions are approximately 55%–70% CH_4 and 30%–45% CO_2, depending on substrate and process conditions (Nashmin Elyasi et al., 2021). The digestate contains stabilised organic matter and plant nutrients (e.g., N, P, K) and can serve as a soil amendment; phosphorus can also be recovered as struvite where appropriate (Pranckietienė et al., 2023; Trotta et al, 2023).

Overall AD performance and stability depend on the structure and activity of the microbial community and on operating conditions such as temperature regime (mesophilic or thermophilic), pH near neutrality, organic loading rate (OLR), hydraulic retention time (HRT), and mixing (Dustin and Hansen, 2011; Kougias and Angelidaki, 2018). Community dynamics are especially important during start-up and transitions, when feedstock and operational changes can shift microbial composition and function (Alcántara-Hernández et al., 2017; Lyu et al., 2024).

Reactor designs and feedstocks. Widely used reactor configurations include continuously stirred tank reactors and plug-flow systems for slurries and manures, as well as high-rate concepts such as induced-bed reactors for specific applications (Kougias and Angelidaki, 2018; Dustin and Hansen, 2011; Cheng et al., 2013). Appropriate feedstock selection and blending are central to stable operation; co-digestion of complementary substrates and the use of pretreatments (mechanical, thermal, biological) are well-documented strategies to enhance hydrolysis, improve biogas yields, and broaden the range of usable wastes (Tabatabaei et al., 2020; Kougias and Angelidaki, 2018).

Sustainable Biogas Production. DOI: https://doi.org/10.1016/B978-0-443-40557-0.00004-4

1.1.1.2 Stages and microbial consortia

AD proceeds through four biologically mediated stages: hydrolysis, acidogenesis, acetogenesis, and methanogenesis. In hydrolysis, polymeric materials (carbohydrates, proteins, lipids) are enzymatically broken down into monomers by hydrolytic bacteria (e.g., *Clostridium*, *Bacillus*) (Kougias and Angelidaki, 2018; Shah et al., 2024). During acidogenesis, fermentative bacteria (e.g., *Clostridium* spp., *Lactobacillus* spp.) convert these monomers to volatile fatty acids (VFAs; e.g., acetate, propionate, butyrate), alcohols, H_2, CO_2, and NH_3, typically under mildly acidic to near-neutral pH (Nashmin Elyasi et al., 2021). Acetogenesis then oxidises higher VFAs and alcohols, often syntrophically with methanogens, yielding mainly acetate, CO_2 and H_2; classic acetogens include *Syntrophomonas* and *Syntrophobacter* spp., and the reactions proceed favourably only when H_2 is continuously scavenged by methanogens (Kougias and Angelidaki, 2018; Wahid et al., 2019). Methanogenesis is the final stage, converting acetate and H_2/CO_2 to CH_4 via two principal guilds: acetoclastic (e.g., Methanothrix syn. *Methanosaeta*, Methanosarcina) and hydrogenotrophic methanogens (e.g., Methanobacterium, Methanoculleus) (Kougias and Angelidaki, 2018; Grohmann et al., 2018). Optimal methanogenic activity generally occurs at mesophilic (~35°C) or thermophilic (~55°C) temperatures and near-neutral pH; substantial acidification inhibits methanogens and destabilises the process (Kougias and Angelidaki, 2018; Shah et al., 2024). Because each guild prepares substrates for the next, balanced cooperation is essential; inhibition in any stage (e.g., slowed methanogenesis with H_2/VFA accumulation) diminishes overall efficiency (Alcántara-Hernández et al., 2017; Wahid et al., 2019).

1.1.1.3 Operating parameters

Key operational levers include temperature regime, HRT, OLR, mixing, and pH/alkalinity control. Digesters are typically run mesophilically or thermophilically; higher temperatures can accelerate kinetics but may narrow stability margins, so the choice is feedstock- and design-dependent (Dustin and Hansen, 2011; Kougias and Angelidaki, 2018; Satpathy, 2017; Matambo and Christian, 2024; Ntaganda et al., 2024; Butt et al., 2025). HRT must be long enough for slow-growing methanogens; too short an HRT risks biomass washout and lower methane yields (Kougias and Angelidaki, 2018; Shah et al., 2024). OLR should match the system's conversion capacity, overloading readily fermentable substrates promotes VFA build-up faster than methanogens can consume, leading to pH drop and upset (Tabatabaei et al., 2020; Kougias and Angelidaki, 2018). Mixing helps prevent stratification and enhances mass transfer, but excessive shear can disrupt beneficial aggregates or granules; practice targets sufficient but not overly vigorous mixing (Dustin and Hansen, 2011; Tabatabaei et al., 2020). Maintaining a near-neutral pH with adequate buffering is crucial to neutralise acids produced upstream and sustain methanogenic activity. Operators often adjust the loading rate or other conditions based on pH and alkalinity trends to keep the microbial ecosystem in balance (Kougias and Angelidaki, 2018; Alcántara-Hernández et al., 2017; Gangagni Rao et al., 2011) (Shah et al., 2024).

1.1.1.4 Feedstocks and co-digestion

A wide variety of organic feedstocks can be processed via AD, including livestock manure, crop residues, food and agro-industrial wastes, sewage sludge, the organic fraction of municipal solid waste, and dedicated energy crops (Mshandete and Parawira, 2009; Kougias and Angelidaki, 2018). Feedstock characteristics, moisture and nutrient content, lignocellulosic fibre fraction, and C:N ratio, strongly influence digestion performance. For example, manures provide nutrients and buffering but are often dilute and yield less gas per unit volume, whereas lipid-rich wastes offer high methane potential yet can inhibit microbes if overloaded due to long-chain fatty acids (Kougias and Angelidaki, 2018; Shah et al., 2024).

Co-digestion, processing complementary substrates together, is a common strategy to improve stability and yields by balancing the overall C:N ratio and micronutrient supply (Mshandete and Parawira, 2009; Kougias and Angelidaki, 2018). Blending nitrogen-rich manures with carbon-rich residues or food waste can mitigate ammonia stress while better utilising fermentable carbon, often resulting in higher methane production and more robust operation than single-substrate digestion (Burg et al., 2023; Huttunen et al., 2014). Many modern agricultural plants are designed as co-digestion facilities (e.g., manure with silage and agro-wastes), combining energy recovery with regional waste management services (Burg et al., 2023).

1.1.1.5 Pretreatment of substrates

For fibrous or otherwise recalcitrant materials, pretreatment can enhance hydrolysis and overall conversion by increasing substrate accessibility to enzymes. Common approaches include mechanical size reduction, thermal treatments, chemical (alkaline/acid/oxidative), and biological options (enzymes or lignocellulose-degrading fungi) (Tabatabaei et al., 2020; Kougias and Angelidaki, 2018). Thermal or alkaline pretreatments of straw/grass-type materials, for instance, can partially disrupt hemicellulose and lignin structures and thereby increase biogas yields (Tabatabaei et al., 2020).

Because pretreatments require additional energy or reagents, their deployment is most justified for highly lignified feedstocks (e.g., crop residues, yard trimmings) when the extra methane produced outweighs the pretreatment cost; conversely, typical manures or source-separated food wastes often do not need intensive pretreatment in full-scale practice (Kougias and Angelidaki, 2018; Tabatabaei et al., 2020). Co-digestion addresses nutrient and C:N imbalances to stabilise the process, while pretreatment tackles substrate recalcitrance, together enabling higher and more reliable biogas production (Kougias and Angelidaki, 2018; Tabatabaei et al., 2020).

1.1.1.6 Reactor designs

A range of reactor configurations has been developed to accommodate different feedstocks and improve throughput. Batch digesters load waste once for the entire retention period and are useful for dry/high-solids digestion, though gas output is inherently unsteady (Kougias and Angelidaki, 2018). For liquid slurries, continuous-flow systems dominate. The most widespread is the continuous stirred-tank reactor (CSTR), a mixed vessel operated at (semi)steady state with (semi)continuous feeding and withdrawal. CSTRs are well suited to slurry manures, sewage sludge and food wastes; they are commonly run under mesophilic conditions with multi-week retention to avoid biomass washout and maintain stability (Kougias and Angelidaki, 2018; Shah et al., 2024). Effective mixing promotes contact between microbes and fresh substrate but, because biomass is not specifically retained, volumetric productivity is moderate (Kougias and Angelidaki, 2018).

To achieve higher loadings and smaller reactor volumes, high-rate anaerobic digesters retain biomass in the vessel so that microbial concentration far exceeds that of the influent. In upflow anaerobic sludge blanket (UASB) reactors, wastewater passes upward through a dense granular sludge bed that self-aggregates and settles, enabling short hydraulic retention for low-solids streams (Kougias and Angelidaki, 2018). Variants such as the expanded granular sludge bed add recirculation to enhance contact, while anaerobic filters (fixed-/packed-bed) grow biofilms on media to prevent biomass washout and tolerate higher organic loading (Kougias and Angelidaki, 2018). Induced bed reactors (IBR) combine media-assisted retention with features of sludge-bed systems; laboratory and pilot studies show stable operation at short HRTs, with thermophilic IBR achieving higher methane production rates than mesophilic under comparable loads (Dustin and Hansen, 2011). For high-solids organic fractions (e.g., source-separated organics or crop residues), dry (high-solids) digestion is implemented in batch "garage-type" units or continuous plug-flow designs suited to stackable substrates (Kougias and Angelidaki, 2018; Akhai, 2023).

1.1.2 Multi-stage systems

Multi-stage AD separates the process across two or more reactors to tailor conditions for each phase. A typical two-stage setup uses a first-stage hydrolysis/acidification reactor (shorter retention, mildly acidic pH ~ 5–6) followed by a methanogenic reactor (longer retention, near-neutral pH). This configuration can increase overall conversion efficiency and stability by isolating fast acid producers from slower methanogens, limiting VFA accumulation in the methanogenic stage and allowing independent control of each unit (Kougias and Angelidaki, 2018; Shah et al., 2024). Multi-stage designs are particularly helpful for hard-to-digest or "souring-prone" substrates such as sugar-rich food wastes, albeit at the cost of higher capital and operational complexity (Kougias and Angelidaki, 2018; Gangagni Rao et al., 2011; Shah et al., 2024).

1.1.2.1 Thermal and process management

Operational innovations continue to raise biogas yields and stability. One example is temperature-phased digestion, in which a thermophilic stage (rapid hydrolysis, pathogen reduction) is followed by a mesophilic stage (enhanced stability and methane conversion) (Kougias and Angelidaki, 2018; Dustin and Hansen, 2011). To maintain thermal conditions, plants commonly integrate heat recovery from combined heat and power (CHP) units or solar heating solutions, both documented in full-scale and pilot contexts (Lohri et al., 2014; Feng et al., 2015). Improvements in mixing (e.g., gas recirculation, external pumped-loop systems) are used to avoid dead zones while limiting energy demand, with practice aiming for sufficient contact without disrupting beneficial aggregates (Kougias and Angelidaki, 2018; Dustin and Hansen, 2011).

Modern AD systems now span from household-scale digesters of a few cubic meters to industrial plants of several thousand cubic meters handling tens of thousands of tons per year, with rising loading rates, biogas productivity, and feedstock flexibility (Kougias and Angelidaki, 2018). Europe has led deployment of advanced configurations, including high-rate and multi-stage systems, supported by policy frameworks (Gustafsson and Anderberg, 2021; Kougias and Angelidaki, 2018), while China has historically focussed on low-cost household and farm-scale digesters and, more recently, large-scale installations for waste management (Cheng et al., 2013; Kougias and Angelidaki, 2018).

1.1.2.2 Biogas utilisation and upgrading

Raw biogas from anaerobic digestion can be used on site for space/water heating, cooking, or CHP generation; in rural contexts, household or farm digesters often supply stoves or lamps, displacing firewood or fossil fuels (Alam, 2016; Andriamanohiarisoamanana et al., 2022; Boissevain et al., 2013). For grid injection and vehicle applications, the gas is upgraded to biomethane by removing CO_2 and trace contaminants (e.g., H_2S, moisture) and increasing methane content. Typical specifications require >95% CH_4 and <3% CO_2, with strict impurity limits, acknowledging jurisdictional variation (Adnan et al., 2019; Lombardi and Francini, 2020; Kapoor et al., 2019; Vilardi et al., 2020). Upgrading enables natural-gas pipeline injection and compression as compressed natural gas (bio-CNG) for transport (Adnan et al., 2019; Lombardi and Francini, 2020).

Commercial technologies include water scrubbing, chemical scrubbing (amine absorption), pressure swing adsorption (PSA), membrane separation, and cryogenic separation, each with characteristic trade-offs in methane recovery, specific energy use, capital/operating cost, footprint, and operational complexity (Kapoor et al., 2019; Lombardi and Francini, 2020; Rodero et al., 2023). Table 1.1 summarises the anaerobic digestion stages, key microbial guilds, and typical operating considerations. Biological upgrading routes are also advancing: hydrogenotrophic methanation (in-situ or ex-situ) converts CO_2 to CH_4 by dosing H_2, photosynthetic upgrading uses microalgae to assimilate CO_2, and electro-assisted configurations (e.g., microbial electrolysis cells) have been demonstrated at pilot scale (Wahid et al., 2019; Nashmin Elyasi et al., 2021; Toledo-Cervantes et al., 2017; Rodero et al., 2023; Cristiani et al., 2021; Contreras et al., 2024).

summarises the anaerobic digestion stages, key microbial guilds, and typical operating considerations (Kougias and Angelidaki, 2018; Alcántara-Hernández et al., 2017; Wahid et al., 2019; Mshandete and Parawira, 2009; Castellan, N., Conti, F., 2019; Graunke and Wilkie, 2008; Shah et al., 2024). Table 1.2 summarises the principal biogas upgrading pathways, process principles, typical efficiencies

Table 1.1 Anaerobic digestion stages, key microbial guilds, and operating considerations.

Stage	Key microbial guilds (examples)	Operating considerations (typical conditions)
Hydrolysis	Hydrolytic bacteria (e.g., *Clostridium* spp., *Bacillus* spp.)	Enzymatic depolymerisation of polymers (carbohydrates, proteins, lipids) to monomers; enhanced by adequate moisture and mixing. Frequently rate-limiting for lignocellulosic wastes. Near-neutral pH (~6–7) and mesophilic/thermophilic temperatures (~30°C–55°C) favour activity. Kougias and Angelidaki (2018); Mshandete and Parawira (2009); Shah et al. (2024)
Acidogenesis	Fermentative (acidogenic) bacteria (e.g., *Clostridia*, *Lactobacillus* spp.)	Rapid conversion of monomers to VFAs, alcohols, CO_2, H_2, and NH_3. Operates under mildly acidic to near-neutral pH (~5–6 tolerated). Low H_2 partial pressure is beneficial for forward flux and especially important for the subsequent acetogenic step. Matambo and Christian (2024); Tundup et al. (2021); Shah et al. (2024)
Acetogenesis	Syntrophic acetogenic bacteria (e.g., *Syntrophomonas*, *Syntrophobacter* spp.) in association with methanogens	Oxidation of higher VFAs and alcohols to acetate, CO_2, and H_2. Thermodynamically feasible only when H_2 is continuously removed by methanogens; thus tight coupling with methanogenesis is required. Near-neutral pH; sensitive to H_2 buildup. Kougias and Angelidaki (2018); Wahid et al. (2019); Alcántara-Hernández et al., 2017)
Methanogenesis	Methanogenic archaea: acetoclastic (e.g., *Methanosaeta*/*Methanothrix*, *Methanosarcina*), hydrogenotrophic (e.g., *Methanobacterium*, *Methanoculleus*)	Final conversion of acetate and H_2/CO_2 to CH_4 and CO_2. Optimal near-neutral pH (~6.8–7.5); strict anaerobes. Mesophilic (~35°C) or thermophilic (~55°C) operation is common. Slow growth necessitates sufficient retention time. Inhibited by low pH, free ammonia, sulphide, or oxygen ingress. Kougias and Angelidaki (2018); Alcántara-Hernández et al. (2017); Shah et al. (2024)

Table 1.2 Biogas upgrading pathways, principles, performance, costs, and challenges.

Upgrading method	Principle	Performance and impact	Cost factors	Key challenges
Water scrubbing	Physical absorption of CO_2 (and partly H_2S) in pressurised water; CH_4 remains in the gas phase.	Simple, no chemical solvent. Typical CH_4 recovery ~90–98%; methane slip commonly ~1–3% unless optimised. Electric demand for pumping/ cooling; overall moderate energy use. Kapoor et al. (2019); Lombardi and Francini (2020)	Moderate CAPEX (column (s), pumps, CO_2 stripper). OPEX from water circulation/cooling; water make-up and CO_2 stripping/vent treatment. Kapoor et al. (2019); Lombardi and Francini (2020)	Water consumption/ management; biofouling or scaling; off-water handling (CO_2-rich stream). Often requires upstream H_2S control for materials protection. Kapoor et al. (2019); Lombardi and Francini (2020)
Chemical scrubbing (amines)	Chemical absorption of CO_2 in amine solvent (e.g., MEA) with thermal regeneration to release CO_2 and recover solvent.	High CO_2 removal; biomethane ≥95%–99% CH_4 with very low CH_4 slip when properly integrated. Thermal duty for regeneration reduces net energy efficiency but product purity is high. Lombardi and Francini (2020); Kapoor et al. (2019)	Higher CAPEX (absorber, stripper, reboiler, heat exchangers). Significant OPEX due to steam/heat for regeneration and solvent make-up. Lombardi and Francini (2020)	Solvent degradation and corrosion; solvent losses/ emissions control; need for heat integration to curb energy use. Lombardi and Francini (2020); Kapoor et al. (2019)

(Continued)

Table 1.2 (Continued)

Upgrading method	Principle	Performance and impact	Cost factors	Key challenges
Pressure swing adsorption (PSA)	Physical adsorption of CO_2 and other impurities on porous adsorbents at elevated pressure; regeneration by depressurisation/purge.	Biomethane typically ~95%–98% CH_4; CH_4 slip ~2%–5% without recycle, lower with off-gas recovery. Electric demand for compression; no solvent/steam. Kapoor et al. (2019); Lombardi and Francini (2020)	Moderate to high CAPEX (multiple vessels, compressors, valves). OPEX dominated by electricity and periodic adsorbent replacement. Kapoor et al. (2019)	Methane loss in purge gas (needs recycle/flare); adsorbent sensitivity to H_2S and moisture, requires gas pre-cleaning; complex sequencing/control. Kapoor et al. (2019); Lombardi and Francini (2020)
Membrane separation	Selective permeation of CO_2 (and H_2O) through polymer membranes under pressure; CH_4 retained on the high-pressure side; usually multi-stage.	Compact, modular, no chemical solvent. Multi-stage systems can reach ≥95% CH_4 recovery; single-stage may show ~5%–15% CH_4 slip. Energy mainly for compression; recycle streams can lower exergy efficiency. Kapoor et al. (2019); Vilardi et al. (2020); Lombardi and Francini (2020)	Modular CAPEX (membrane skids, compressors). OPEX: electricity for compression and periodic membrane replacement. Kapoor et al. (2019); Lombardi and Francini (2020)	Fouling/aging by H_2S, VOCs, particulates, pre-cleaning required. Trade-off between purity and recovery; higher pressures raise energy cost. Kapoor et al. (2019); Lombardi and Francini (2020)
Cryogenic separation	Low-temperature/pressure processing to condense/ partially liquefy CO_2 and other components while retaining CH_4; optional LCO_2 and liquefied	Very high CH_4 purity (~99% +), potential for liquid products. No chemical solvent. High electricity demand for refrigeration, typically the least energy-efficient unless waste cold	High CAPEX (refrigeration trains, cryogenic equipment) and expertise requirements. OPEX substantial due to cooling duty. Lombardi and Francini (2020)	Operational complexity (icing/solid CO_2 control), stringent materials and safety, generally economical at larger scales or when liquid products

	biomethane (LBG) production.	is available. CH_4 slip low with tight containment. Lombardi and Francini (2020); Kapoor et al. (2019)		justify costs. Lombardi and Francini (2020)
Biological upgrading	Microbial conversion of CO_2 to CH_4 via hydrogenotrophic methanogens using dosed H_2 (in-situ or ex-situ bioreactors); alternatively, photosynthetic upgrading by microalgae to assimilate CO_2.	When H_2 is adequately supplied and transferred, ex-/in-situ methanation can approach ≥98%–99% CH_4 with very low inherent CH_4 slip; energy input lies mainly in H_2 production. Environmental performance depends on the H_2 source (ideally surplus renewable power). Nashmin Elyasi et al. (2021); Wahid et al. (2019); Toledo-Cervantes et al. (2017)	CAPEX: low for in-situ retrofit (H_2 injection/ safety), higher for ex-situ reactors and gas-liquid contacting; OPEX dominated by H_2 cost. Nashmin Elyasi et al. (2021)	Gas-liquid mass transfer limits for H_2; culture stability under variable H_2 supply; strict anaerobiosis (avoid O_2 ingress); technology maturity and scale-up; economics hinge on low-cost renewable H_2. Wahid et al. (2019); Nashmin Elyasi et al. (2021)

(methane recovery and energy use), indicative costs, and key operational challenges, drawing on comparative assessments by Lombardi and Francini (2020), Kapoor et al. (2019), Adnan et al. (2019), and Vilardi et al. (2020).

Notes: *Methanosaeta* is also referred to as *Methanothrix* in the literature; both names are used in engineered AD contexts (Kougias and Angelidaki, 2018). Numeric ranges are typical, not prescriptive, and may vary with substrate, configuration, and scale (Kougias and Angelidaki, 2018; Shah et al., 2024).

Notes: Performance figures represent typical ranges consolidated from comparative reviews; exact specifications and outcomes vary with raw-gas composition, plant scale, integration, and local grid/vehicle standards.

Biogas upgrading is critical for producing biomethane that complies with natural-gas grid specifications or transport-fuel standards. Across technologies, there are inherent trade-offs among methane recovery, energy/exergy efficiency and cost (Vilardi et al., 2020; Lombardi and Francini, 2020; Kapoor et al., 2019). For example, amine scrubbing attains very high methane purity with minimal slip, but requires substantial heat for solvent regeneration; membrane systems are comparatively simple and modular, yet single-stage designs may sacrifice a few percent of methane unless configured in multiple stages (Lombardi and Francini, 2020; Kapoor et al., 2019; Vilardi et al., 2020). Water scrubbing and PSA are established, relatively moderate-cost options, but their effluent streams (CO_2-laden water or PSA off-gas) must be managed to minimise methane emissions (Kapoor et al., 2019; Lombardi and Francini, 2020; Adnan et al., 2019).

Biological methanation offers an upgrading route with potentially lower greenhouse-gas footprints when sustainable H_2 is available, effectively storing surplus renewable electricity as methane, although challenges remain in gas-liquid mass transfer, process control and reliable hydrogen supply (Nashmin Elyasi et al., 2021; Wahid et al., 2019). In practice, technology selection depends on the scale and end-use. Smaller plants often adopt membrane or PSA systems for simplicity. Larger installations may favour amine or water scrubbing for efficiency, while cryogenic upgrading is usually considered only when liquefied biomethane is required (Kapoor et al., 2019; Lombardi and Francini, 2020). Ongoing innovations seek to improve these trade-offs, for example, integrating waste heat for amine regeneration or deploying membrane_PSA hybrids, to increase net energy gains and reduce methane slip (Vilardi et al., 2020; Lombardi and Francini, 2020). Table 1.2 summarises the principal pathways, typical performance, costs, and key operational challenges.

1.1.3 Historical context and evolution of biogas technology

The production of flammable gas from organic decay has been recognised for centuries, but its controlled use emerged only after systematic scientific investigation.

Early scientific discoveries. Anecdotal reports of "combustible air" over swamps predate formal science (Kougias and Angelidaki, 2018). In 1776, Alessandro Volta collected gas from Lake Maggiore sediments and demonstrated its flammability, linking marsh emissions to a distinct combustible gas, what would later be identified as methane (Kougias and Angelidaki, 2018). In 1808, Humphry Davy showed that methane arises from the anaerobic fermentation of cattle manure, providing early experimental evidence that biological processes generate the gas (Kougias and Angelidaki, 2018; Sharma and Sukumaran, 1980). By the mid-19th century, Louis Pasteur and other scientists had firmly established the microbial basis of biogas formation, paving the way for controlled, engineered applications (Kougias and Angelidaki, 2018; Sharma and Sukumaran, 1980).

1.1.3.1 Pioneering practical uses

Documented practical use of biogas dates to the 19th century. In 1859, an anaerobic digester was reportedly constructed at a leper colony in Bombay (Mumbai), India, to capture gas from septic tanks for lighting, an early demonstration that sanitary wastes could be converted to useful energy (Kougias and Angelidaki, 2018). In 1895, the city of Exeter (UK) installed a recovery system to utilise gas from sewage treatment for street lighting, one of the first municipal applications (Kougias and Angelidaki, 2018). By the turn of the 20th century, several European treatment facilities were experimenting with anaerobic sludge digestion to stabilise waste and reduce odours, capturing biogas as a useful by-product (Kougias and Angelidaki, 2018).

Between 1906 and 1910, foundational scientific advances improved engineering predictability: Buswell formulated stoichiometric relationships that allowed estimation of methane yields from substrate composition and contemporaries identified methanogens as the organisms responsible for methane formation (Kougias and Angelidaki, 2018). Through the 1930s, Buswell and colleagues further isolated methane-producing microorganisms and developed design equations, laying the basis for modern AD process engineering (Kougias and Angelidaki, 2018). Although intermittent farm-scale digesters appeared in Europe during the interwar years, deployment remained limited (Kougias and Angelidaki, 2018).

1.1.3.2 Mid-20th-century development

After World War II, anaerobic digestion became established for municipal and agricultural waste treatment. From the 1950s_1960s, many wastewater plants adopted anaerobic sludge digestion to stabilise biosolids and reduce odours, commonly using the gas on-site for heating and electricity generation (Kougias and Angelidaki, 2018). In rural settings, small-scale digesters spread in Asia. India initiated household and community biogas units by the 1940s–1950s, leading to simple floating-drum designs (KVIC model) for manure-fed household systems intended to supply cooking fuel and fertiliser (Sharma and Sukumaran, 1980; Bhat et al., 2001). China introduced domestic biogas systems in the 1950s, with rapid uptake during the 1970s; early digesters were typically below-ground brick or concrete tanks, unheated, supplying single households with cooking gas and slurry fertiliser (Cheng et al., 2013; Chen et al., 1983). By the late 1970s, China had built millions of family-sized biogas digesters under national rural energy programmes. Variability in construction quality and limited user training during that boom led to subsequent programme improvements (Cheng et al., 2013; Chen et al., 1983).

1.1.3.3 1970s energy crisis and renewed interest

The global oil shocks of 1973 and 1979 catalysed renewed interest in AD as an indigenous energy option and a waste-management tool. In India, national and NGO-led dissemination programmes in the early 1980s promoted household and community digesters to supply cooking fuel and fertiliser; tens of thousands of small plants (typically a few cubic metres) were installed, although long-term utilisation was uneven due to maintenance needs and seasonal gas shortages (Bhat et al., 2001; Sivagnanam et al., 1984). In China, government-backed rural energy campaigns accelerated uptake from the 1950s through the 1970s, resulting in millions of domestic digesters by the late 1970s/early 1980s; typical units were below-ground brick or concrete tanks supplying single households (Cheng et al., 2013; Chen et al., 1983). Such large-scale efforts established biogas as a practical solution for rural energy access and sanitation across Asia.

In industrialised regions, policy concerns around waste, odour and nutrient emissions drove interest in larger-scale, centralised systems. Denmark pioneered centralised co-digestion plants in the mid-1980s, aggregating manures from multiple farms with organic industrial wastes to reduce environmental impacts while producing energy and sanitised fertiliser; subsequent deployment was supported by policy and cooperation among stakeholders (Nielsen, 2022; Gustafsson and Anderberg, 2021). Across Europe, biogas development increasingly aligned with environmental and energy policies, and agricultural plants evolved into regional hubs that integrate waste treatment with energy and nutrient recycling (Nielsen, 2022; Güsewell et al., 2021; Burg et al., 2023).

Beyond Asia, Sub-Saharan Africa saw pilot and early dissemination efforts from the 1970s–1980s onward, but adoption was often modest due to upfront costs, limited awareness and maintenance challenges; subsequent studies in Ghana and Ethiopia underline the importance of user support, financing and fit-for-context designs for sustained operation (Mshandete and Parawira, 2009; Osei-Marfo et al., 2018; Shallo et al., 2020).

1.1.3.4 Industrial expansion and modernisation (1990s–present)

From the 1990s onward, biogas moved decisively into full-scale industrial applications, propelled by environmental regulation and renewable-energy support. In Europe, waste and energy policies, landfill diversion, nutrient and odour controls, and market instruments such as feed-in tariffs and investment aid, created strong incentives for AD across municipal, agricultural and industrial streams (Nielsen, 2022; Güsewell et al., 2021). Denmark's experience with centralised co-digestion illustrates how policy, cooperative ownership and environmental drivers (e.g. nitrogen and ammonia mitigation) supported deployment and operational learning (Nielsen, 2022). Europe also became an early leader in biomethane upgrading and grid injection, with water/amine scrubbing, PSA, membranes and cryogenic systems adopted according to scale, gas quality requirements and integration options (Kougias and Angelidaki, 2018; Lombardi and Francini, 2020; Rodero et al., 2023; Wannasek et al., 2022; Saito et al., 2014).

In Asia, China's biogas sector diversified from millions of household units toward a mix including medium- and large-scale digesters serving livestock operations, agro-industries and municipal organics, an evolution driven by environmental management and energy goals (Cheng et al., 2013; Kougias and Angelidaki, 2018; Chen et al., 1983). India's long-running dissemination initiatives continued to support rural household plants while urban AD expanded for organic wastes and industrial effluents (Bhat et al., 2001). Beyond these two countries, national programmes in South Asia (e.g., Bangladesh, Nepal) targeted households and small farms, focusing on adoption drivers, financing and willingness-to-pay, and highlighting the importance of context-appropriate designs and user support (Sarker et al., 2020; Li, 2000; Alam, 2016; Thapa et al., 2021).

Across regions, modernisation has emphasised: (1) integration of AD into waste- and energy-system planning; (2) scalability from decentralised digesters to large centralised facilities; (3) process intensification (higher loadings, improved gas upgrading) to enhance energy yields; and (4) policy alignment to balance environmental performance (emissions control, nutrient recycling) with economic viability (Gustafsson and Anderberg, 2021; Burg et al., 2023; Lombardi and Francini, 2020; Vilardi et al., 2020; Güsewell et al., 2021).

1.1.3.5 Global perspective (21st century)

In the 21st century, biogas has come to be recognised not only as a rural energy option but as a multipurpose technology for renewable power, waste treatment, greenhouse-gas mitigation and nutrient recycling, consistent with circular-bioeconomy objectives (Gustafsson and Anderberg, 2021; Burg et al., 2023; Kougias and Angelidaki, 2018; Tundup et al., 2021; Al-Smairan et al., 2015). Modern plants increasingly co-produce energy and value-added outputs, including nutrient products (e.g., struvite, ammonium salts), soil amendments/compost, and, in upgrading chains, captured CO_2 suitable for industrial uses (Trotta et al., 2023; Lombardi and Francini, 2020; Burg et al., 2023). Deployment patterns reflect context: higher-income regions favour larger, centralised digesters with advanced process control and upgrading, while lower-income regions continue to adopt simpler, small-scale units for cooking fuel and sanitation, with uptake shaped by household economics and adoption drivers (Kelebe et al., 2017; Bhat et al., 2001; Osei-Marfo et al., 2018; Shallo et al., 2020). Historically strong adopters include European countries (policy-driven waste and energy integration) and Asian programmes in China and India spanning household to large-scale facilities (Kougias and Angelidaki, 2018; Cheng et al., 2013; Chen et al., 1983; Bhat et al., 2001).

1.1.3.6 Modern innovations and trends

Recent developments target process efficiency, flexibility, and system integration. Operational strategies such as thermophilic operation, two-phase (multi-stage) digestion, and high-rate reactor concepts are used to intensify conversion, enhance hygienisation and shorten retention times (Kougias and Angelidaki, 2018; Dustin and Hansen, 2011). European operators have piloted flexible operation, for example, intermittent or load-following feeding and expanded gas storage, to better match power output to grid demand (Barz et al., 2021; Gustafsson and Anderberg, 2021). Integration into biorefinery schemes is advancing, combining AD with algae cultivation, composting or thermochemical steps to valorise all streams (Burg et al., 2023; Kougias and Angelidaki, 2018). Nutrient recovery from digestate is gaining ground, with optimised struvite precipitation and related routes producing marketable fertilisers and addressing regional nutrient surpluses (Trotta et al., 2023; Pranckietienė et al., 2023). Policy instruments are also evolving, from fixed tariffs to broader incentives linked to waste diversion and carbon performance, which reshapes project economics and encourages higher efficiency and lower emissions (Gustafsson and Anderberg, 2021; Lombardi and Francini, 2020). Collectively, these trends reinforce biogas as a versatile platform: a single infrastructure that treats wastes, generates renewable energy and recycles nutrients, aligning with circular-economy and decarbonisation goals (Burg et al., 2023; Kougias and Angelidaki, 2018). Some researchers even propose 'anaerobic digestion without biogas' concepts that shift the focus to higher-value biochemical products instead of methane, underscoring the versatility of AD as a resource recovery platform (Kleerebezem et al., 2015). A recent study demonstrated that optimising digestate handling can improve both environmental and economic outcomes in food-waste biogas systems, exemplifying such synergy (Zheng et al., 2025).

1.1.4 The role of biogas in sustainable waste management

Biogas systems lie at the nexus of waste management and renewable energy, offering a pathway to turn organic wastes into valuable outputs while mitigating environmental impacts (Kougias and Angelidaki, 2018; Pranckietienė et al., 2023; Šál and Nováková, 2019; Huttunen et al., 2014). AD is a key component of integrated waste-management strategies that prioritise resource recovery and pollution reduction (Kougias and Angelidaki, 2018; Burg et al., 2023).

Waste diversion and emissions mitigation: One of the primary benefits of anaerobic digestion is the diversion of organic waste from landfills and uncontrolled dumping. Biodegradable wastes that are landfilled decompose anaerobically in an uncontrolled manner, producing methane if not captured; methane's 100-year GWP is on the order of 28–34× CO_2 (Huttunen et al., 2014). Channeling organics into sealed digesters captures methane for energy use and directly reduces emissions (Huttunen et al., 2014). Life-cycle assessments consistently show lower net GHG emissions for AD of organics versus landfilling, even after energy inputs, because recovered methane displaces fossil fuels (Kougias and Angelidaki, 2018; Huttunen et al., 2014). For example, treating ~1 t of food waste via AD can avoid ~0.5–1.0 t CO_2-eq relative to landfilling, depending on capture rates and energy substitution (Akhai, 2023). Diverting wet organics to AD also reduces leachate and local air-quality issues (odours, VOCs) at landfills (Huttunen et al., 2014). Many countries incentivise AD for source-separated municipal organics and agricultural residues as a waste-management strategy (Adnan et al., 2019); in the EU, landfill bans for organics have driven adoption of AD and composting (Lombardi and Francini, 2020). In developing cities, community digesters for food and market wastes cut open dumping and improve sanitation (Akhai, 2023). Biogas systems convert would-be waste emissions into a renewable resource, directly supporting climate mitigation goals by capturing methane and offsetting fossil fuel use (Kougias and Angelidaki, 2018).

AD stabilises organic matter, reducing odour and pathogen loads and yielding a more sanitised digestate, an outcome that addresses public health concerns in waste management (Kougias and Angelidaki, 2018; Lohri et al., 2014; Huttunen et al., 2014). For example, untreated manures can emit ammonia and odours and harbour pathogens when spread on fields, whereas digestion both produces energy and, in mesophilic/thermophilic regimes, attenuates pathogens and odour emissions (Kougias and Angelidaki, 2018; Nielsen, 2022; Huttunen et al., 2014). AD functions as a waste-treatment technology that mitigates environmental impacts (greenhouse gases, leachate, odour) while recovering value, consistent with the waste-management hierarchy's emphasis on recovery over disposal (Akhai, 2023).

Renewable energy generation. Biogas is a versatile renewable energy carrier. Combusted in boilers or engines, biogas supplies heat and electricity with a lower life-cycle carbon footprint than fossil fuels. Its CO_2 emissions are biogenic, reflecting short-cycle carbon from recent photosynthesis (Kougias and Angelidaki, 2018). In rural areas lacking reliable grid power, biogas can provide decentralised energy for cooking, lighting or mechanical work, reducing reliance on firewood/charcoal (Gwavuya et al., 2012) and improving indoor air quality by displacing smoky biomass stoves (Bhat et al., 2001). At larger scales, biogas-fuelled CHP plants export electricity while supplying useful heat (e.g., building heat, drying), with supportive policies in Europe facilitating grid integration (Lombardi and Francini, 2020).

A growing application is upgrading biogas to biomethane and injecting it into natural-gas grids or using it as vehicle fuel. This extends fossil-fuel displacement to sectors that use pipeline gas or CNG. Several countries operate biomethane injection and vehicle-fuel programmes, treating upgraded biogas as a renewable substitute for natural gas (Adnan et al., 2019; Lombardi and Francini, 2020; Kapoor et al., 2019). Using biomethane in vehicles (bio-CNG or bio-LNG (liquefied natural gas)) can reduce life-cycle GHG emissions, particularly for heavy-duty fleets, and lower selected tailpipe pollutants relative to diesel, contingent on fuel quality and after-treatment (Kapoor et al., 2019; Lombardi and Francini, 2020). The energy yield from AD can be substantial at the plant scale: for example, ~100–150 m³ biogas per tonne of food waste (~60–90 m³ CH_4, ~2–3 GJ LHV depending on composition) (Akhai, 2023). At system level, biomethane can provide a material share of renewable gas where suitable waste streams and policy support exist (Kougias and Angelidaki, 2018; Lombardi and Francini, 2020).

Circular economy and nutrient recycling. Beyond energy, AD closes nutrient loops by returning organic-waste nutrients to soils. Digestate retains most of the N, P, K and micronutrients present in the feedstock and can function as a biofertiliser when applied appropriately (Huttunen et al., 2014). Field studies report agronomic benefits (e.g., nutrient supply and soil improvements) from digestate use (Pranckietienė et al., 2023). Targeted nutrient recovery, notably struvite precipitation, can produce concentrated P fertilisers and improve nutrient management (Trotta et al., 2023). In integrated systems, AD plants operate as regional hubs that valorise wastes into energy and nutrient products, aligning with circular-bioeconomy goals and reducing reliance on imported mineral fertilisers (Burg et al., 2023; Huttunen et al., 2014).

Nutrient management and environmental safeguards. Nutrient recycling via digestate must be managed carefully. Over-application or poorly timed application can lead to nitrate and phosphate losses via leaching or runoff (Trotta et al., 2023; Huttunen et al., 2014). Digestate typically contains a higher share of ammonium-N than raw manure, making N more plant-available but also more prone to volatilisation or loss if not promptly incorporated (Pranckietienė et al., 2023; Huttunen et al., 2014). Sustainable practice includes applying to crop nutrient demand, timing to crop uptake, using low-emission spreading (injection/banding rather than broadcasting), and covered storage to limit ammonia emissions (Pranckietienė et al., 2023; Huttunen et al., 2014). Many jurisdictions regulate nutrient applications to protect water quality; when managed well, evidence indicates digestate can substitute a substantial fraction of synthetic fertilisers without adverse environmental effects (Burg et al., 2023; Huttunen et al., 2014).

Resource recovery and co-products. Modern plants increasingly valorise all streams: biogas for energy, digestate for fertiliser, and concentrated CO_2 from upgrading for local uses (e.g., greenhouses), depending on market context (Lombardi and Francini, 2020; Adnan et al., 2019; Burg et al., 2023). Digestate can be separated into solid (OM/P-rich) and liquid (N/K-rich) fractions; downstream recovery includes struvite precipitation to produce a granular P fertiliser (Trotta et al., 2023) and ammonia recovery (e.g., stripping or membrane concentration) to make ammonium products (Tabatabaei et al., 2020). Material uses have also been demonstrated (e.g., incorporating fermentation residues into brick blocks) (Šál and Nováková, 2019). Bioproduct routes include solid-state fermentation of residues to cultivate *Bacillus thuringiensis* (biopesticide) (Zhang et al., 2013) and broader biorefinery concepts for added-value chemicals (Tabatabaei et al., 2020). These diverse pathways demonstrate that AD is not only an energy process but also a resource-recovery platform aligned with circular-economy goals (Burg et al., 2023).

Environmental and site-specific considerations. While biogas systems offer clear environmental benefits, net impacts depend on local conditions and management. LCAs indicate AD generally reduces GHG emissions, fossil-energy use and nutrient losses when well integrated; poor operation (e.g., digestate spills, methane leakage from storage) can erode these gains (Tiwari et al., 2024; Huttunen et al., 2014). Accordingly, plants employ gas-tight covers, monitoring and flare/back-up systems to minimise fugitive emissions (Akhai, 2023; Lombardi and Francini, 2020). The comparative advantage of AD is context-dependent: composting may be simpler for some solid streams but forgoes energy recovery and can exhibit higher net GHGs relative to AD with methane capture (Adnan et al., 2019; Huttunen et al., 2014). For very dilute streams, aerobic options may be more practical. In many cases, a combined approach, AD for energy followed by composting or tailored land application of digestate, yields best overall outcomes (Burg et al., 2023; Bourdin et al., 2020). Siting and logistics also matter: transport distances for feedstocks/digestate influence footprints and social acceptance; programmes that align with local policy goals and stakeholder preferences perform better (Gustafsson and Anderberg, 2021; Bourdin et al., 2020). Optimally, locally integrated systems treat regional wastes, use energy locally or via existing grids, and return nutrients to nearby soils, closing loops and limiting transport (Burg et al., 2023).

Socioeconomic impacts and adoption challenges. Biogas projects can deliver energy-cost savings and co-benefits in rural and peri-urban areas by displacing firewood, kerosene or LPG, improving indoor air quality and freeing time otherwise spent on fuel collection (Gwavuya et al., 2012; Bhat et al., 2001). Evidence from Indonesia confirms these co-benefits at the smallholder level: a case study of farms in Bali and Lombok reports energy-cost savings, improved manure management and agricultural productivity gains following biodigester adoption (Smith and Culas 2025). In institutions (schools, hospitals, prisons), on-site digestion improves hygiene and reduces fuel purchases (Lohri et al., 2014). Financial performance is context-specific: payback depends on capital cost, subsidies, utilisation rate and displaced-fuel prices; analyses emphasise the need for robust O&M and financing frameworks (Yousuf et al., 2016; Gwavuya et al., 2012; Andriamanohiarisoamanana et al., 2022). Adoption studies highlight the roles of credit access, training/after-sales service, and fit-for-context designs; national programmes that build local supply chains (technicians, masons, service providers) tend to achieve more durable uptake (Kelebe et al., 2017; Sarker et al., 2020; Uhunamure et al., 2021). When these enabling conditions are in place, AD links waste services, energy access and nutrient recycling, delivering environmental and socioeconomic value.

Adoption barriers. Despite clear benefits, uptake can be constrained by upfront capital costs and limited access to credit/micro-finance, which deter lower-income households from investing in digesters without subsidies (Kelebe et al., 2017; Osei-Marfo et al., 2018). Successful operation also requires a reliable feedstock supply; households with insufficient livestock or organic waste may not generate enough gas to justify costs (Shallo et al., 2020; Osei-Marfo et al., 2018). Community/cooperative plants can pool feedstocks but add governance and cost-sharing complexity (Uhunamure et al., 2021; Situmeang et al., 2022). Sociocultural acceptance matters as well: perceptions about digestate handling, stove preferences, and gas safety can hinder adoption, underscoring the role of education and extension services (Rasimphi et al., 2024; Bhat et al., 2001; Osei-Marfo et al., 2018).

Operational challenges. Household systems require regular feeding, basic mixing, and appliance maintenance; lack of labour, loss of livestock, or weak after-sales support can lead to non-use or abandonment (Shallo et al., 2020; Osei-Marfo et al., 2018). Building local repair capacity and spare-parts supply chains (valves, piping, burners) is critical for long-term functionality (Kelebe et al., 2017; Osei-Marfo et al., 2018). In institutional settings (schools, prisons), biogas improves sanitation and reduces fuel costs when roles and routines are defined and staff are trained (Lohri et al., 2014); consistent operation and feeding strategies also improve techno-economic performance (Glivin et al., 2018).

Large-scale deployment barriers. Centralised plants face high capital costs, permitting complexity (waste handling/energy generation), and, at times, local opposition over odour and traffic; early engagement and odour control are critical for social acceptance (Ishii et al., 2024; Bourdin et al., 2020). Project viability often hinges on policy support (e.g., tariffs/credits, carbon incentives, waste-management instruments); where support is weak or inconsistent, investment uncertainty slows adoption (Gustafsson and Anderberg, 2021; Tembo et al., 2023). As externalities (e.g., landfill levies, carbon pricing) are internalised, AD economics improve (Gustafsson and Anderberg, 2021).

Summary. Biogas delivers combined waste, energy and nutrient benefits, but realising its potential requires fit-for-purpose policy frameworks (capital support, price signals for renewable gas, favourable gate fees), capacity building and community engagement (Gustafsson and Anderberg, 2021; Bhat et al., 2001). Programmes pairing technology deployment with training, finance and stakeholder collaboration have shown durable environmental and socioeconomic gains across regions (Burg et al., 2023; Osei-Marfo et al., 2018).

Digestate management and valorisation

A critical link to sustainable waste management is sound digestate management. Applying the 4Rs (reduce, reuse, recycle, recover) helps maximise resource value and minimise residuals (Akhai, 2023). Practical routes include:

- Volume reduction via dewatering/drying to lower transport burdens; liquid/solid separation enables targeted use (Akhai, 2023; Huttunen et al., 2014).
- Direct land application as a biofertiliser, with agronomic practice (rate, timing, incorporation, low-emission spreading) to limit ammonia loss and leaching (Pranckietienė et al., 2023; Huttunen et al., 2014).
- Nutrient recovery: struvite precipitation from liquid digestate to produce a granular P fertiliser (Trotta et al., 2023); ammonia recovery/concentration (e.g., stripping or membrane-based concentration) to generate ammonium products where viable (Tabatabaei et al., 2020).
- Further valorisation: post-digestion composting for handling and stability (Huttunen et al., 2014); material uses such as incorporating fermentation residues into brick blocks (Šál and Nováková, 2019); bioproducts via solid-state fermentation to cultivate *Bacillus thuringiensis* biopesticides (Zhang et al., 2013).

These pathways show that AD serves as a resource-recovery platform, producing energy while also returning nutrients and creating value-added products within circular bioeconomy systems (Burg et al., 2023).

Environmental and site-specific safeguards. Even with land application, safeguards are essential. Facilities may polish the liquid fraction in nature-based units (e.g., constructed wetlands/biofilters) to remove residual nutrients and organics, and deploy phytomanagement (willows/grasses) on spreading areas to take up excess nutrients, measures aligned with "green principles" for safe digestate handling (Rabapane, 2024; Akhai, 2023; Huttunen et al., 2014). Regulators commonly require nutrient-management plans and recognise digestate as a regulated fertiliser/soil conditioner subject to quality criteria; compliance depends on source control (clean feedstocks) and, where required, sanitisation/pasteurisation to ensure pathogen reduction (Gustafsson and Anderberg, 2021; Huttunen et al., 2014). Stakeholder engagement is likewise important: farmers need agronomic guidance and acceptance of digestate use, and early communication with local communities helps address odour/traffic concerns and build social licence (Gustafsson and Anderberg, 2021; Bhat et al., 2001).

Conclusion. Biogas technology converts waste liabilities into renewable energy and recycled nutrients, supporting climate mitigation, clean energy access and sustainable waste services. Realising these benefits at scale requires fit-for-purpose regulation, investment and training, together with community participation; numerous programmes demonstrate that, under such conditions, AD delivers durable environmental and socioeconomic gains (Kougias and Angelidaki, 2018; Burg et al., 2023; Lohri et al., 2014; Osei-Marfo et al., 2018) (Table 1.3).

Table 1.3 Summary of anaerobic digestion (AD).

Section	Focus	Actionable takeaways	Indicative cues/metrics	References
Fundamentals	Process and microbiology	Four stages (hydrolysis → acidogenesis → acetogenesis → methanogenesis); stable CH_4 formation needs tight syntrophy (bacteria + archaea). Match substrate complexity to retention strategy (e.g., fats → LCFA handling).	pH ~6.8–7.5; meso ~35°C–38°C/thermo ~50°C–55°C.	Shah et al. (2024), Kougias and Angelidaki (2018)
	Stability levers	Balance C:N; keep alkalinity reserve; ramp OLR slowly; monitor VFAs and ammonia; couple HRT with biodegradability.	VFA:alkalinity ratio (keep low); watch free NH_3 at high pH/T.	Kougias and Angelidaki (2018), Alcántara-Hernández et al. (2017), Shah et al. (2024)
	Feedstocks and enhancement	Co-digestion to complement nutrients/trace metals; pretreat recalcitrants (mechanical/thermal/chemical/biological) to push rates/yields.	Look for improved biodegradability and specific CH_4 yield uplift post-pretreatment.	Mshandete and Parawira (2009), Tabatabaei et al. (2020), Kougias and Angelidaki (2018)
	Reactor configurations	Wet wastes → CSTR; low-solids/high-rate → UASB, fixed/packed beds, induced-bed; high-solids → dry/plug-flow; souring-prone or tight control → multi-stage.	Pair OLR targets with biomass retention; shorten HRT only with good granulation/retention.	Kougias and Angelidaki (2018), Dustin and Hansen (2011), Akhai (2023), Hennig et al. (2018)

(Continued)

Table 1.3 (Continued)

Section	Focus	Actionable takeaways	Indicative cues/metrics	References
Development and diffusion	Early science and use	From Volta's methane to biological understanding (Davy/Béchamp); late-1800s sewage-gas recovery (India/England) seeded engineered digesters.	—	Kougias and Angelidaki (2018), Sharma and Sukumaran (1980)
	20th-century programmes	Massive roll-outs in India/China; reliability hinges on training/ maintenance; 1970s oil shocks → global push; EU policy integrated AD into waste/energy.	Prioritise service chains (spares, O&M, user training).	Bhat et al. (2001), Chen et al. (1983), Cheng et al. (2013), Gustafsson and Anderberg (2021), Kougias and Angelidaki (2018)
	Contemporary circularity	Scale spans household → centralised hubs; targets: energy efficiency + nutrient recovery + biomethane for grid/transport.	Track energy use per m^3 biogas; nutrient recovery rates (e.g., P via struvite).	Burg et al. (2023), Trotta et al. (2023), Lombardi and Francini (2020), Adnan et al. (2019)
Utilisation and upgrading	End-uses	Heat/CHP/cooking from raw biogas; choose route by local heat sinks and tariffs.	CHP electrical η typically lower than heat η; match to on-site demand profile.	Lombardi and Francini (2020), Adnan et al. (2019)
	Upgrading options	Physical/chemical (water/amine scrubbing, PSA, membranes, cryogenic) vs. biological methanation ($H_2 + CO_2 \rightarrow CH_4$). Trade-offs: CH_4 slip, exergy, CAPEX/OPEX, complexity.	Biomethane ≥ grid spec; minimise CH_4 slip; consider H_2 availability for bio-methanation.	Lombardi and Francini (2020), Kapoor et al. (2019), Nashmin Elyasi et al. (2021), Wahid et al. (2019), Vilardi et al. (2020)

Environmental and resource outcomes	Waste-to-value and climate	Diverts organics from dump/landfill; controlled capture reduces fugitive CH_4/odour/pathogens vs. landfill.	Compare GHG intensity vs. baseline landfill/ open dumping.	Kougias and Angelidaki (2018), Akhai (2023)
	Renewable energy role	Decentral heat/power; biomethane displaces fossil gas in grids/vehicles, supporting climate/energy targets.	Report net CO_2-eq per kWh or per Nm^3 biomethane.	Lombardi and Francini (2020), Huttunen et al. (2014)
	Nutrient recycling	Digestate returns N/P/C to soils; agronomic practice (rate/timing/ incorporation) limits losses/leaching.	Monitor soil N/P status; adjust application windows.	Pranckietienė et al. (2023), Huttunen et al. (2014)
	Multiple circular outputs	Beyond energy: bio-fertilisers, purified CO_2, struvite; design plants as resource-recovery hubs.	KPIs for P recovery %, CO_2 purity/use, fertiliser displacement.	Trotta et al. (2023), Lombardi and Francini (2020), Burg et al. (2023)
Socioeconomic performance	People and places	Lower household energy cost, rural jobs, sanitation benefits; barriers (upfront cost, maintenance, awareness) addressed via credit/ subsidies, training, engagement.	Track adoption/uptime; LCOE/LCOG vs. alternatives; user-satisfaction/health outcomes.	Yousuf et al. (2016), Lohri et al. (2014), Osei-Marfo et al. (2018), Kelebe et al. (2017)

AI disclosure

During the preparation of this work the author(s) used DEEPL, ChatGPT, Grammarly, Mendeley in order to Translation, language correction, bibliography management. After using this tool/service, the author(s) reviewed and edited the content as needed and take(s) full responsibility for the content of the publication.

References

Adnan, A.I., Ong, M.Y., Nomanbhay, S., Chew, K.W., Show, P.L., 2019. Technologies for biogas upgrading to biomethane: a review. Bioengineering 6 (4). Available at https://doi.org/10.3390/bioengineering6040092.

Akhai, S., 2023. Biogas plants for sustainable municipal waste management: a brief review. In: Srivastava, P., Ramteke, D., Bedyal, A., Gupta, M., Sandhu, J. (Eds.), Handbook of Research on Safe Disposal Methods of Municipal Solid Wastes for a Sustainable Environment. IGI Global, Hershey, PA, pp. 162–179. Available at https://doi.org/10.4018/978-1-6684-8117-2.ch012.

Al-Smairan, M., Al-Harahsheh, S., Al-Khazaleh, H., 2015. Biomass energy utilisation in northeast Badia of Jordan. Research Journal of Applied Sciences, Engineering and Technology 10 (11), 1322–1329. Available at https://doi.org/10.19026/rjaset.10.1829.

Alam, M., 2016. Taming the renewables: Actors' innovation in improving the utilisation of biogas for daily use in agricultural setting. International Journal of Renewable Energy Development. 5 (1), 57–64. Available at https://doi.org/10.14710/ijred.5.1.57-64.

Alcántara-Hernández, R.J., Taş, N., Carlos-Pinedo, S., Durán-Moreno, A., Falcón, L.I., 2017. Microbial dynamics in anaerobic digestion reactors for treating organic urban residues during the start-up process. Letters in Applied Microbiology 64 (6), 438–445. Available at https://doi.org/10.1111/lam.12734.

Andriamanohiarisoamanana, F.J., Randrianantoandro, T.N., Ranaivoarisoa, H.F., Kono, H., Yoshida, G., Ihara, I., et al., 2022. Integration of biogas technology into livestock farming: Study on farmers' willingness to pay for biodigesters in Madagascar. Biomass and Bioenergy 164. Available at https://doi.org/10.1016/j.biombioe.2022.106557.

Barz, M., Wesenfeld, H., Laß-Seyoum, A., Meibohm, A., Knist, S., 2021. Increasing flexibility of biogas plants through the application of innovative concepts. In: Koukios, E., Sacio-Szymańska, A. (Eds.), Bio Futures: Foreseeing and Exploring the Bioeconomy. Springer International Publishing, Cham, pp. 435–459. Available at https://doi.org/10.1007/978-3-030-64969-2_20.

Bhat, P.R., Chanakya, H.N., Ravindranath, N.H., 2001. Biogas plant dissemination: Success story of Sirsi, India. Energy for Sustainable Development 5 (1), 39–46. Available at https://doi.org/10.1016/S0973-0826(09)60019-3.

Boissevain, B.L., Zemke, P.E., Wood, B.D., 2013. Improving energy utilisation in anaerobic digester power systems. Biological Engineering Transactions 6 (1), 43–56. Available at https://doi.org/10.13031/2013.42632.

Bourdin, S., Raulin, F., Josset, C., 2020. On the (un)successful deployment of renewable energies: Territorial context matters. A conceptual framework and an empirical analysis of biogas projects. Energy Studies Review 24 (1), 4088. Available at https://doi.org/10.15173/ESR.V24I1.4088.

Burg, V., Rolli, C., Schnorf, V., Scharfy, D., Anspach, V., Bowman, G., 2023. Agricultural biogas plants as a hub to foster circular economy and bioenergy: An assessment using substance and energy flow analysis. Resources, Conservation and Recycling 190, 106770. Available at https://doi.org/10.1016/j.resconrec.2022.106770.

Butt, S., Shujat, A., Arshad, S., Zahra, T., Mehmood, A., 2025. Biogas development and usage current perspectives and opportunities, future outcomes and prospects. Pakistan Journal of Scientific and Industrial Research Series B: Biological Sciences 68 (2), 223–228.

Castellan, N., Conti, F., 2019. Molecular biotechnology to improve biofuel production from biomass. Proceedings of the 27th European Biomass Conference and Exhibition (EUBCE), Lisbon, Portugal 951–957. Available at https://doi.org/10.5071/27thEUBCE2019-2CV.6.24.

Chen, G.-Q., Liu, K.-X., Chen, M., 1983. A survey of rural biogas fermentative technology in China. In: Strub, A.S., Chartier, P., Schleser, G. (Eds.), Energy from Biomass: 2nd E.C. Conference, Berlin. Applied Science Publishers, London, pp. 1072–1084 20-23 Sept 1982.

Cheng, S., Li, Z., Mang, H.P., Huba, E.M., 2013. A review of prefabricated biogas digesters in China. Renewable and Sustainable Energy Reviews 8, 738–748. Available at https://doi.org/10.1016/j.rser.2013.08.030.

Contreras, C.A., Palacios, O.A., de-Bashan, L.E., Choix, F.J., 2024. Microalga growth-promoting bacteria as strategy to improve CO_2 removal from biogas. Bioenergy Research 17 (4), 2082–2099. Available at https://doi.org/10.1007/s12155-024-10800-0.

Cristiani, L., Zeppilli, M., Villano, M., Majone, M., 2021. Role of the organic loading rate and the electrodes' potential control strategy on the performance of a micro pilot tubular microbial electrolysis cell for biogas upgrading. Chemical Engineering Journal 426, 131909. Available at https://doi.org/10.1016/j.cej.2021.131909.

Dustin, J.S., Hansen, C.L., 2011. Performance of the induced bed reactor anaerobic digester at mesophilic and thermophilic temperatures. Transactions of the ASABE 54 (2), 729–740. Available at https://doi.org/10.13031/2013.36476.

Feng, R., Li, J., Li, X., Yang, J., 2015. Thermal performance of over-ground household biogas production system heated by solar energy. Nongye Gongcheng Xuebao/Transactions of the Chinese Society of Agricultural Engineering 31 (15), 196–200. Available at https://doi.org/10.11975/j.issn.1002-6819.2015.15.027.

Gangagni Rao, A., Surya Prakash, S., Joseph, J., Rajashekhara Reddy, A., Sarma, P.N., 2011. Multi stage high rate biomethanation of poultry litter with self mixed anaerobic digester. Bioresource Technology 102 (2), 729–735. Available at https://doi.org/10.1016/j.biortech.2010.08.069.

Glivin, G., Edwin, M., Sekhar, S.J., 2018. Techno-economic studies on the influences of nonuniform feeding in the biogas plants of educational institutions. Environmental Progress and Sustainable Energy 37 (6), 2156–2164. Available at https://doi.org/10.1002/ep.12892.

Graunke, R.E., Wilkie, A.C., 2008. Research and solutions: AASHE student award-winning paper: converting food waste to biogas. Sustainability (United States 1 (6), 391–394. Available at https://doi.org/10.1089/SUS.2008.9914.

Grohmann, A., Vainshtein, Y., Euchner, E., Grumaz, C., Bryniok, D., Rabus, R., et al., 2018. Genetic repertoires of anaerobic microbiomes driving generation of biogas. Biotechnology for biofuels 11 (1). Available at https://doi.org/10.1186/s13068-018-1258-x.

Gustafsson, M., Anderberg, S., 2021. Dimensions and characteristics of biogas policies – Modelling the European policy landscape. Renewable and Sustainable Energy Reviews 135, 110200. Available at https://doi.org/10.1016/j.rser.2020.110200.

Gwavuya, S.G., Abele, S., Barfuss, I., Zeller, M., Müller, J., 2012. Household energy economics in rural Ethiopia: a cost-benefit analysis of biogas energy. Renewable Energy 48, 202–209. Available at https://doi.org/10.1016/j.renene.2012.04.042.

Güsewell, J., Scherzinger, K., Holstenkamp, L., Vincent, L., Eltrop, L., 2021. Extending the operation of existing biogas plants: which follow-up concepts and plants will prevail? Frontiers in Energy Research 9. Available at https://doi.org/10.3389/fenrg.2021.719697.

Hennig, S., Wenzel, M., Haas, C., Hoffmann, A., Weber, J., Rödel, G., et al., 2018. New approaches in bioprocess-control: consortium guidance by synthetic cell-cell communication based on fungal pheromones. Engineering in Life Sciences 18 (6), 387–400. Available at https://doi.org/10.1002/elsc.201700181.

Huttunen, S., Manninen, K., Leskinen, P., 2014. Combining biogas LCA reviews with stakeholder interviews to analyse life cycle impacts at a practical level. Journal of Cleaner Production 80, 5–16. Available at https://doi.org/10.1016/j.jclepro.2014.05.081.

Ishii, K., Yamanaka, M., Ochiai, S., Sato, M., 2024. Multicriteria evaluation of biogas plants receiving food waste from a broad area considering different spatial system boundaries and stakeholders. Journal of Solid Waste Technology and Management 50 (2), 512–526. Available at https://doi.org/10.5276/jswtm/iswmaw/502/2024.512.

Kapoor, R., Ghosh, P., Kumar, M., Vijay, V.K., 2019. Evaluation of biogas upgrading technologies and future perspectives: a review. Environmental Science and Pollution Research 26 (12), 11631–11661. Available at https://doi.org/10.1007/s11356-019-04767-1.

Kelebe, H.E., Ayimut, K.M., Berhe, G.H., Hintsa, K., 2017. Determinants for adoption decision of small scale biogas technology by rural households in Tigray, Ethiopia. Energy Economics 66, 272–278. Available at https://doi.org/10.1016/j.eneco.2017.06.022.

Kleerebezem, R., Joosse, B., Rozendal, R., Van Loosdrecht, M.C.M., 2015. Anaerobic digestion without biogas? Reviews in Environmental Science and Biotechnology 14 (4), 787–801. Available at https://doi.org/10.1007/s11157-015-9374-6.

Kougias, P.G., Angelidaki, I., 2018. Biogas and its opportunities—a review. Frontiers of Environmental Science and Engineering 12 (3). Available at https://doi.org/10.1007/s11783-018-1037-8.

Li, Y., 2000. Case study 19: Dissemination of biogas digester technology. In: Metz, B., Davidson, O., Martens, J.-W., van Rooijen, S., Van Wie McGrory, L. (Eds.), Methodological and Technological Issues in Technology Transfer. Cambridge University Press, Cambridge, UK.

Lohri, C.R., Gauthier, M., Oppliger, A., Zurbrügg, C., 2014. Ensuring appropriateness of biogas sanitation systems for prisons: analysis from Rwanda, Nepal and the Philippines. In: Bolay, J.C., Hostettler, S., Hazboun, E. (Eds.), Technologies for Sustainable Development. Springer, Cham, pp. 57–72. Available at https://doi.org/10.1007/978-3-319-00639-0_6.

Lombardi, L., Francini, G., 2020. Techno-economic and environmental assessment of the main biogas upgrading technologies. Renewable Energy 156, 440–458. Available at https://doi.org/10.1016/j.renene.2020.04.083.

Lyu, X., Nuhu, M., Candry, P., Wolfanger, J., Betenbaugh, M., Saldivar, A., et al., 2024. Top-down and bottom-up microbiome engineering approaches to enable biomanufacturing from waste biomass. Journal of Industrial Microbiology and Biotechnology 51. Available at https://doi.org/10.1093/jimb/kuae025.

Matambo, T.S., Christian, R., 2024. Innovations in the Global Biogas industry: Applications of Green Principles. Woodhead Publishing, pp. 1–534. Available at 10.1016/C2023-0-00619-7.

Mshandete, A.M., Parawira, W., 2009. Biogas technology research in selected sub-saharan African countries—a review. African Journal of Biotechnology 8 (2), 116–125.

Nashmin Elyasi, S., He, L., Tsapekos, P., Rafiee, S., Khoshnevisan, B., Carbajales-Dale, M., et al., 2021. Could biological biogas upgrading be a sustainable substitution for water scrubbing technology? A case study in Denmark. Energy Conversion and Management 245, 114550. Available at https://doi.org/10.1016/j.enconman.2021.114550.

Nielsen, M.B., 2022. Identifying challenges and drivers for deployment of centralized biogas plants in Denmark. Sustainability (Switzerland) 14 (13). Available at https://doi.org/10.3390/su14138021.

Ntaganda, J., Gasore, G., Mukashyaka, A., Salvador Tamele, B.Z., Twahirwa, E., Mukamugema, J., et al., 2024. Enhancing Biogas Production and Use by Remote Data Acquisition and Analysis-Household Biogas Use in Rwanda as Case for the Study. 2024 IEEE PES/IAS PowerAfrica, Johannesburg, South Africa. IEEE, pp. 01–07. Available at 10.1109/PowerAfrica61624.2024.10759380.

Osei-Marfo, M., Awuah, E., de Vries, N.K., 2018. Biogas technology diffusion and shortfalls in the central and greater Accra regions of Ghana. Water Practice and Technology 13 (4), 932–946. Available at https://doi.org/10.2166/wpt.2018.100.

Pilarski, K., Pilarska, A.A., Dach, J., 2025. Biogas as renewable energy source: a brief overview. Journal of Ecological Engineering 26 (7), 408–416. Available at https://doi.org/10.12911/22998993/203376.

Pranckietienė, I., Navickas, K., Venslauskas, K., Jodaugienė, D., Buivydas, E., Žalys, B., et al., 2023. The effect of digestate from liquid cow manure on spring wheat chlorophyll content, soil properties, and risk of leaching. Agronomy 13 (3), 626. Available at https://doi.org/10.3390/agronomy13030626.

Rabapane, K.J., 2024. The merits of applying green principles in handling potential biogas digestate hazardous chemicals and pathogenicity. In: Matambo, T.S., Christian, R. (Eds.), Innovations in the Global Biogas Industry: Applications of Green Principles. Woodhead Publishing, pp. 423–453. Available at https://doi.org/10.1016/B978-0-443-22372-3.00016-9.

Rasimphi, T., Kilonzo, B., Tinarwo, D., Nyamukondiwa, P., 2024. Challenges, opportunities, and possible interventions in the biogas sector in rural areas of Limpopo, South Africa. Energy Strategy Reviews 56, 101562. Available at https://doi.org/10.1016/j.esr.2024.101562.

Rodero, Md.R., Ángeles, R., García-Depraect, O., Lebrero, R., Muñoz, R., 2023. Chapter 5 - Recent advances on photosynthetic biogas upgrading to biomethane. In: Yousuf, A., Melville, L. (Eds.), Biogas to Biomethane: Engineering, Production, Sustainability. Woodhead Publishing, pp. 117–140. Available at https://doi.org/10.1016/B978-0-443-18479-6.00010-7.

Saito, A., Osaka, N., Yamagishi, T., 2014. Development of biogas upgrading system with membrane separators. International Gas Union Research Conference (IGRC 2014) Proceedings, Vol. 3, Copenhagen, Denmark, 17-19 Sept 2014. pp. 2662–2673.

Šál, J., Nováková, P., 2019. Utilization of fermentation residues in the production of a brick block. IOP Conference Series: Earth and Environmental Science 290 (1), 012034. Available at https://doi.org/10.1088/1755-1315/290/1/012034.

Sarker, S.A., Wang, S., Adnan, K.M.M., Sattar, M.N., 2020. Economic feasibility and determinants of biogas technology adoption: evidence from Bangladesh. Renewable and Sustainable Energy Reviews 123. Available at https://doi.org/10.1016/j.rser.2020.109766.

Satpathy, P., 2017. Recent developments in biogas technology. In: Vico, A., Artemio, N. (Eds.), Biogas: Production, Applications and Global Developments. Nova Science Publishers, Hauppauge, NY, pp. 209–239.

Shah, P., Gurumurthy, J., Segaran, G., Sathiavelu, M., 2024. Chapter 9 - Anaerobic fermentation for biogas production. In: Jeguirim, M., Zorpas, A.A. (Eds.), Advances in Biofuels Production, Optimization and Applications. Elsevier, pp. 165–183. Available at https://doi.org/10.1016/B978-0-323-95076-3.00013-2.

Shallo, L., Ayele, M., Sime, G., 2020. Determinants of biogas technology adoption in southern Ethiopia. Energy, Sustainability and Society 10 (1), 1. Available at https://doi.org/10.1186/s13705-019-0236-x.

Sharma, M., Sukumaran, K.P., 1980. BIOGAS technology - a renewable energy source for the future. Chemical Age of India 31 (12).

Situmeang, R., Mazancová, J., Roubík, H., 2022. Technological, economic, social and environmental barriers to adoption of small-scale biogas plants: case of Indonesia. Energies 15 (14), 5105. Available at https://doi.org/10.3390/en15145105.

Sivagnanam, R.V.R., Sastri, M.V.C., Sathyamurthy, M.R., 1984. Barriers to the growth of biogas systems in India. Energy Management (New Delhi) 8 (4), 271–278.

Smith, A., Culas, R.J., 2025. The economic and agricultural impacts of biogas technology on small-scale farms in Indonesia: a case study in Bali and Lombok. In: Gorawala, P., Mandhatri, S. (Eds.), Agricultural Research Updates: Volume 49. Nova Science Publishers, Hauppauge, NY, pp. 1–48. Available at https://doi.org/10.52305/JXIX3014.

Tabatabaei, M., Aghbashlo, M., Valijanian, E., Kazemi Shariat Panahi, H., Nizami, A.S., Ghanavati, H., et al., 2020. A comprehensive review on recent biological innovations to improve biogas production, part 2: mainstream and downstream strategies. Renewable Energy 146, 1392–1407. Available at https://doi.org/10.1016/j.renene.2019.07.047.

Tembo, A., Rahman, M.M., Jerin, T., 2023. Barriers to development and adoption of biogas in Mokambo peri-urban of Mufulira, Zambia: how does local government fail to provide renewable energy? Biofuels 14 (6), 583–594. Available at https://doi.org/10.1080/17597269.2022.2156055.

Thapa, S., Morrison, M., Parton, K.A., 2021. Willingness to pay for domestic biogas plants and distributing carbon revenues to influence their purchase: a case study in Nepal. Energy Policy 158, 112521. Available at https://doi.org/10.1016/j.enpol.2021.112521.

Tiwari, A., Niveditha, S.V., Mohan, C., Bendi, A., 2024. Waste management from agricultural field: sustainable practices and framework. Exploring Waste Management in Sustainable Development Contexts. IGI Global, Hershey, PA, pp. 154–169. Available at 10.4018/979-8-3693-4264-0.ch011.

Toledo-Cervantes, A., Estrada, J.M., Lebrero, R., Muñoz, R., 2017. A comparative analysis of biogas upgrading technologies: photosynthetic vs physical/chemical processes. Algal Research 25, 237–243. Available at https://doi.org/10.1016/j.algal.2017.05.006.

Trotta, S., Adani, F., Fedele, M., Salvatori, M., 2023. Nitrogen and phosphorus recovery from cow digestate by struvite precipitation: process optimization to maximize phosphorus recovery. Results in Engineering 20, 101478. Available at https://doi.org/10.1016/j.rineng.2023.101478.

Tundup, S., Selvam, S.M., Roshini, P.S., Kumar, A., Sahoo, A., Paramasivan, B., 2021. Evaluating the scientific contributions of biogas technology on rural development through scientometric analysis. Environmental Technology & Innovation 24, 101879. Available at https://doi.org/10.1016/j.eti.2021.101879.

Uhunamure, S.E., Nethengwe, N.S., Tinarwo, D., 2021. Development of a comprehensive conceptual framework for biogas technology adoption in South Africa. Resources 10 (8), 76. Available at https://doi.org/10.3390/resources10080076.

Vilardi, G., Bassano, C., Deiana, P., Verdone, N., 2020. Exergy and energy analysis of three biogas upgrading processes. Energy Conversion and Management 224, 113323. Available at https://doi.org/10.1016/j.enconman.2020.113323.

Wahid, R., Mulat, D.G., Gaby, J.C., Horn, S.J., 2019. Effects of H_2:CO_2 ratio and H_2 supply fluctuation on methane content and microbial community composition during in-situ biological biogas upgrading. Biotechnology for Biofuels 12 (1). Available at https://doi.org/10.1186/s13068-019-1443-6.

Wannasek, L., May, D., Tyris, D., Amon, T., 2022. Design of an on-farm biomethane upgrading plant with hybrid compression and filling station. VDI Berichte 2022 (2406), 259–266. Available at https://doi.org/10.51202/9783181024065-259.

Yousuf, A., Khan, M.R., Pirozzi, D., Ab Wahid, Z., 2016. Financial sustainability of biogas technology: barriers, opportunities, and solutions. Energy Sources, Part B: Economics, Planning and Policy 11 (9), 841–848. Available at https://doi.org/10.1080/15567249.2016.1148084.

Zhang, W., Gong, A., Qiu, L., Yao, R., 2013. Solid state fermentation of biogas residues for production of Bacillus thuringiensis based bio-pesticide. Nongye Gongcheng Xuebao/Transactions of the Chinese Society of Agricultural Engineering 29 (8), 212–217. Available at https://doi.org/10.3969/j.issn.1002-6819.2013.08.025.

Zheng, L., Ma, X., Lyu, Y., Pan, H., Zhang, X., 2025. Promoting synergy among environmental and economic performances in food waste treatment by optimizing biogas residue management practices. Bioresource Technology 419, 132101. Available at https://doi.org/10.1016/j.biortech.2025.132101.

Waste to energy: principles of biogas production

2

2.1 Waste to energy: principles of biogas production

Biogas production from organic waste occurs via a multi-step anaerobic digestion (AD) process, namely hydrolysis, acidogenesis, acetogenesis, and methanogenesis (Atelge et al., 2020; Ometto et al., 2019; Tabatabaei et al., 2020). A wide range of organic substrates can be converted, including animal manures, municipal solid waste, agricultural residues, sewage sludge, agro-industrial effluents, and even algal biomass (Banks and Heaven, 2013; Getahun et al., 2014; Lo et al., 2019; Pavi et al., 2017). The achievable methane yield depends strongly on substrate characteristics (e.g., C/N ratio, lignin content, moisture), process parameters (pH, temperature, organic loading rate [OLR], retention time), and any pretreatment or co-digestion strategies employed; yields are further shaped by inhibitory compounds such as ammonia and sulphates, and by accumulation of volatile fatty acids (VFAs) (Chen et al., 2019; Mirmohamadsadeghi et al., 2021; Pastor-Poquet et al., 2019; Schnürer and Nordberg, 2008; Triolo et al., 2011; Wang et al., 2009; Xu et al., 2018). Optimising biogas production therefore requires balancing these factors and, increasingly, leveraging emerging tools such as machine learning and targeted bioaugmentation to enhance yield and process stability (Adeleke et al., 2025; Basak et al., 2025; Song et al., 2024; Town and Dumonceaux, 2016; Zhu et al., 2018; Li et al., 2024). A comprehensive discussion of AD processes, products (biogas, digestate) and applications is provided by Nizami (2011).

2.1.1 Types of organic waste suitable for biogas production

A diverse array of organic wastes can serve as AD feedstocks. Major categories include animal manures and slurries, the organic fraction of municipal solid waste (OFMSW) including food/kitchen waste, agricultural residues, various agro-industrial by-products (e.g., from food processing or slaughterhouses), sewage sludge from wastewater treatment, and algal biomass (Atelge et al., 2020; Banks and Heaven, 2013; Getahun et al., 2014; Lo et al., 2019; Pavi et al., 2017). Even among industrial wastewaters, anaerobic treatability can vary significantly. An early study using an upflow anaerobic filter revealed substantial differences in digestion performance across different organic effluents (Swamy and Kotaiah, 1996). Each substrate type has characteristic properties that affect its biogas potential and digestion behaviour, including total solids (TS) and volatile solids (VS), elemental composition (especially C/N), levels of lipids, proteins, and fibers (e.g., lignin), and the presence of potential inhibitors such as ammonia or sulphates (Pastor-Poquet et al., 2019; Qiang, Li and Pei, 2018; Triolo et al., 2011; Xu et al., 2018). Co-digestion, combining multiple waste types, is often employed to balance

Sustainable Biogas Production. DOI: https://doi.org/10.1016/B978-0-443-40557-0.00006-8

nutrients and mitigate inhibition, thereby improving overall performance (Ahmad et al., 2024; Pavi et al., 2017; Shahbaz et al., 2019; Shrestha et al., 2023; Zheng et al., 2022). Table 2.1 provides indicative ranges of biogas yields and typical methane content for different waste categories (Atelge et al., 2020; Banks and Heaven, 2013; Naik et al., 2014; Pavi et al., 2017) (see e.g., Table 2.1).

Animal manures (from cattle, pigs, poultry, etc.) are widely employed as AD substrates owing to their availability and the imperative to manage manures in an environmentally sound manner (Atelge et al., 2020; Banks and Heaven, 2013). Although they generally contain less readily biodegradable organic carbon than food wastes, they still yield substantial methane. Typical BMP ranges under mesophilic conditions are ~0.10–0.30 m^3 CH_4/kg VS for cattle slurries and ~0.20–0.40 m^3 CH_4/kg VS for swine, with poultry manures reaching ~0.40–0.45 m^3 CH_4/kg VS under optimised conditions (Atelge et al., 2020; Banks and Heaven, 2013; Lo et al., 2019). Manure C/N typically lies around 10–25; co-digestion with higher-carbon co-substrates is often used to adjust cattle and pig slurries towards ~20, whereas poultry manures, rich in nitrogen, mainly as uric acid, can exhibit C/N < 10 and pose an ammonia-inhibition risk (Atelge et al., 2020; Banks and Heaven, 2013; Shrestha et al., 2023). Slurries commonly contain ~5%–15% TS (cattle/pig), which reduces VS per unit volume but also dilutes potential inhibitors (Banks and Heaven, 2013). A key limitation, especially for cattle manure with bedding and undigested forage, is the recalcitrance of lignocellulosic fibres; hydrolysis becomes rate-limiting, and modest gains are achievable via mechanical size reduction or mild chemical pretreatment (Triolo et al., 2011; Lindmark et al., 2012; Amnuaycheewa et al., 2016). Typical biogas compositions are ~55%–65% CH_4 for cattle and ~60%–70% for swine, while poultry manures often yield 60%–80% CH_4 due to higher protein/fat contents (Banks and Heaven, 2013; Lo et al., 2019). Elevated ammonia released during protein degradation is the principal constraint for poultry manures; total ammonia nitrogen (TAN) levels on the order of ~3–5 g/L inhibit many methanogens, and successful operation often relies on dilution or co-digestion, pH management to shift NH_3 to NH_4^+, and acclimation towards syntrophic acetate oxidation pathways that tolerate high-ammonia conditions (Schnürer and Nordberg, 2008; Weng et al., 2024). Animal manures represent robust, year-round substrates that typically require nutrient balancing yet deliver moderate methane yields and produce nutrient-rich digestate suitable for fertiliser use (Atelge et al., 2020; Banks and Heaven, 2013). For instance, methods have been developed to recover solid fertilisers from the liquid digestate, further enhancing the resource value of AD by-products (Cotabarren and Piña, 2021).

2.1.1.1 Municipal solid waste and food waste

The organic fraction of municipal solid waste, including source-separated food scraps, yard trimmings and other biodegradable household fractions, is an attractive AD feedstock because it simultaneously diverts waste from landfill and generates renewable energy (Atelge et al., 2020; Banks and Heaven, 2013; Pavi et al., 2017). Source-separated food waste is highly degradable; typical mesophilic BMP values are on the order of 0.30–0.50 m^3 CH_4/kg VS, with richer fractions (e.g., lipid- or

Table 2.1 Indicative methane yield (BMP) and methane content for various organic waste substrates (batch tests, mesophilic).

Waste category	Specific methane yield (m^3 CH_4/kg VS)[a]	Typical CH_4 content (%)	Notes (conditions, key factors)	References
Animal manure - cattle	0.10–0.30	55–65	Lower end for high-fibre diets; manure often dilute; moderate C/N (~20); lignin slows hydrolysis.	Banks and Heaven (2013), Naik et al. (2014), Triolo et al. (2011), Lo et al. (2019)
Animal manure - swine	0.20–0.35	60–70	Typically higher VS than cattle; moderate protein, ammonia risk generally manageable.	Banks and Heaven (2013), Lo et al. (2019), Schnürer and Nordberg (2008)
Animal manure - poultry	0.25–0.45	60–70	High protein → higher ammonia potential; dilution/co-digestion often needed to avoid NH_3 inhibition.	Banks and Heaven (2013), Atelge et al. (2020), Schnürer and Nordberg (2008), Shrestha et al. (2023)
OFMSW (food and green waste)	0.30–0.50	50–65	Readily digestible; risk of rapid VFA build-up if buffering is insufficient; remove contaminants in preprocessing.	Atelge et al. (2020), Pavi et al. (2017), Wang et al. (2009), Banks and Heaven (2013),
Kitchen waste (source-separated)	0.40–0.60	55–70	Often lipid-rich; high yields but LCFA accumulation possible; co-digest with manures/high-C wastes to stabilise pH.	Pavi et al. (2017), Atelge et al. (2020), Shahbaz et al. (2019), Shrestha et al. (2023), Banks and Heaven (2013)
Agricultural residues (straw)	0.10–0.25	50–60	High lignocellulose (higher lignin → lower BMP); pretreatment (milling/alkali/biological) improves yields; co-digestion for nutrient balance.	Triolo et al. (2011), Amnuaycheewa et al. (2016), Guan et al. (2018), Lindmark et al. (2012), Banks and Heaven (2013)

(Continued)

Table 2.1 (Continued)

Waste category	Specific methane yield (m^3 CH_4/kg VS)[a]	Typical CH_4 content (%)	Notes (conditions, key factors)	References
Agro-industrial wastewater	0.50–0.94	70–80	High-COD, readily biodegradable effluents (e.g., dairy/distillery/slaughterhouse); lipid-rich streams can approach ~0.9–1.0 m^3 CH_4/kg VS; staged feeding may help stability; slaughterhouse wastewater ~0.94 m^3 CH_4/kg VS.	Atelge et al. (2020), Banks and Heaven (2013), Çetinkaya and Yetilmezsoy (2019)
Sewage sludge (mixed primary + WAS)	0.20–0.30	60–70	Primary sludge more degradable than WAS; mesophilic HRT ~20 d common; micronutrients influence methanogens.	Banks and Heaven (2013), Ometto et al. (2019)
Algae (micro-/macroalgae)	0.14–0.30	50–65	Strongly species/pretreatment dependent; microalgal proteins can raise ammonia; co-digestion or thermal pretreatment often beneficial.	Atelge et al. (2020), Ometto et al. (2019)

[a] Typical biochemical methane potential (BMP) ranges reported in the literature for each category (STP). Actual yields depend on inoculum quality, substrate variability, organic loading rate (OLR) and hydraulic retention time (HRT), mixing, and other conditions. 1 m^3 CH_4/kg VS=1000 mL CH_4/g VS.

sugar-rich wastes such as grease trap residues) sometimes approaching ~0.55–0.60 m^3 CH_4/kg VS, whereas mixed MSW with paper/soiled organics tends towards lower values (~0.15–0.30 m^3 CH_4/kg VS) (Atelge et al., 2020; Banks and Heaven, 2013; Pavi et al., 2017). Food wastes frequently contain high moisture and substantial lipid fractions; while lipids have high methane potential, rapid acidogenesis can lead to long-chain fatty acid accumulation and pH depression if over-fed, inhibiting methanogenesis (Pavi et al., 2017; Shrestha et al., 2023; Wang et al., 2009). Co-digestion is therefore widely applied: blending food waste with manure or wastewater sludge provides alkalinity and a more diverse microbiome, improving buffering and overall stability; gains in methane output have been reported when mixture ratios and substrate-to-inoculum ratios are appropriately selected (Shahbaz et al., 2019; Zheng et al., 2022). As-collected MSW typically contains plastics, metals and glass that are inert but operationally problematic; effective preprocessing (screening, depackaging, shredding/pulping) is standard practice to deliver a cleaner organic feed (Atelge et al., 2020; Banks and Heaven, 2013). Specialised preprocessing systems are available to eliminate such inert contaminants from the organic waste stream, preventing operational issues (Finsterwalder and Energy, 2008). With good source separation or effective preprocessing, biogas from OFMSW commonly contains ~55%–65% CH_4, although composition and yield remain composition-dependent; for example, a tropical-climate study reported MSW BMP spanning ~0.17–0.32 m^3 CH_4/kg VS depending on waste composition (Getahun et al., 2014). Although OFMSW and food wastes offer high biogas yields, stable operation requires careful feed management and often benefits from co-digestion and nutrient/alkalinity supplementation (Atelge et al., 2020; Pavi et al., 2017).

2.1.1.2 Agricultural crop residues

Agricultural residues such as cereal straws, corn stover, sugarcane bagasse, rice husks, and related lignocellulosic biomasses are abundant AD feedstocks, yet their conversion is constrained by slow biodegradation due to the protective lignin-carbohydrate matrix surrounding cellulose and hemicellulose (Mirmohamadsadeghi et al., 2021; Tabatabaei et al., 2020). These constraints result in lower biogas yields and slower digestion rates than those of readily degradable wastes. For example, untreated wheat or similar cereal straws commonly yield on the order of ~ 0.10-0.20 m^3 CH_4/kg VS under mesophilic BMP conditions, reflecting incomplete breakdown of structural polymers (Ozbayram et al., 2017; Triolo et al., 2011). Lignin is recalcitrant under standard AD, and methane potential inversely correlates with lignin content; Triolo et al. (2011) proposed a composition-based algorithm that captures this dependence.

Pretreatment is therefore a key enabler for lignocellulosic residues. Mechanical size reduction/intensive defibration can markedly increase hydrolysis and methane yield (Lindmark et al., 2012), while physicochemical routes such as steam explosion, ammonia fibre expansion and alkaline soaking/disruption further open up structure and partially delignify the substrate, improving subsequent enzymatic attack by the

AD microbiome (Amnuaycheewa et al., 2016; Mirmohamadsadeghi et al., 2021; Tabatabaei et al., 2020). Alkaline or combined thermo-alkaline schemes and related variants have reported substantial gains in straw BMP, albeit with case-dependent performance and the need to balance chemical/energy inputs against net energy recovery (Amnuaycheewa et al., 2016; Dhavaleswarapu, Hoysall, and Srinivasaiah, 2025; Guan et al., 2018; Boulamanti et al., 2013). In practice, co-digestion is a complementary, low-complexity strategy: carbon-rich residues are frequently mixed with manure or other nitrogen- and micronutrient-bearing substrates to supply buffering, moisture and trace elements, improving stability and overall methane productivity (Banks and Heaven, 2013; Shrestha et al., 2023; Xu et al., 2018; Zheng et al., 2022).

2.1.1.3 Agro-industrial by-products and wastewaters

Many agro-industrial operations generate concentrated, readily biodegradable streams that are well-suited to AD, for example dairy effluents, fruit and vegetable processing residues (e.g., pomace), brewery/distillery stillage, slaughterhouse by-products and wastewaters, poultry manure slurries from egg farms, or palm-oil mill effluent. Owing to their high VS and biodegradability, these substrates can deliver high methane yields when managed appropriately (Atelge et al., 2020; Banks and Heaven, 2013). Livestock and poultry slaughterhouse residues are among the most energy-dense: batch BMP values around ~0.60–0.65 m^3 CH_4/kg VS with ~60%–70% CH_4 in the biogas are commonly reported for mixed slaughterhouse by-products (Atelge et al., 2020; Banks and Heaven, 2013). Fat-rich wastewaters from rendering or edible-oil processing can approach or even exceed ~0.90 m^3 CH_4/kg VS under favorable conditions, reflecting the high calorific value of lipids (Atelge et al., 2020; Banks and Heaven, 2013). These advantages come with operational risks: protein- and fat-rich feeds promote ammonia release and long-chain fatty acid (LCFA) accumulation, respectively, which can inhibit methanogenesis if loading and buffering are not controlled; phased digestion and careful feeding strategies are therefore recommended (Banks and Heaven, 2013; Pavi et al., 2017). For high-COD industrial wastewaters, UASB configurations are widely applied; performance enhancements with conductive/adsorptive additives have also been demonstrated, for instance, biochar addition increased methane production in a UASB treating high-strength plant juice (Ometto et al., 2019; Kassaye et al., 2025).

A further consideration is sulphate. Several agro-industrial streams (e.g., molasses distillery slops or certain fermentation broths) contain sulphate or other external electron acceptors. Under such conditions, sulphate-reducing bacteria compete with methanogens, diverting substrate to H_2S rather than CH_4 (Isa, Grusenmeyer and Verstraete, 1986; Chen et al., 2019). Maintaining a high COD/SO_4^{2-} ratio (commonly >10) favors methanogenesis; downstream sulphide control via iron dosing (e.g., $FeCl_2$/$FeCl_3$) precipitates H_2S as FeS (note: this removes sulphide, not sulphate) and mitigates toxicity and corrosion (Qiang, Li and Pei, 2018; Ometto et al., 2019). In practice, many agro-industrial residues are co-digested, fruit/vegetable cannery wastes with municipal sludge, or slaughterhouse streams with manures or OFMSW, to dilute inhibitors, supply missing nutrients, and stabilise pH/alkalinity. Both lab- and pilot-scale studies report improved biogas output from such pairings when mixture ratios are properly controlled (Pavi et al., 2017; Zheng et al., 2022; Pulgarín-Muñoz et al., 2025; Shrestha et al., 2023). In co-digestion of plant biomass with manure, the dominant substrate can dictate the microbial pathway usage and overall performance, an effect observed in mixing Pennisetum grass with livestock waste (Zhong et al., 2024). Agro-industrial by-products can rank among the highest-yielding and most impactful waste-to-energy feedstocks, provided that adequate preprocessing (e.g., screening and depackaging), careful loading control, and sulphide management (for sulphate-rich wastes) are in place (Atelge et al., 2020; Banks and Heaven, 2013).

2.1.1.4 Sewage sludge

Sewage sludge from municipal wastewater treatment is a mature AD feedstock, conventionally stabilized in large digesters to recover energy (Banks and Heaven, 2013; Ometto et al., 2019). Two sludge types are typically distinguished: primary sludge, which is richer in readily degradable organics (fats, proteins, carbohydrates) and therefore more methanogenic, and waste activated sludge (WAS), which is dominated by microbial biomass and extracellular polymers and is comparatively recalcitrant (Banks and Heaven, 2013; Ometto et al., 2019). When co-mingled (primary + WAS), the mixture is characteristically nitrogen-rich with a low C/N (~6–10), and its mesophilic BMP commonly falls around 0.20–0.30 m^3 CH_4/kg VS; the resulting biogas typically contains ~60%–70% CH_4 (Banks and Heaven, 2013; Ometto et al., 2019). These yields are lower than for many food wastes because a substantial share of the easily degradable organics has already been oxidised during upstream wastewater treatment, while WAS contains biomass with robust cell walls and EPS that degrade slowly (Banks and Heaven, 2013; Ometto et al., 2019).

To improve biodegradability, facilities increasingly apply pretreatments to WAS, most commonly thermal hydrolysis (high-temperature/pressure) and, in some plants, ultrasonic or chemical routes, to rupture cell structures, increase solubilisation and enhance downstream methane generation; the magnitude of improvement is site- and process-dependent (Ometto et al., 2019). Co-digestion is also widely practiced: co-feeding food waste or FOG with sewage sludge raises the organic loading and often improves C/N balance and buffering, provided mixture ratios are controlled to avoid acidification or ammonia stress (Banks and Heaven, 2013; Pavi et al., 2017). Empirically, stable performance with increased methane production and distinct microbial shifts has been reported for sludge-food-waste co-digestion at relatively low C/N, and pilot-scale work with OFMSW-sludge mixtures confirms that appropriate co-substrate ratios enhance gas output (Zheng et al., 2022; Pulgarín-Muñoz et al., 2025). Because sludge digesters can accumulate elevated ammonium, acclimation towards high-ammonia-tolerant pathways (e.g., syntrophic acetate oxidation with hydrogenotrophic methanogenesis) and careful pH/alkalinity control are important to maintain stability (Schnürer and Nordberg, 2008; Weng et al., 2024).

2.1.1.5 Algae and aquatic biomass

Algal biomass, both microalgae (cultivated in wastewater or dedicated ponds) and macroalgae (seaweeds, floating macrophytes), is a prospective AD feedstock that avoids competition with food crops and can couple energy recovery with nutrient recycling from wastewater (Atelge et al., 2020; Ometto et al., 2019). Despite these advantages, the biodegradability of algal biomass is highly variable. Many microalgae have robust cell walls that hinder hydrolysis, and several macroalgae contain recalcitrant or inhibitory compounds that depress conversion rates (Atelge et al., 2020; Ometto et al., 2019). Reported mesophilic BMPs for macroalgal feedstocks typically fall around ~0.14–0.30 m^3 CH_4/kg VS, with biogas compositions commonly ~50%–65% CH_4, depending on species and pretreatment (Atelge et al., 2020; Ometto et al., 2019).

Microalgae cultivated on nutrient-rich waters can be protein-rich; upon digestion this favours ammonia release and, at elevated TAN/free NH_3, can inhibit methanogens unless loading and buffering are carefully managed (Schnürer and Nordberg, 2008; Weng et al., 2024). Practical mitigation strategies include co-digestion with carbon-rich co-substrates to improve C/N and buffering, and the application of cell-disruptive pretreatments (mechanical, thermal, enzymatic) to accelerate hydrolysis (Atelge et al., 2020; Ometto et al., 2019; Xu et al., 2018). Aquatic biomass case studies also show process intensification options; for example, biochar addition improved methane production from a high-strength plant juice in UASB operation (Kassaye et al., 2025). For marine macroalgae, sulphate-bearing polysaccharides can promote sulphide formation via sulphate-reducing bacteria, competing with methanogenesis; maintaining a high COD/SO_4^{2-} ratio and precipitating sulphide with iron salts are standard controls (Isa, Grusenmeyer, and Verstraete, 1986; Chen et al., 2019; Qiang, Li, and Pei, 2018; Ometto et al., 2019). Although algae and other aquatic biomasses are feasible AD substrates, they typically require pretreatment or co-digestion to achieve stable, competitive methane yields (Atelge et al., 2020; Ometto et al., 2019).

2.1.1.6 Co-digestion strategies and nutrient balance

Co-digestion, the simultaneous digestion of complementary feedstocks, is a core means of intensifying AD performance. Operators can tune the overall C/N ratio to a desirable range (~15–25) by blending wastes with contrasting compositions. This practice maintains alkalinity and dilutes inhibitory constituents, thereby improving process stability and methane productivity (Atelge et al., 2020; Pavi et al., 2017; Xu et al., 2018). A classic pairing is food waste with manure or municipal sludge: food waste supplies readily fermentable carbon but limited buffering, whereas manure/sludge contributes nitrogen, alkalinity, and a more diverse inoculum. Appropriately chosen mixture ratios increase methane yields and dampen acidification compared with mono-digestion of either stream (Banks and Heaven, 2013; Shahbaz et al., 2019; Shrestha et al., 2023; Zheng et al., 2022). On farms, co-digesting energy-crop residues or other carbon-rich materials with animal slurries is widely adopted to optimise nutrient balance and loading (Banks and Heaven, 2013; Naik et al., 2014).

Determining an effective blend is case-specific. Recent studies use mixture-design experiments and modeling to locate operating "sweet spots" that avoid both nutrient limitation and ammonia stress; pilot-scale work on sludge-OFMSW systems, for example, identifies an optimum sludge: OFMSW window beyond which performance declines (Kelif Ibro, Ramayya Ancha, and Beyene Lemma, 2024; Pulgarín-Muñoz et al., 2025; Zheng et al., 2022). In practice, keeping high-nitrogen fractions moderated, via dilution, staged feeding, or co-substrates with higher C/N, helps prevent free-ammonia inhibition, while sufficient alkalinity and gradual loading mitigate VFA accumulation (Banks and Heaven, 2013; Shrestha et al., 2023; Wang et al., 2009). Start-up and batch testing further benefit from a moderate substrate-to-inoculum (S/I) ratio; S/I on the order of 0.5–1.0 (VS basis) is frequently associated with stable performance and robust methane yields, and the same principle informs feeding rates and digestate recycle in continuous operation (Habib, Torii, and Mol, 2024; Shahbaz et al., 2019).

Co-digestion can also dilute inhibitors. Blending a sulphate-rich stream with a higher-COD co-substrate sustains substrate availability for methanogens and raises the overall COD/SO_4^{2-} ratio, conditions that suppress sulphate-reducing competition and favor methanogenesis (Isa, Grusenmeyer, and Verstraete, 1986; Qiang, Li, and Pei, 2018; Chen et al., 2019). Adding a small fraction of fatty wastes (e.g., FOG) to a largely carbohydrate-based feed can boost methane output, provided the bulk feedstock buffers against LCFA accumulation and acidification (Banks and Heaven, 2013; Pavi et al., 2017; Atelge et al., 2020). Across numerous trials, co-digestion improves stability and methane productivity while fostering more robust microbial communities (Atelge et al., 2020; Shrestha et al., 2023; Xu et al., 2018; Zheng et al., 2022).

Microbial community responses. Amplicon-based surveys consistently show that co-digestion tends to increase microbial diversity and enrich functional guilds that stabilise carbon flow. In lignocellulose-rich systems, efficient acetate and hydrogen converters (e.g., *Methanosaeta/Methanosarcina* and *Methanoculleus*) are key, with their relative dominance shaped by ammonia and loading conditions (Jensen et al., 2021; Weng et al., 2024; Schnürer and Nordberg, 2008). In sludge-food-waste co-digestion specifically, distinct community shifts, including increased representation of versatile *Methanosarcina* at lower C/N, have been linked to improved stability (Zheng et al., 2022).

Practical considerations. Successful co-digestion hinges on adequate mixing and feed handling (often with separate preprocessing lines) and, where needed, trace-metal supplementation. If all inputs are poor in key cofactors (e.g., Co, Ni), targeted dosing can enhance methanogenic activity; conversely, micronutrient-rich wastes such as manures or sludges can supply these to carbon-rich residues (Banks and Heaven, 2013; Li et al., 2022; Pavi et al., 2017). Owing to these synergistic benefits, many commercial plants operate deliberately as co-digestion facilities combining manures, organic wastes and, in some regions, crop residues (Banks and Heaven, 2013; Atelge et al., 2020; Santhappan et al., 2024).

2.2 Anaerobic digestion process explained

Anaerobic digestion is a microbially mediated conversion of complex organic matter to biogas through four interlinked stages, hydrolysis, acidogenesis, acetogenesis and methanogenesis. In single-stage digesters these transformations proceed concurrently in a mixed consortium, but the stagewise view remains useful for diagnosis and optimisation (Atelge et al., 2020; Ometto et al., 2019). Optimising biogas output must often be balanced against efficiency and cost constraints (Nakashima and de Oliveira 2021).

2.2.1 *Hydrolysis*

In the initial step, extracellular enzymes depolymerise particulate substrates: cellulases/hemicellulases target plant fibres, proteases cleave proteins, and lipases split triglycerides to long-chain fatty acids and glycerol. Kinetically, the hydrolysis rate is governed by the activity and stability of these enzyme classes; tuning pH and temperature and, where warranted, dosing exogenous enzymes can substantially accelerate substrate depolymerisation (Singh, 2025). The activity is dominated by hydrolytic/fermentative bacteria, frequently Firmicutes, with clostridial and ruminococcal lineages prominent on lignocellulosic feeds (Atelge et al., 2020; Basak et al., 2022; Jensen et al., 2021). Hydrolysis is commonly rate-limiting for solid, fibre-rich wastes because lignin and fibre architecture restrict enzyme access; methane potential correlates inversely with lignin content (Jensen et al., 2021; Mirmohamadsadeghi et al., 2021; Triolo et al., 2011). Accordingly, measures that increase substrate accessibility, mechanical comminution/defibration, thermal/alkaline or organic-acid pretreatments, or biologically oriented strategies, can accelerate hydrolysis and raise BMP (Amnuaycheewa et al., 2016; Lindmark et al., 2012; Mirmohamadsadeghi et al., 2021). Targeted bioaugmentation with potent hydrolysers has also proved effective: additions of cellulolytic consortia or *Clostridium thermocellum* increased fibre breakdown and methane formation on agricultural residues (Ecem Öner et al., 2018; Ozbayram et al., 2017). Good hydrodynamic contact between solids and the active biomass through adequate mixing further supports hydrolytic turnover and downstream stability (Atelge et al., 2020; Ometto et al., 2019). Comparable gains have been reported in sludge-based systems, where coupling mild pretreatment with bioaugmentation by cellulolytic cultures accelerated lignocellulose breakdown and increased methane production (Hu et al., 2016). Direct addition of hydrolytic enzymes has shown improved initial breakdown of lignocellulosic waste in batch AD tests (Quiñones et al., 2012).

2.2.2 *Acidogenesis*

During acidogenesis, the soluble monomers released by hydrolysis are fermented by a broad consortium of facultative and obligate anaerobes into volatile fatty acids (VFAs; primarily acetate, propionate and butyrate), alcohols (e.g., ethanol), ketones, CO_2, and H_2 (Atelge et al., 2020; Ometto et al., 2019). Typical acidogens include clostridial lineages and related fermenters frequently detected in AD consortia. This step is kinetically rapid on simple substrates and can outpace downstream conversions, so insufficient buffering or excessive loading leads to VFA accumulation and pH decline that inhibit methanogenesis (Pavi et al., 2017; Wang et al., 2009). Hydrogen-producing acidogens are pivotal because H_2 partial pressure exerts thermodynamic control over subsequent pathways: lower H_2 favours more oxidised products (acetate), whereas elevated H_2 shifts fermentation towards reduced products such as propionate and butyrate (Ometto et al., 2019; Weng et al., 2024). Maintaining adequate alkalinity and coupling to efficient H_2 consumers downstream are therefore essential for avoiding acidification and for sustaining flux from acidogenesis to methanogenesis (Atelge et al., 2020; Ometto et al., 2019).

2.2.3 Acetogenesis

Acetogenesis channels higher-order acids and alcohols produced in acidogenesis into acetate, CO_2 and H_2, the direct precursors for methanogenesis. Two functional groups are involved. (1) Obligate proton-reducing (syntrophic) acetogens oxidise VFAs near the thermodynamic limit and require tight coupling to H_2-consuming partners: *Syntrophobacter* spp. oxidise propionate, and *Syntrophomonas* spp. oxidise butyrate, but only when methanogens (or, undesirably, sulphate reducers) keep H_2 at very low levels (Weng et al., 2024; Chen et al., 2019). Disruption of this coupling elevates H_2, stalls fatty-acid oxidation, and drives propionate/butyrate accumulation, a hallmark of overloaded or imbalanced digesters (Weng et al., 2024; Wang et al., 2009). (2) Homoacetogens reduce CO_2 with H_2 via the Wood–Ljungdahl pathway or ferment selected substrates to acetate; in mixed AD communities they usually play a secondary role because methanogens are stronger H_2/CO_2 competitors, but homoacetogenesis can contribute additional acetate under specific constraints (e.g., inhibitory stress or atypical pH) (Ometto et al., 2019; Atelge et al., 2020). In steady operation, a major share of carbon is funnelled to acetate that is subsequently cleaved by acetoclastic methanogens, with the remainder converted via the hydrogenotrophic route from H_2/CO_2, hence the centrality of efficient syntrophic H_2 removal for maintaining acetogenic flux (Banks and Heaven, 2013; Ometto et al., 2019; Weng et al., 2024).

2.2.4 Methanogenesis

The terminal step of AD is catalysed by methanogenic archaea through two principal routes: (1) the acetoclastic cleavage of acetate to CH_4 and CO_2, and (2) the hydrogenotrophic reduction of CO_2 with H_2 ($CO_2 + 4H_2 \rightarrow CH_4 + 2H_2O$). A third, methylotrophic pathway (from methanol or methylamines) can operate when such substrates are available, but is generally minor in municipal and agricultural wastes (Banks and Heaven, 2013; Ometto et al., 2019). In practice, acetoclastic taxa such as *Methanosaeta* (syn. *Methanothrix*) and *Methanosarcina* and hydrogenotrophic genera including *Methanobacterium*, *Methanospirillum*, *Methanoculleus*, and *Methanobrevibacter* constitute the core methanogenic guilds in digesters (Banks and Heaven, 2013; Weng et al., 2024; Zheng et al., 2022). Under mesophilic, low-acetate conditions, *Methanosaeta* often predominates and its filamentous growth can support granule structure, whereas *Methanosarcina*, able to use both acetate and H_2/CO_2, grows faster and tends to dominate at elevated acetate, albeit sometimes with less stable operation (Banks and Heaven, 2013; Zheng et al., 2022).

Temperature and ammonia strongly modulate pathway use. At thermophilic conditions and/or under elevated ammonium/free-ammonia stress, acetoclastic activity is frequently curtailed and carbon flux shifts towards syntrophic acetate oxidation (SAO) coupled to hydrogenotrophic methanogenesis; this SAO-hydrogenotrophic route is also selected under sustained high-ammonia mesophilic operation (Schnürer and Nordberg, 2008; Ometto et al., 2019; Weng et al., 2024). Hydrogenotrophic lineages (e.g. *Methanoculleus*, *Methanobacterium*) typically show greater tolerance to ammonia and higher temperatures than obligate acetoclasts, explaining their dominance in such regimes (Schnürer and Nordberg, 2008; Weng et al., 2024).

Methanogens exhibit comparatively slow growth (doubling times of days), making methanogenesis the kinetic bottleneck that dictates the hydraulic/solids retention times required for stable operation. The step is also pH-sensitive, with optima near neutrality (~7.0–7.5); inhibition of methanogens leads to VFA accumulation from upstream fermentation, further depressing pH and triggering failure if not corrected (Banks and Heaven, 2013; Ometto et al., 2019; Wang et al., 2009).

2.2.5 Microbial consortia and syntrophy

AD performance emerges from a tightly integrated consortium in which bacteria and archaea exchange metabolites and reducing equivalents. Syntrophic interactions are most conspicuous at the interface between acetogenesis and methanogenesis: obligately proton-reducing acetogens oxidise VFAs only when partnered with fast H_2 consumers, typically hydrogenotrophic methanogens, which depress H_2 partial pressure to render otherwise endergonic reactions feasible (Banks and Heaven, 2013; Ometto et al., 2019; Weng et al., 2024). These partnerships are often manifested in close physical association (flocs/granules) that shorten diffusion paths for intermediates. Beyond classical interspecies H_2 transfer, direct interspecies electron transfer (DIET) has been proposed in several digester types; conductive appendages and redox-active particles can mediate electron flow more efficiently than dissolved H_2. Reviews highlight DIET/EET as an emerging lever for stability under stress (e.g., ammonia, organic overloading), and the targeted addition of conductive materials (e.g., biochar) has improved methane formation in high-strength systems (James et al., 2025; Kassaye et al., 2025).

Competition with sulphate-reducing bacteria (SRB) is a further syntrophic boundary condition when sulphate is present. SRB outcompete methanogens at low COD/SO_4^{2-}, diverting electron flow to H_2S; raising the COD/SO_4^{2-} ratio and/or precipitating sulphide with iron salts mitigates inhibition and re-balances carbon towards methanogenesis (Isa, Grusenmeyer, and Verstraete, 1986; Qiang, Li, and Pei, 2018; Chen et al., 2019). In practice, operators track VFA profiles and biogas composition as process "vital signs," because a disturbance in any guild (e.g., methanogens) is rapidly reflected in intermediate build-up and gas quality (Banks and

Heaven, 2013; Pavi et al., 2017). A practical complement is the specific methanogenic activity (SMA) assay, which rapidly gauges the methanogenic potential of the biomass and helps diagnose capacity limitations (Rojas, Netto, and Zaiat, 2008). High-throughput sequencing surveys consistently resolve a recurrent core: clostridial/ruminococcal fermenters during hydrolysis-acidogenesis; *Syntrophomonas/Syntrophobacter* during fatty-acid oxidation; and acetoclastic (*Methanosaeta*, *Methanosarcina*) plus hydrogenotrophic (*Methanobacterium*, *Methanoculleus*, *Methanospirillum*, *Methanobrevibacter*) archaea during methanogenesis, with community structure tuned by feedstock and constraints such as ammonia (Jensen et al., 2021; Weng et al., 2024; Zheng et al., 2022). Co-digestion generally increases functional diversity and resilience relative to mono-digestion, aligning with observed gains in stability and methane productivity (Shrestha et al., 2023; Zheng et al., 2022).

2.2.6 Acetoclastic versus hydrogenotrophic pathways

Under mesophilic, moderate-ammonia operation, acetoclastic methanogenesis, mediated by obligate acetoclasts such as *Methanosaeta* or by the versatile *Methanosarcina*, typically contributes the majority of methane formation, with the remainder derived from the hydrogenotrophic reduction of CO_2 (Banks and Heaven, 2013; Ometto et al., 2019). Environmental stressors re-partition this balance. Elevated ammonium/free NH_3 and/or thermophilic temperature depress acetoclastic activity and select for the SAO route coupled to hydrogenotrophic methanogenesis; *Methanoculleus* and related hydrogenotrophs often increase in dominance under these regimes (Schnürer and Nordberg, 2008; Weng et al., 2024). The SAO shift is well documented in high-ammonia systems, where acetate is oxidised to $H_2 + CO_2$ by SAO bacteria and subsequently reduced to CH_4 by hydrogenotrophic archaea; concomitantly, acetoclastic *Methanosaeta* declines (Schnürer and Nordberg, 2008; Weng et al., 2024; Alsouleman, 2019). Process-wise, maintaining conditions that preserve acetoclasts (e.g., moderated ammonia via dilution/co-digestion and adequate buffering) minimises retention-time penalties, while ensuring robust hydrogenotrophic capacity prevents VFA accumulation during transients (Banks and Heaven, 2013; Shrestha et al., 2023). Stable-isotope approaches now provide in situ diagnostics: ^{13}C-based fractionation/labelling of CH_4/CO_2 can resolve the relative contributions of acetate cleavage vs. CO_2 reduction and reveal pathway shifts with loading, temperature, or inhibitors (Rodrigues Oliveira et al., 2024).

Pretreatment and emerging strategies. Before feeding wastes into AD, a range of pretreatments can be applied to increase digestibility; the choice is substrate-specific and must consider net energy balance.

Mechanical pretreatment. Shredding, grinding, maceration and related disintegration methods reduce particle size and rupture structures, increasing surface area and enzyme accessibility (Atelge et al., 2020; Ometto et al., 2019). For fibre-rich crops, intensive size reduction can yield sizeable gains, for example, ley-crop silage showed ~40%–60% higher methane after mechanical defibration/comminution (Lindmark et al., 2012). Complementary approaches include high-shear/extrusion, liquid-shear maceration, ultrasound and even freeze-thaw; these can enhance solubilisation and hydrolysis, although effectiveness is case-dependent (Atelge et al., 2020; Ometto et al., 2019). Because mechanical steps can be energy-intensive, specific energy input must remain well below the extra energy recovered from biogas; sustainability assessments emphasise optimizing pretreatment intensity and heat/power integration (Banks and Heaven, 2013; Boulamanti et al., 2013). For lignocellulosics, pairing physical size reduction with mild chemical assistance (e.g., organic acids) further improves accessibility (Amnuaycheewa et al., 2016; Guan et al., 2018).

Thermal pretreatment. Heating substrates in the presence of moisture accelerates hydrolysis and solubilisation. Moderate heating (~70–100°C) can improve hygienisation and protein/starch disintegration, providing modest biogas gains in some applications (Ometto et al., 2019). High-severity options such as steam explosion (~160–200°C followed by rapid depressurisation) disrupt hemicellulose and alter lignin structure, thereby increasing subsequent biomethanation rates from straw and other lignocellulosics (Mirmohamadsadeghi et al., 2021; Tabatabaei et al., 2020). For sewage sludge, commercial thermal-hydrolysis processes (e.g., ~165°C for ~20 min prior to digestion) are widely deployed, enabling higher loading rates and substantial gas-yield increases alongside pathogen reduction (Ometto et al., 2019). Thermo-alkaline variants and alkaline-assisted schemes (e.g., NaOH or combined reagents) enhance solubilisation and methane yields across diverse feedstocks, although benefits must be weighed against reagent and heat demands (Dhavaleswarapu, Hoysall and Srinivasaiah, 2025; Guan et al., 2018; Amnuaycheewa et al., 2016). Integrating waste heat and right-sizing pretreatment severity are central to maintaining a positive energy balance (Boulamanti et al., 2013; Banks and Heaven, 2013). Pre-treating straw with ammonia (ammoniation) in a high-solids AD process improved its biodegradation but also selected for ammonia-tolerant microbial communities (Tang et al., 2022).

2.2.7 Chemical pretreatment

Alkaline agents (e.g., NaOH, KOH, lime) swell fibres, cleave ester/ether linkages and increase cellulose accessibility, effects that are particularly valuable on high-lignin residues. Across lignocellulosics, alkaline or thermo-alkaline schemes often raise solubilisation and BMP by tens of percent, with the magnitude depending on severity and substrate (Mirmohamadsadeghi et al., 2021; Tabatabaei et al., 2020; Guan et al., 2018; Dhavaleswarapu, Hoysall and Srinivasaiah, 2025). Dilute-acid routes (e.g., H_2SO_4 at elevated temperature) can release fermentable sugars but may concomitantly form furans and phenolics, so they are used more in bioethanol pretreatment than in AD unless followed by careful conditioning (Mirmohamadsadeghi et al., 2021; Tabatabaei et al., 2020). Oxidative options (ozone, H_2O_2 and related) have been investigated for delignification, although reagent costs and side-product management are constraints (Mirmohamadsadeghi et al., 2021). Recent work has explored stronger oxidants and synergies: a mild potassium ferrate treatment combined with ultrasound improved methane yield from nut shells by ~30% (Şenol, 2026). Integrating waste heat and right-sizing pretreatment severity are central to maintaining a positive energy balance (Boulamanti et al., 2013; Banks and Heaven, 2013).

2.2.8 Biological pretreatment

Microbial/enzymatic routes pre-condition substrates prior to digestion. Partial aerobic steps (e.g., ensiling) can soften fibres, while selective delignification by white-rot fungi (e.g., *Trametes*, *Pleurotus*) leaves carbohydrates more accessible, albeit at the expense of multi-day residence times (Mirmohamadsadeghi et al., 2021; Tabatabaei et al., 2020). Biological pretreatments with specific fungi have shown success on certain substrates: treating spent mushroom compost with Trichoderma viride spores significantly increased its methane yield (Zhu et al., 2025). Exogenous enzymes (cellulases, hemicellulases, pectinases) applied before or at start-up can accelerate hydrolysis; for instance, enzyme blends increased the initial degradation rate and methane yield from tomato residues (Martínez et al., 2025). Cost remains a key limitation, motivating interest in in-situ enzyme generation and enzyme-producing bioaugmentation (Ecem Öner et al., 2018). Targeted bioaugmentation has also been used to address specific bottlenecks: inoculation with *Clostridium thermocellum* or rumen-derived cellulolytic consortia enhances fibre conversion on agricultural residues (Ecem Öner et al., 2018; Ozbayram et al., 2017), and, in a stalled manure digester with VFA accumulation, augmentation with an enriched acetogen/methanogen culture relieved the imbalance and restored methane production (Town and Dumonceaux, 2016). Advanced methods like metaproteomic analysis have been proposed to identify key cellulolytic enzymes in AD consortia, potentially guiding more effective bioaugmentation strategies (Villalobos Solis et al., 2022; Li et al., 2022). Effect persistence depends on ecological fit; without a niche advantage, introduced strains may wane, so bioaugmentation is most effective when matched to a clearly diagnosed limitation (Mirmohamadsadeghi et al., 2021; Tabatabaei et al., 2020). In one approach, enzymes themselves are produced from waste biomass: Silva et al. (2019) extracted cellulases from passion fruit peel residues and used the hydrolysate to boost biogas yields.

2.2.9 Advanced and combined strategies

Recent work increasingly combines pretreatment modes and uses data-driven tools to tune severity to the substrate. Sequential, multimodal schemes, mechanical, chemical, and thermal, have delivered biomethane gains surpassing single-method controls on agricultural feedstocks (Austen et al., 2025; Mirmohamadsadeghi et al., 2021). Optimisation can be guided by machine learning: multimodal ML has been used to predict yields from NaOH-pretreated biomass and to identify optimal pretreatment conditions (Adeleke et al., 2025), while separate studies correlate pretreatment variables (e.g., temperature, alkali dose, residence time) with methane outcomes and rank their importance, informing scale-up and customisation to feedstock (Song et al., 2024).

An additional emerging route is in-situ biogas upgrading via hydrogen addition (biological methanation): external H_2, for example, from surplus renewable power via electrolysis, is injected and converted with CO_2 to CH_4 by hydrogenotrophic methanogens, enriching biogas and storing renewable H_2 as methane (Abdelrahman et al., 2025; Holl et al., 2023). Design considerations include H_2 mass transfer/mixing, gas–liquid contact, and avoidance of H_2 slip; pilot- and reactor-scale trials report promising methane enrichment alongside these engineering challenges (Holl et al., 2023; Abdelrahman et al., 2025).

AD performance hinges on the four stages and their microbial consortia. For solids, hydrolysis is often rate-limiting, making pretreatment and targeted bioaugmentation key levers. Stability depends on syntrophy to methanogenesis; maintaining acetoclastic and hydrogenotrophic routes prevents VFA buildup. In practice, co-digestion, trace-nutrient tuning, and adsorbent/conductive amendments stabilise tough feeds, while high-throughput monitoring and ML-guided control enable adaptive operation. Integrating $H_2 \rightarrow CH_4$ conversion broadens flexibility (Jensen et al., 2021; Li et al., 2022; Kassaye et al., 2025; Song et al., 2024; Abdelrahman et al., 2025). For overviews and recent advances, see (Nizami, 2011; Ramachandran, Singh and Singh, 2023).

Four principal stages proceed in a coordinated microbiome: (1) Hydrolysis, extracellular enzymes (cellulases/hemicellulases, proteases, lipases) from hydrolytic bacteria depolymerise complex substrates (carbohydrates, proteins, lipids) to soluble monomers (sugars, amino acids, long-chain fatty acids). (2) Acidogenesis, fermentative bacteria convert monomers into volatile fatty acids (VFAs; e.g., acetate, propionate, butyrate), alcohols, CO_2 and H_2. (3) Acetogenesis, obligately proton-reducing (syntrophic) bacteria oxidise higher VFAs and alcohols to acetate while releasing H_2/CO_2; reactions proceed only at low H_2 maintained by partner microbes. (4) Methanogenesis, archaea produce CH_4 either by acetoclastic cleavage of acetate or by hydrogenotrophic reduction of CO_2 with H_2. Syntrophic partnerships are essential: acid oxidisers (e.g., *Syntrophomonas*, *Syntrophobacter*) and hydrogenotrophic methanogens (e.g., *Methanoculleus*, *Methanobacterium*) operate in tight association; direct interspecies electron transfer (DIET) via conductive pili or particles may complement dissolved H_2 transfer. The figure also highlights common stressors (ammonia, NH_3, released from protein; sulphide, H_2S, from sulphate reduction) and operational countermeasures (co-digestion for C/N and alkalinity balance; Fe^{2+} dosing to bind sulphide). Carbon flow is indicated by arrows: polymers → monomers → VFAs/alcohols → acetate + H_2/CO_2 → CH_4 + CO_2. Representative taxa can be annotated per stage (e.g., *Clostridium* spp. in hydrolysis/acidogenesis; *Syntrophomonas* in acetogenesis; *Methanosaeta*/*Methanosarcina* and *Methanobacterium* in methanogenesis).

2.3 Factors influencing biogas yield

AD performance (biogas output, CH_4%) is driven by interrelated substrate, microbiological/inhibitory, and operational factors (Atelge et al., 2020; Banks and Heaven, 2013; Ometto et al., 2019). Key drivers include feedstock composition, temperature, and retention time (Dobre et al., 2014). This section outlines control/mitigation strategies, lists typical operating ranges, and summarises major inhibitors and countermeasures (Tables 2.2 and 2.3).

2.3.1 *Substrate factors—composition and biodegradability*

The intrinsic properties of the feedstock strongly influence biogas yield. One primary factor is the biodegradable VS fraction: substrates rich in simple carbohydrates, proteins, and lipids convert more completely than those high in lignin or ash (Atelge et al., 2020; Banks and Heaven, 2013). For example, with the same VS (~80%), waste frying oil (mostly triglycerides) can approach the theoretical methane potential for lipids (~1.0 m^3 CH_4/kg VS), whereas untreated cereal straws typically yield ~0.10–0.30 m^3 CH_4/kg VS unless pretreated (Banks and Heaven, 2013; Triolo et al., 2011; Guan et al., 2018; Lindmark et al., 2012). Lignin content negatively correlates with BMP; Triolo et al. (2011) quantified an inverse relationship and proposed a composition-based BMP algorithm (Triolo et al., 2011; Mirmohamadsadeghi et al., 2021). Hence crop residues, wood wastes and many paper wastes, being lignin-rich, are inherently lower-yielding per VS added; co-digesting with low-lignin substrates (e.g., manures, food waste) raises the overall degradable fraction (Banks and Heaven, 2013; Pavi et al., 2017; Xu et al., 2018; Shrestha et al., 2023).

2.3.2 *Carbohydrates versus proteins versus lipids*

Lipids have the highest theoretical methane potential (~1.014 m^3 CH_4/kg VS for pure fats) but hydrolyse to LCFAs that can adsorb to biomass and inhibit if overfed; LCFA degradation proceeds via syntrophic β-oxidation with methanogens (Banks and Heaven, 2013; Pavi et al., 2017; Atelge et al., 2020). Moderating lipid share (~10%–30% of VS), co-digesting with carbohydrate-rich feeds, and using sorbents can mitigate LCFA inhibition; biochar addition has improved methane production in high-strength systems (Pavi et al., 2017; Kassaye et al., 2025; Atelge et al., 2020). Proteins (~0.5 m^3 CH_4/kg VS) release ammonia upon deamination; rising TAN/free NH_3 inhibits acetoclastic methanogens and can shift pathways towards SAO + hydrogenotrophic methanogenesis (Schnürer and Nordberg, 2008; Weng et al., 2024; Banks and Heaven, 2013). Operationally, maintaining bulk pH near 7.0–7.2 and mesophilic temperature lowers the free-NH_3 fraction at a given TAN; inhibition commonly emerges once free NH_3 reaches on the order of ~80–150+ mg N/L (system-dependent), while a few-tens mg N/L typically show little effect (Schnürer and Nordberg, 2008; Weng et al., 2024). Where nitrogen loads are high (e.g., poultry manure, slaughter residues), co-digestion/dilution and, in extreme cases, ammonia removal (e.g., stripping/struvite) are applied (Banks and Heaven, 2013; Shrestha et al., 2023). The choice of primary carbon source in the feedstock has been shown to influence the composition of methanogenic archaea and the methane production capacity of the system (An et al., 2025).

Table 2.2 Common inhibitors in AD and countermeasures (single-stage CSTRs; mesophilic).

Inhibitor (source)	Inhibitory threshold	Effects on AD	Mitigation strategies	References
Volatile fatty acids (VFAs), acetate, propionate (product of overload/ imbalance)	Acetate: about 2000 mg/ L may suppress methanogens; Propionate: around 900 mg/ L is strongly inhibitory. Total VFA above about 3000 mg/ L often correlates with pH < 6.5.	Buffer exhaustion → pH drop; direct inhibition of methanogenesis (propionate is particularly problematic as methanogens do not use it directly). Lower CH_4% and risk of souring.	Increase buffering (e.g., $NaHCO_3$; NH_4OH where safe) to keep pH ~ 7; reduce/stop feeding (cut OLR) until VFAs are consumed; ensure adequate mixing; recycle effluent for dilution/inoculum; in two-stage AD, purge VFA in stage 1; for chronic propionate, consider bioaugmentation with propionate-oxidising consortia.	Wang et al. (2009), Banks and Heaven (2013), Ometto et al. (2019), Town and Dumonceaux (2016)
Ammonia (NH_3/ NH_4^+), high-protein wastes (manure, meat, recirculated digestate)	Free NH_3-N: often inhibitory around 150–200 mg/ L (mesophilic, pH ~ 7). Roughly corresponds to TAN ~3–4 g/ L at 35°C, pH 7. In thermophilic or pH > 7.5, lower TAN can inhibit (higher NH_3 fraction). Acclimated systems may tolerate TAN ~ 5–7 g/ L via SAO shift.	NH_3 diffuses into cells → proton imbalance; inhibits methanogenic enzymes; acetate and H_2 accumulate; community shifts towards SAO + hydrogenotrophic methanogenesis (acetoclast decline, Methanoculleus-type increase).	Co-digest to raise C/N; dilute with water/low-N feed; ammonia stripping (air/ vacuum at elevated pH/T; recover as ammonium salt); hold pH 6.8–7.2 and prefer mesophilic T to minimise NH_3 fraction; lengthen HRT for adaptation; bioaugment with ammonia-tolerant SAO + hydrogenotrophic consortia.	Schnürer and Nordberg (2008), Banks and Heaven (2013), Ometto et al. (2019), Weng et al. (2024), Town and Dumonceaux (2016)

Sulphide (H_2S)/ sulphate (SO_4^{2-}), sulphate-bearing feeds, S-rich proteins	Dissolved H_2S: about 50 mg/ L onset of inhibition; about 200 mg/ L strong inhibition. Gas phase: target below about 1000 ppm (equipment/ corrosion). COD/SO_4^{2-} ratio: below about 10, SRB tend to outcompete methanogens.	H_2S is toxic (enzyme inhibition; precipitates essential metals like Co/Ni); SRB competition diverts COD to sulphide and generates corrosive H_2S.	Dose Fe^{2+}/Fe^{3+} to precipitate sulphide as FeS; maintain high COD/SO_4^{2-} ratio; limit sulphate in feed; consider two-stage operation (sulphate reduction then methanogenesis); polish biogas (iron oxide, activated carbon, biofilter); carefully controlled microaeration in some units.	Banks and Heaven (2013), Isa, Grusenmeyer and Verstraete (1986), Qiang, Li and Pei (2018), Ometto et al. (2019)
Long-chain fatty acids (LCFAs), FOG, lipid-rich wastes (dairy, slaughter-house)	Total LCFA above about 1–2 g/ L can coat biomass and inhibit; > ~30% of influent COD as lipids is often problematic without adaptation.	LCFAs adsorb to cells/ granules, block transport/ enzymes; foaming, sludge flotation, granule disintegration; β-oxidation intermediates accumulate; CH_4 drops as acids are not fully converted.	Co-digest with carbohydrate/ fibre-rich substrates to dilute/adsorb LCFAs; limit lipid loading and feed slowly; add adsorbents (biochar, activated carbon) to buffer LCFA concentration and provide attachment sites; maintain mixing; pre-hydrolyse fats or use an acidogenic pre-stage; use acclimated inoculum enriched in LCFA-oxidising consortia (e.g., Syntrophomonas-type).	Pavi et al. (2017), Banks and Heaven (2013), Atelge et al. (2020), Shrestha et al. (2023)

(*Continued*)

Table 2.2 (Continued)

Inhibitor (source)	Inhibitory threshold	Effects on AD	Mitigation strategies	References
Heavy metals (Zn, Cu, Cr, etc.), industrial wastewater, feed additives, contamination	Soluble Cu^{2+}/Zn^{2+} above about 10–20 mg/ L can inhibit methanogens; Cr^{6+} inhibitory at much lower levels (about 1–5 mg/ L). Bioavailability (dissolved fraction) is key; many metals precipitate in AD.	Enzyme inactivation and microbial toxicity; methanogens often most sensitive. Sulfide/hydroxide precipitation and biosorption reduce bioavailability in situ.	Promote precipitation/ sequestration: encourage metal sulphide formation (endogenous sulphide or dosing iron while maintaining reducing conditions); upstream pretreatment of industrial streams; cautious use of chelants; add competing/ precipitating ions (e.g., phosphate) where appropriate; increase inoculum-to-feed; in persistent cases, select metal-tolerant consortia.	Banks and Heaven (2013), Ometto et al. (2019)

Table 2.3 Typical operating parameter ranges for stable anaerobic digestion (single-stage, CSTR-type). Values are approximate optima; different designs or special conditions may allow departures.

Parameter	Typical Range/Optimum	Notes (effects, key factors)	References
Temperature	Mesophilic: 30–40°C (optimum ~ 37°C); Thermophilic: 50–60°C (optimum ~ 55°C)	Psychrophilic (< 25°C) greatly reduces rates. Thermophilic gives higher kinetics/throughput but needs heating and is less tolerant to shocks; avoid fluctuations > ±1–2°C.	Banks and Heaven (2013), Ometto et al. (2019), Atelge et al. (2020)
pH	Overall pH 6.8–7.5 (optimum ~ 7.0)	Methanogens inhibited if pH < 6.5 or > 8.0. Unchecked acidogenesis trends to pH ~5–6, maintain alkalinity. Two-phase systems allow lower pH in the acidogenic stage and ~7 in the methanogenic stage.	Banks and Heaven (2013), Ometto et al. (2019), Wang et al. (2009)
Organic Loading Rate (OLR)	CSTR (wet): 1–4 kg VS m^{-3} d^{-1}; occasionally up to ~6 with readily biodegradable soluble feeds and robust buffering	OLR depends on biodegradability and mixing; too high → VFA accumulation/souring. Higher OLRs (> 10 kg VS m^{-3} d^{-1}) are typical of high-rate reactors (e.g., UASB/EGSB), not CSTR.	Banks and Heaven (2013), Ometto et al. (2019), Pavi et al. (2017), Pastor-Poquet et al. (2019)
Hydraulic Retention Time (HRT)	Mesophilic: 15–30 d; Thermophilic: often 12–20 d for readily degradable slurries, 20–40 d used in practice for stability; 50–80 d if high solids or inhibition present	HRT must accommodate slow-growing methanogens. Biomass retention (granules/media) raises SRT and permits lower HRT.	Banks and Heaven (2013), Ometto et al. (2019), Pastor-Poquet et al. (2019)
Total Ammonia Nitrogen (TAN)	< 2 g NH_4-N L^{-1} (little/no inhibition); 2–5 g L^{-1} (moderate, controllable); > 6 g L^{-1} (strong inhibition likely)	Inhibition mainly from free NH_3, which increases with pH and temperature. At pH ~7 and 35 °C: 2 g TAN L^{-1} ~ 10–20 mg NH_3-N L^{-1}; 5 g TAN L^{-1} ~ 30–50 mg NH_3-N L^{-1}. Inhibitory ranges often start around ≥ 80–150 mg NH_3-N L^{-1}; acclimation raises tolerance.	Schnürer and Nordberg (2008), Weng et al. (2024), Ometto et al. (2019)
Free ammonia (NH_3-N)	Keep typically < 80–150 mg L^{-1} (onset of inhibition); acclimated consortia may tolerate ~200–300 mg L^{-1}	Direct inhibitor of methanogens; rises with pH and temperature, control via pH ~7.0–7.2, temperature, dilution/co-digestion, or stripping.	Schnürer and Nordberg (2008), Weng et al. (2024), Ometto et al. (2019)

(Continued)

Table 2.3 (Continued)

Parameter	Typical Range/Optimum	Notes (effects, key factors)	References
Alkalinity (as $CaCO_3$)	2500–5000 mg L^{-1} for pH stability	Provides buffering against VFA. In practice maintain VFA/alkalinity ratio < 0.3–0.4; below about 1500 mg L^{-1} alkalinity increases risk of rapid pH drop during acid pulses.	Banks and Heaven (2013), Ometto et al. (2019)
Volatile Fatty Acids (VFA)	Steady-state total VFA typically ≤ 0.5–1.5 g L^{-1} (as acetic)	Rising VFA and propionate accumulation indicate overload or inhibition; manage by lowering OLR, increasing alkalinity, improving mixing/trace metals. VFA/alkalinity < 0.3–0.4 is a useful operational target.	Wang et al. (2009), Banks and Heaven (2013), Pastor-Poquet et al. (2019)
Sulphide/H_2S	Dissolved H_2S: ~50 mg L^{-1} onset of inhibition; ~200 mg L^{-1} strong inhibition. Gas phase target: below about 1000 ppm (equipment/corrosion).	Sulphate-reducing bacteria (SRB) generate sulphide and can outcompete methanogens; manage via iron salts/air dosing/stripping and substrate sulphate control.	Banks and Heaven (2013), Ometto et al. (2019), Isa et al. (1986)
Sulphate competition (COD/SO_4^{2-})	Below about 10 (COD/SO_4^{2-}), SRB tend to outcompete methanogens	Low COD/SO_4^{2-} drives electrons to sulphate reduction, lowering methane; adjust by co-substrates with higher COD or sulphate removal.	Banks and Heaven (2013), Isa et al. (1986); Qiang et al. (2018)
Mixing	Continuous or intermittent gentle mixing (CSTRs typically mixed most of the time)	Prevents stratification/scum and ensures microbe-substrate contact. Over-mixing may disrupt flocs/syntrophy. High-solids systems often use slower or discontinuous mixing to preserve structure.	Banks and Heaven (2013), Ometto et al. (2019), Pastor-Poquet et al. (2019)
Total solids (TS)	Wet CSTR feed: ~5%–12% TS; high-solids ("dry") batch AD: ~20%–40% TS	TS drives viscosity/mixing energy; high TS often paired with batch, plug-flow or percolate systems rather than fully mixed CSTRs.	Banks and Heaven (2013), Pastor-Poquet et al. (2019)
Nutrients/Trace metals	Overall C/N ~20–30:1; essential trace elements (Co, Ni, Fe, Se, Mo) at μg L^{-1}-low mg L^{-1}	Off-optimal C/N causes ammonia toxicity (too low C/N) or nutrient limitation (too high C/N); co-digestion adjusts C/N. Trace-metal shortages (e.g., Co, Ni for methanogenic cofactors) can limit gas; targeted supplementation can restore activity.	Banks and Heaven (2013), Li et al. (2022), Xu et al. (2018)

2.3.3 Substrate factors - composition and biodegradability

Carbohydrates (polysaccharides) typically yield around 0.35–0.45 m^3 CH_4/kg VS (glucose theoretical ~0.373 m^3/kg) (Banks and Heaven, 2013; Ometto et al., 2019). They ferment quickly (especially simple sugars and starch); hence carbohydrate-rich wastes (e.g., bakery or fruit waste) can trigger an "acid crash" if buffering is insufficient, blending with higher-N, well-buffered substrates (e.g., sludge or manure) mitigates this (Pavi et al., 2017; Zheng et al., 2022; Wang et al., 2009). Lignocellulosic carbohydrates require hydrolysis as a slow step; once hydrolysed, monomers ferment readily. Particle size matters: smaller particles offer more surface area, grinding fibrous feeds to <1 cm (ideally a few mm) improves conversion; very fine milling can be energetically uneconomic unless applied to recalcitrant feeds (Lindmark et al., 2012; Amnuaycheewa et al., 2016; Boulamanti et al., 2013).

Contaminants and toxins in feed. Besides core composition, toxicants reduce yields. Potential inhibitors include heavy metals, antibiotics, disinfectants, and pesticides; low levels of certain metals act as micronutrients (Co, Ni, Fe), but excess becomes inhibitory (Banks and Heaven, 2013; Atelge et al., 2020; Li et al., 2022). In manures, Cu/Zn can accumulate; many dissolved metals precipitate as sulphides in AD, which naturally mitigates toxicity (Isa, Grusenmeyer, and Verstraete, 1986; Ometto et al., 2019). Antibiotic residues in manures may transiently suppress sub-populations, but diverse AD consortia often attenuate moderate inhibitory inputs. Large shock loads of sanitisers or solvents can still crash digesters, so new waste streams should be screened and BMP assays performed when necessary (Atelge et al., 2020; Banks and Heaven, 2013). Sorbents such as biochar can help manage inhibitory organics/LCFAs in high-strength feeds (Kassaye et al., 2025).

Volatile fatty acids: Moderate levels (<~1000 mg L^{-1} total) are normal intermediates; high VFA causes pH drop and further methanogenesis inhibition. Propionate is often most problematic; concentrations around ~0.9 g L^{-1} have been associated with marked yield declines, whereas acetate is tolerated at higher concentrations if pH is controlled (Wang et al., 2009; Ometto et al., 2019; Banks and Heaven, 2013; Weng et al., 2024). A high propionate/acetate ratio is a recognised early warning of souring (Banks and Heaven, 2013; Ometto et al., 2019). Operators respond by cutting OLR, adding alkalinity, improving mixing, and allowing methanogens to catch up (Banks and Heaven, 2013; Wang et al., 2009).

Ammonia: Protein-rich feeds raise TAN; free NH_3 (favoured by high pH/temperature) selectively inhibits acetoclastic methanogens and can shift pathways towards SAO + hydrogenotrophic methanogenesis (Schnürer and Nordberg, 2008; Weng et al., 2024; Banks and Heaven, 2013). Control includes co-digestion/dilution, pH management (~7.0–7.2), and, in high-N cases, ammonia removal (Banks and Heaven, 2013).

Sulphide/sulphate: In sulphate-bearing streams, SRB compete with methanogens and generate H_2S; maintaining a high COD/SO_4^{2-} and precipitating sulphide with iron salts are standard countermeasures (Isa, Grusenmeyer, and Verstraete, 1986; Qiang, Li, and Pei, 2018; Chen et al., 2019).

LCFAs: Lipid-rich feeds have high potential but LCFAs can adsorb to biomass and inhibit; staged/gradual feeding and co-digestion with carbohydrate-rich substrates, sometimes with sorbents (e.g., biochar), improve stability (Pavi et al., 2017; Atelge et al., 2020; Kassaye et al., 2025).

Ammonia (NH_3/NH_4^+): Ammonia at high free concentrations is an inhibitor, affecting methanogens by diffusing into cells and causing proton imbalance and potassium deficiency; cellulolytic bacteria may also be inhibited but typically to a lesser extent (Schnürer and Nordberg, 2008; Weng et al., 2024). When TAN is high, maintaining a slightly lower pH (~6.8–7.2) and mesophilic temperature helps keep free NH_3 below toxic levels (Banks and Heaven, 2013; Schnürer and Nordberg, 2008). If ammonia is chronically high (e.g., ≥5 g/L TAN) because the feed is inherently N-rich, options include dilution with water or low-N co-substrate, ammonia stripping (air or vacuum at elevated pH/temperature with acid capture to recover ammonium salts), or fostering SAO by community adaptation/bioaugmentation (Banks and Heaven, 2013; Ometto et al., 2019; Schnürer and Nordberg, 2008). Bioaugmentation with acetate-oxidising/methanogenic enrichments has been shown to relieve VFA accumulation in stalled digesters and can support operation when acetoclastic routes are impaired (Town and Dumonceaux, 2016; Weng et al., 2024). Over time, communities can acclimate: long-term high-ammonia operation often selects syntrophic acetate oxidisers working with hydrogenotrophic methanogens (frequently *Methanoculleus*), enabling stable, albeit slower, conversion (Schnürer and Nordberg, 2008; Weng et al., 2024).

Sulphide (H_2S) and sulphates: Sulphate in feed promotes activity of sulphate-reducing bacteria (SRB), which compete with methanogens and generate sulphide that is toxic to many enzymes and also corrosive; management is essential for sulphur-rich wastes (Isa, Grusenmeyer and Verstraete, 1986; Banks and Heaven, 2013). The common approach is dosing iron salts (e.g., $FeCl_2$) to precipitate sulphide as FeS, thereby reducing dissolved sulphide and lowering H_2S in biogas; keeping H_2S low protects equipment and microbial activity (Banks and Heaven, 2013; Qiang, Li, and Pei, 2018; Ometto et al., 2019). Carefully controlled microaeration of the gas headspace is also used in some units to biologically oxidise a portion of H_2S to elemental sulphur (Banks and Heaven, 2013). For sulphate-rich streams, two-stage AD can be advantageous: initial sulphide management followed by a methanogenic stage with reduced sulphate load (Banks and Heaven, 2013; Ometto et al., 2019). At moderate sulphate, coexistence of SRB and methanogens is possible and can influence intermediates via H_2 control; increasing sulphate shifts community responses and can suppress methanogenesis if COD/SO_4^{2-} is too low (Chen et al., 2019). Maintaining a high COD/SO_4^{2-} ratio (commonly >10) and precipitating sulphide with Fe^{2+} are practical rules of thumb to favor methanogenesis (Isa, Grusenmeyer, and Verstraete, 1986; Qiang, Li, and Pei, 2018).

Long-chain fatty acids. LCFAs (oleic, palmitic, etc.) can form a film on microbial aggregates and are slowly degraded by β-oxidising bacteria (which themselves are often inhibited at high LCFA) (Banks and Heaven, 2013; Pavi et al., 2017; Atelge et al., 2020). If lipids constitute a large fraction of the feedstock, digesters often exhibit an acclimation period during which LCFA-degraders increase; a sudden dump of fats can cause immediate upset (floating grease layers, foaming, low gas via methanogen inhibition) (Banks and Heaven, 2013; Pavi et al., 2017). To mitigate LCFA issues: (1) avoid overloading, keep the lipid share moderate (~10%–30% of VS/COD) unless specially adapted (Banks and Heaven, 2013; Pavi et al., 2017; Atelge et al., 2020); (2) co-feed fibrous material so fibres adsorb LCFAs and provide biofilm surface (Banks and Heaven, 2013); (3) add biochar/activated carbon to adsorb hydrophobic LCFAs and provide attachment sites, biochar additions have improved methane yield and stability in high-strength systems (Atelge et al., 2020; Kassaye et al., 2025); (4) maintain mixing to avoid oil stratification; a short pre-hydrolysis tank for very fatty wastes (with inoculum) can smooth acid formation before feeding the main digester (Banks and Heaven, 2013; Ometto et al., 2019).

Operational factors—OLR, HRT, temperature, mixing. The way the digester is run has a large impact on yields. OLR: a low OLR under-utilises capacity but is very stable; very high OLR risks VFA accumulation beyond methanogenic capacity and lowers gas yield. There is typically an optimal OLR that maximises volumetric production without instability (Banks and Heaven, 2013; Ometto et al., 2019; Pavi et al., 2017). For example, small-scale digesters showed a sweet spot around ~2–3 kg VS/m^3/d; beyond this, gas per added VS declined and variability increased (Naik et al., 2014). Feeding regime matters: continuous or semi-continuous feeding evens metabolic load, whereas large periodic pulses can cause transient VFA spikes (Banks and Heaven, 2013; Ometto et al., 2019; Pavi et al., 2017). If an overload is detected (rising VFAs, dropping alkalinity), operators cut back OLR (skip/dilute feeds) to let methanogens catch up and recover pH (Wang et al., 2009; Banks and Heaven, 2013).

Hydraulic/solids retention time (HRT/SRT): This determines how long the substrate and microbes stay in the reactor. For CSTRs digesting slowly biodegradable stuff (like manure with straw), an HRT of at least ~20 days is needed to reach reasonable VS reductions (Banks and Heaven, 2013; Ometto et al., 2019; Pastor-Poquet et al., 2019). Shorter HRT leads to washout of slow-growing methanogens and undegraded solids leaving, thus lower overall methane extraction from feed (Banks and Heaven, 2013; Ometto et al., 2019). Many high-rate systems achieve high throughput by decoupling SRT from HRT—for example, UASB reactors retain biomass granules so that methanogens are not washed out even if HRT is only a day or two (used for liquid wastes like ethanol distillery waste) (Ometto et al., 2019; Zhu et al., 2018). But for solid wastes in complete mix digesters, HRT equals SRT (Ometto et al., 2019). A general performance indicator is VS destruction percentage; if HRT is too short, VS destruction will be low (lots of energy left in digestate). For example, a food waste digester might destroy 80% of VS at 30 d HRT, but only 60% at 15 d HRT—meaning less methane extracted (Banks and Heaven, 2013; Pavi et al., 2017). Temperature ties in: thermophilic digesters can often use ~20% shorter HRT for similar VS destruction, but in practice they might keep HRT similar to mesophilic to get even more VS destruction (since thermophilic can achieve slightly higher extent due to faster hydrolysis) (Atelge et al., 2020; Ometto et al., 2019). For high-solids batch digesters (garage type), retention time is the batch duration, often 30–60 days with percolate recirculation to evenly distribute inoculum and nutrients (Pastor-Poquet et al., 2019; Atelge et al., 2020). If a digester faces chronic VFA issues, sometimes increasing HRT (slowing down throughput) can help because it effectively lowers OLR and gives more time for consumption of intermediates (Wang et al., 2009; Banks and Heaven, 2013).

Temperature stability is critical in both mesophilic and thermophilic AD. Rapid temperature shifts can inhibit sensitive microbial groups, particularly methanogens. Therefore, stable heating and insulation are essential for consistent yields (Banks and Heaven, 2013; Ometto et al., 2019). Seasonal temperature drops are a known cause of digester upsets in systems without proper heating (e.g., small-scale digesters) (Naik et al., 2014). Each temperature range has pros and cons: mesophilic digesters are more stable and often have better effluent quality, whereas thermophilic units produce gas faster (approximate Q_{10} effect) and achieve pathogen kill but can be more prone to foaming/imbalance and require more heating energy input (Atelge et al., 2020; Ometto et al., 2019; Banks and Heaven, 2013). Some facilities opt for a two-stage temperature: for example, first stage thermophilic for rapid breakdown and pathogen kill, second stage mesophilic for polishing and stability (Ometto et al., 2019). For example, alternating-temperature operation increased methane yield for corn straw digestion by leveraging distinct microbial communities at different phases (He et al., 2023). This temperature-phased anaerobic digestion (TPAD) approach marries advantages of both ranges (Ometto et al., 2019; He et al., 2023).

Mixing and mass transfer. Proper mixing ensures that substrates, microbes and by-products (e.g., H_2, VFAs) are evenly distributed; stratification can cause local build-ups of acids or ammonia that inhibit nearby microbes. In high-solids systems, mixing prevents dead zones and dry pockets, but the mixing intensity must be carefully controlled. Excessive mixing can shear microbial flocs and disturb syntrophic pairings, as many methanogens occur in aggregates or granules with partner bacteria that are disrupted by extreme turbulence (Banks and Heaven, 2013; Ometto et al., 2019; Weng et al., 2024). To minimise shear, CSTRs often employ continuous gentle mixing or intermittent mixing regimes (e.g., short mixing periods each hour). Some studies report stability benefits of intermittent mixing in manure digesters, likely via moderating acidification transients, but practices are site-specific (Banks and Heaven, 2013; Pastor-Poquet et al., 2019). Mixing also helps release biogas bubbles trapped in sludge (reducing floating layers) and limits grit sedimentation that would otherwise reduce active volume; energy demand for mixing is typically lower than for heating, so over-mixing wastes energy without benefit (Banks and Heaven, 2013; Ometto et al., 2019).

Start-up and inoculation. Start-up strategy strongly affects how quickly design performance is reached. Using an active inoculum from a similar feedstock shortens the lag and avoids a prolonged acidogenic phase with low methane (Banks and Heaven, 2013; Ometto et al., 2019). In BMP/start-up contexts, a substrate-to-inoculum (S/I) ratio around 0.5–1.0 (VS basis) is often recommended to prevent initial VFA spikes and support stable methanogenesis; too high S/I (excess substrate) increases the risk of souring (Shahbaz et al., 2019; Habib, Torii, and Mol, 2024). For continuous systems, operators typically begin at a low OLR and ramp gradually over weeks to allow acclimation; rushing start-up commonly leads to VFA accumulation and recovery downtime (Banks and Heaven, 2013; Pavi et al., 2017).

2.3.4 Mitigation of inhibitors and process upsets

To handle VFA accumulation: increase mixing, add alkalinity (e.g., bicarbonate; or ammonia where safe), and reduce or pause feeding; effluent recycle can dilute acids and seed active biomass. Once VFAs decline and pH normalises, feeding can resume. Many facilities track VFA/alkalinity (and, where available, on-line VFA) to detect acidification early (Banks and Heaven, 2013; Ometto et al., 2019; Wang et al., 2009; Pavi et al., 2017). Bioaugmentation with targeted consortia has also relieved acetate accumulation in stalled digesters (Town and Dumonceaux, 2016).

To counter ammonia inhibition: in addition to dilution or stripping, gradually acclimating the microbial community can select ammonia-tolerant communities that rely more on SAO coupled to hydrogenotrophic methanogenesis; such high-ammonia operation has been documented under mesophilic conditions (Schnürer and Nordberg, 2008; Weng et al., 2024). Bioaugmentation with adapted communities can aid recovery after ammonia shocks (Town and Dumonceaux, 2016). As a simpler chemical measure, slightly lowering the pH (e.g., ~7.5→7.1–7.2) reduces free NH_3 at a given TAN (Banks and Heaven, 2013; Schnürer and Nordberg, 2008).

To mitigate sulphide/H_2S: dose iron salts (e.g., $FeCl_2$) to precipitate sulphide as FeS and suppress dissolved/biogas H_2S; controlled microaeration and external polishing (e.g., iron oxide or activated carbon) are also used. Keeping a high COD/SO_4^{2-} ratio and removing sulphide favor methanogenesis over SRB competition (Isa, Grusenmeyer and Verstraete, 1986; Qiang, Li and Pei, 2018; Chen et al., 2019; Banks and Heaven, 2013; Ometto et al., 2019).

To overcome LCFA inhibition: co-digest with fibrous substrates to sorb LCFAs and provide attachment surfaces; add sorbents such as biochar/activated carbon to buffer LCFA concentrations; feed fats slowly and maintain effective mixing (Banks and Heaven, 2013; Pavi et al., 2017; Atelge et al., 2020; Kassaye et al., 2025; Ometto et al., 2019).

Trace element supplementation. If a digester feed is micronutrient-poor, targeted dosing can markedly improve gas production and stability by restoring key methanogenic enzymes. In sewage-sludge AD, adding Fe^0 together with Co and Ni altered the microbial community and increased methane yield; Ni and Co supported methanogens (Ni in cofactor F_{430}; Co in corrinoids), while Fe^0 served as an effective supplement (Li et al., 2022). Numerous plants use commercial trace-element mixes (e.g., Co, Ni, Fe, Se, Mo) to stabilise mono-substrate operations such as food-waste or energy-crop digestion when deficiencies are diagnosed (Banks and Heaven, 2013; Li et al., 2022).

Process control and monitoring. Modern AD facilities routinely track pH, alkalinity, VFAs, biogas flow and CH_4/CO_2 (and sometimes H_2). Maintaining near-neutral pH via the carbonate/bicarbonate system is central; operators often monitor VFA/alkalinity and the propionate/acetate ratio as early-warning indicators of overload or ammonia stress (Banks and Heaven, 2013; Ometto et al., 2019; Wang et al., 2009). Gas composition trends (e.g., falling CH_4, rising CO_2, or detectable H_2) flag syntrophic imbalance or methanogen inhibition (Banks and Heaven, 2013; Ometto et al., 2019; Weng et al., 2024).

Process staging. Splitting AD into an acidification stage and a methanogenic stage allows each to be optimised (lower pH/high OLR upstream; controlled pH/OLR downstream). In thermophilic systems treating SS-OFMSW, two-stage operation delivered a more stable start-up than single-stage digestion (Ghanimeh et al., 2020; Ometto et al., 2019).

High yields are achieved by balancing the microbial ecosystem with the feeding regime: substrate composition sets theoretical potential, but operational control and timely mitigation of inhibitors determine realised conversion. Keeping conditions within favourable ranges and addressing inhibitory build-up proactively enables consistent recovery of a large fraction of BMP (Banks and Heaven, 2013; Pavi et al., 2017; Ometto et al., 2019). Emerging tools, ML models to predict performance or optimise pretreatment/feeding, and conductive/carbon additives that support interspecies electron transfer, are further improving robustness (Song et al., 2024; Adeleke et al., 2025; James et al., 2025; Kassaye et al., 2025). At the energy-systems level, AD facilities can operate as 'programmable' renewables, dispatchable within constraints and readily integrated with storage and demand response, thus adding flexibility value to the grid (Nogueira Nakashima and de Oliveira, 2024). Among biological options, controlled enzymatic or microbial pretreatments can reduce lignocellulose recalcitrance and improve hydrolysis efficiency (Amnuaycheewa et al., 2016; Mirmohamadsadeghi et al., 2021; Tabatabaei et al., 2020).

Every AD system is unique, and the specific inhibitory thresholds can differ based on acclimation and interactions (Schnürer and Nordberg, 2008; Weng et al., 2024; Banks and Heaven, 2013). For instance, moderate ammonia may be more tolerable if

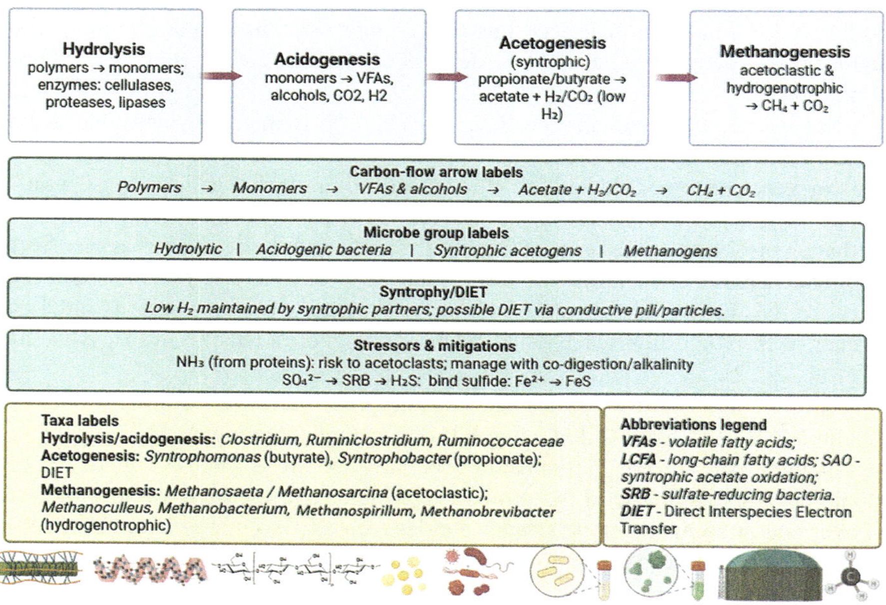

Figure 2.1 Schematic of the anaerobic digestion process.

VFAs are low, but combined high ammonia + high VFA is particularly stressful (Wang et al., 2009; Schnürer and Nordberg, 2008; Weng et al., 2024). A digester might handle ~3 g/L of sulphate if COD is abundant, but not if COD is low, i.e., performance depends strongly on maintaining a high COD/SO_4^{2-} ratio (Isa, Grusenmeyer, and Verstraete, 1986; Qiang, Li, and Pei, 2018; Chen et al., 2019). In practice, operators consider the combined loading of stressors (Banks and Heaven, 2013; Ometto et al., 2019). Often, solving one issue (e.g., removing H_2S by iron addition) simultaneously improves other aspects (e.g., reduces sulphide toxicity, improves overall microbial functionality and trace-metal bioavailability) (Banks and Heaven, 2013; Isa, Grusenmeyer, and Verstraete, 1986; Ometto et al., 2019; Li et al., 2022). As shown, many mitigations revolve around control of environment (pH, temperature, mixing) and feed management (co-digestion ratios, phased feeding) (Banks and Heaven, 2013; Ometto et al., 2019; Pavi et al., 2017; Pastor-Poquet et al., 2019). Applying these strategies allows modern digesters to achieve high conversion efficiencies and remain stable even with "difficult" feedstocks, thereby maximising renewable energy production in the form of biogas (Atelge et al., 2020; Pavi et al., 2017; Ometto et al., 2019) (Fig. 2.1).

AI disclosure

During the preparation of this work the author(s) used DEEPL, ChatGPT, Grammarly, Mendeley in order to Translation, language correction, bibliography management. After using this tool/service, the author(s) reviewed and edited the content as needed and take(s) full responsibility for the content of the publication.

References

Abdelrahman, A.M., Abdelrazec, N., AlSayed, A., Kakar, F.L., Muller, C., Bell, K.Y., et al., 2025. From process fundamentals to engineering perspectives: a technical review on in-situ biogas upgrading via the hydrogenotrophic methanogenesis pathway. Biomass and Bioenergy 200. Available at https://doi.org/10.1016/j.biombioe.2025.108043.

Adeleke, O., Olatunji, K.O., Madyira, D.M., Jen, T.C., 2025. Application of multimodal machine learning-based analysis for the biomethane yields of NaOH-pretreated biomass. Scientific Reports 15, 24372. Available at https://doi.org/10.1038/s41598-025-09527-5.

Ahmad, R.M., Javied, S., Aslam, A., Alamri, S., Zaman, Q.uz, Hassan, A., et al., 2024. Optimizing biogas production and digestive stability through waste co-digestion. Sustainability 16 (7), 3045. Available at https://doi.org/10.3390/su16073045.

Alsouleman, K., 2019. Effect of increasing amounts of ammonium nitrogen induced by consecutive mixture of poultry manure and cattle slurry on the microbial community during thermophilic anaerobic digestion. Journal of Microbiology and Biotechnology 29 (12), 1993–2005. Available at https://doi.org/10.4014/jmb.1909.09023.

Amnuaycheewa, P., Hengaroonprasan, R., Rattanaporn, K., Kirdponpattara, S., Cheenkachorn, K., Sriariyanun, M., 2016. Enhancing enzymatic hydrolysis and biogas production from rice straw by pretreatment with organic acids. Industrial Crops and Products 87, 247–254. Available at https://doi.org/10.1016/j.indcrop.2016.04.069.

An, M.M., Lin, X.H., Liang, R.N., Zhao, G.Z., 2025. Effect of carbon source in methanogenic communities regulation and methane production capacity. The Editorial Office of Biotechnology Bulletin, China Biotechnology Bulletin 41 (6), 355–366. Available at https://doi.org/10.13560/j.cnki.biotech.bull.1985.2024-1180.

Atelge, M.R., Krisa, D., Kumar, G., Eskicioglu, C., Nguyen, D.D., Chang, S.W., et al., 2020. Biogas production from organic waste: recent progress and perspectives. Waste and Biomass Valorization 11 (3), 1019–1040. Available at https://doi.org/10.1007/s12649-018-00546-0.

Austen, N., Theaker, H., Tenci, N.A., Beesley, A., Thompson, I.P., 2025. A combination of mechanical, chemical, and thermal pretreatments of agricultural feedstocks enhances biomethane yields in advanced anaerobic digestion. Bioenergy Research 18 (1). Available at https://doi.org/10.1007/s12155-025-10846-8.

Banks, C.J., Heaven, S., 2013. 6 - Optimisation of biogas yields from anaerobic digestion by feedstock type. In: Wellinger, A., Murphy, J., Baxter, D. (Eds.), The Biogas Handbook: Science, Production and Applications. Woodhead Publishing, pp. 131–165. Available at https://doi.org/10.1533/9780857097415.1.131.

Basak, B., Ahn, Y., Kumar, R., Hwang, J.H., Kim, K.H., Jeon, B.H., 2022. Lignocellulolytic microbiomes for augmenting lignocellulose degradation in anaerobic digestion. Trends in Microbiology 30 (1), 6–9. Available at https://doi.org/10.1016/j.tim.2021.09.006.

Basak, B., Kumar, R., Tanpure, R.S., Mishra, A., Tripathy, S.K., Chakrabortty, S., et al., 2025. Roles of engineered lignocellulolytic microbiota in bioaugmenting lignocellulose biomethanation. Renewable and Sustainable Energy Reviews 7. Available at https://doi.org/10.1016/j.rser.2024.114913.

Boulamanti, A.K., Donida Maglio, S., Giuntoli, J., Agostini, A., 2013. Influence of different practices on biogas sustainability. Biomass and Bioenergy 53, 149–161. Available at https://doi.org/10.1016/j.biombioe.2013.02.020.

Çetinkaya, A.Y., Yetilmezsoy, K., 2019. Evaluation of anaerobic biodegradability potential and comparative kinetics of different agro-industrial substrates using a new hybrid computational coding scheme. Journal of Cleaner Production 238, 117921. Available at https://doi.org/10.1016/j.jclepro.2019.117921.

Chen, J., Wade, M.J., Dolfing, J., Soyer, O.S., 2019. Increasing sulfate levels show a differential impact on synthetic communities comprising different methanogens and a sulfate reducer. Journal of the Royal Society Interface 16 (154), 20190129. Available at https://doi.org/10.1098/rsif.2019.0129.

Cotabarren, I., Piña, J., 2021. Production of solid fertilizers from anaerobic digestion liquid residue. In: Gorawala, P., Mandhatri, S. (Eds.), Agricultural Research Updates: Volume 37. Nova Science Publishers, Hauppauge, NY, pp. 15–55. Available at https://doi.org/10.52305/ZLVV3716.

Dhavaleswarapu, R.K., Hoysall, C.N., Srinivasaiah, D., 2025. Thermal-alkaline pretreatment of different biomass to understand the interplay of composition of biomass, biogas yields, rates and mass balance. Bioresource Technology Reports 30. Available at https://doi.org/10.1016/j.biteb.2025.102162.

Dobre, P., Nicolae, F., Matei, F., 2014. Main factors affecting biogas production: an overview. Romanian Biotechnological Letters 19 (3), 9283–9296.

Ecem Öner, B., Akyol, Ç., Bozan, M., Ince, O., Aydin, S., Ince, B., 2018. Bioaugmentation with Clostridium thermocellum to enhance the anaerobic biodegradation of lignocellulosic agricultural residues. Bioresource Technology 249, 620–625. Available at https://doi.org/10.1016/j.biortech.2017.10.040.

Finsterwalder T., Energy Y. 2008 11 2008/11 Air and Waste Management Association - 27th Annual International Conference on Thermal Treatment Technologies 2008 747 754 Germany Biogas process: Removing solid contaminations from organic waste streams 2.

Getahun, T., Gebrehiwot, M., Ambelu, A., Van Gerven, T., Van Der Bruggen, B., 2014. The potential of biogas production from municipal solid waste in a tropical climate. Environmental monitoring and assessment 186 (7), 4637–4646. Available at https://doi.org/10.1007/s10661-014-3727-4.

Ghanimeh, S., Al-Sanioura, D., Saikaly, P.E., El-Fadel, M., 2020. Comparison of single-stage and two-stage thermophilic anaerobic digestion of ss-ofmsw during the start-up phase. Waste and Biomass Valorization 11 (12), 6709–6716. Available at https://doi.org/10.1007/s12649-019-00891-8.

Guan, R., Li, X., Wachemo, A.C., Yuan, H., Liu, Y., Zou, D., et al., 2018. Enhancing anaerobic digestion performance and degradation of lignocellulosic components of rice straw by combined biological and chemical pretreatment. Science of the Total Environment 637-638, 9–17. Available at https://doi.org/10.1016/j.scitotenv.2018.04.366.

Habib, S.S., Torii, S., Mol, K.S., 2024. New methodologies for the optimization of operational parameters of bio gas power plants: a review. materials and energy research center. Japan Journal of Renewable Energy and Environment 11 (4), 9–27. Available at https://doi.org/10.30501/jree.2024.444163.1840.

He, H., Wang, Z., Yan, J., Wang, W., Zhu, J., Chen, J., et al., 2023. Enhanced biomethane generation from the anaerobic digestion of wilted corn straw via control in mesophilic and thermophilic temperature intervals. Fuel 349, 128616. Available at https://doi.org/10.1016/j.fuel.2023.128616.

Holl, E., Oskina, A., Baier, U., Lemmer, A., 2023. Optimization of thermodynamic parameters of the biological hydrogen methanation in a trickle-bed reactor for the conditioning of biogas to biomethane. Energies 16 (12), 4720. Available at https://doi.org/10.3390/en16124720.

Hu, Y., Hao, X., Wang, J., Cao, Y., 2016. Enhancing anaerobic digestion of lignocellulosic materials in excess sludge by bioaugmentation and pre-treatment. Waste Management 49, 55–63. Available at https://doi.org/10.1016/j.wasman.2015.12.006.

Isa, Z., Grusenmeyer, S., Verstraete, W., 1986. Sulfate reduction relative to methane production in high-rate anaerobic digestion: technical aspects. Applied and Environmental Microbiology 51 (3), 572–579. Available at https://doi.org/10.1128/aem.51.3.572-579.1986.

James, A., Li, M., Mazarji, M., Li, A., Li, Y., Pan, J., 2025. Coupling electron bifurcation and interspecies electron transfer to mitigate ammonia and acids inhibition. Renewable and Sustainable Energy Reviews 210, 115166. Available at https://doi.org/10.1016/j.rser.2024.115166.

Jensen, M.B., de Jonge, N., Dolriis, M.D., Kragelund, C., Fischer, C.H., Eskesen, M.R., et al., 2021. Cellulolytic and xylanolytic microbial communities associated with lignocellulose-rich wheat straw degradation in anaerobic digestion. Frontiers in Microbiology 12. Available at https://doi.org/10.3389/fmicb.2021.645174.

Kassaye, G., Gabbiye, N., Getahun, E., Akizuki, Si, Bantie, Z., Bhatia, P., et al., 2025. Enhancing biogas production from water hyacinth juice through biochar addition in an upflow anaerobic sludge blanket reactor. Journal of Chemical Technology and Biotechnology 100 (11), 2289–2298. Available at https://doi.org/10.1002/jctb.70039.

Kelif Ibro, M., Ramayya Ancha, V., Beyene Lemma, D., 2024. Biogas Production optimization in the anaerobic codigestion process: a critical review on process parameters modeling and simulation tools. Journal of Chemistry 2024, 4599371. Available at https://doi.org/10.1155/2024/4599371.

Li, C., Lü, F., Peng, W., He, P., Zhang, H., 2024. Efficacy of bioaugmentation with nondomesticated mixed microbial consortia under ammonia inhibition in anaerobic digestion. Bioresource Technology 391, 129954. Available at https://doi.org/10.1016/j.biortech.2023.129954.

Li, J., Lei, Y., Pu, X., Liu, Y., Mei, Z., Tang, Y., 2022. Improving biomethane fermentation through trace elements-driven microbial changes: different effects of Fe0 combined with Co/Ni. Process Biochemistry 121, 197–206. Available at https://doi.org/10.1016/j.procbio.2022.07.004.

Lindmark, J., Leksell, N., Schnürer, A., Thorin, E., 2012. Effects of mechanical pre-treatment on the biogas yield from ley crop silage. Applied Energy 97, 498–502. Available at https://doi.org/10.1016/j.apenergy.2011.12.066.

Lo, F.C., Lin, K.L., Pai, T.Y., Yang, W.G., Tzeng, Y.M., Chiu, H.Y., et al., 2019. Biogas production from most agricultural organic wastes by anaerobic digestion in Taiwan. Environmental Progress and Sustainable Energy 38 (6), e13242. Available at https://doi.org/10.1002/ep.13242.

Martínez, M.E., Restovic, F., Urrego, F., Fuentes, D., Ramos, C., 2025. Evaluation of enzymatic pretreatment strategies to enhance anaerobic digestion of tomato waste. Biofuels, Bioproducts and Biorefining 19 (6), 1677–1689. Available at https://doi.org/10.1002/bbb.2767.

Mirmohamadsadeghi, S., Karimi, K., Azarbaijani, R., Parsa Yeganeh, L., Angelidaki, I., Nizami, A.S., et al., 2021. Pretreatment of lignocelluloses for enhanced biogas production: A review on influencing mechanisms and the importance of microbial diversity. Renewable and Sustainable Energy Reviews 135, 110173. Available at https://doi.org/10.1016/j.rser.2020.110173.

Naik, L., Gebreegziabher, Z., Tumwesige, V., Balana, B.B., Mwirigi, J., Austin, G., 2014. Factors determining the stability and productivity of small scale anaerobic digesters. Biomass and Bioenergy 70, 51–57. Available at https://doi.org/10.1016/j.biombioe.2014.01.055.

Nakashima, R.N., de Oliveira Junior, S., 2021. Trade-offs between productivity, efficiency and costs of biogas plants for agricultural wastes. ECOS 2021 - 34th International Conference on Efficiency, Cost, Optimisation, Simulation and Environmental Impact of Energy Systems. ECOS 2021 Program Organizers, pp. 128–139. Available at https://doi.org/10.52202/062738-0012.

Nizami, A.S., 2011. Anaerobic digestion: processes, products, and applications. In: Caruana, D.J., Olsen, A.E. (Eds.), Potential of Activated Sludge Utilization. Ireland Nova Science Publishers, Inc, Commack, New York, pp. 133–147.

Nogueira Nakashima, R., de Oliveira Junior, S., 2024. Chapter 6 - Biomass: Digestion. Small Scale Power Generation Handbook: Towards Distributed Energy Systems. Elsevier, Denmark, pp. 127–141. Available at https://doi.org/10.1016/B978-0-12-821672-9.00007-1.

Ometto, F., Karlsson, A., Ejlertsson, J., Björn, A.V., Shakeri, S.Y., 2019. 5 - Anaerobic digestion: an engineered biological process. In: Materazzi, M., Ugo Foscolo, P. (Eds.), Substitute Natural Gas from Waste: Technical Assessment and Industrial Applications of Biochemical and Thermochemical Processes. Academic Press, pp. 63–74. Available at https://doi.org/10.1016/B978-0-12-815554-7.00005-2.

Ozbayram, E.G., Kleinsteuber, S., Nikolausz, M., Ince, B., Ince, O., 2017. Effect of bioaugmentation by cellulolytic bacteria enriched from sheep rumen on methane production from wheat straw. Anaerobe 46, 122–130. Available at https://doi.org/10.1016/j.anaerobe.2017.03.013.

Pastor-Poquet, V., Papirio, S., Trably, E., Rintala, J., Escudié, R., Esposito, G., 2019. High-solids anaerobic digestion requires a trade-off between total solids, inoculum-to-substrate ratio and ammonia inhibition. International Journal of Environmental Science and Technology 16 (11), 7011–7024. Available at https://doi.org/10.1007/s13762-019-02264-z.

Pavi, S., Kramer, L.E., Gomes, L.P., Miranda, L.A.S., 2017. Biogas production from co-digestion of organic fraction of municipal solid waste and fruit and vegetable waste. Bioresource Technology 228, 362–367. Available at https://doi.org/10.1016/j.biortech.2017.01.003.

Pulgarín-Muñoz, C.E., Saldarriaga-Molina, J.C., Castro-Valencia, J.C., Correa-Ochoa, M.A., Echeverry-Ruiz, J.D., 2025. Effect of the co-substrate ratio on the anaerobic co-digestion of sewage sludge and the organic fraction of municipal solid waste at pilot scale. International Biodeterioration and Biodegradation 198, 106008. Available at https://doi.org/10.1016/j.ibiod.2025.106008.

Qiang, H., Li, Y.Y., Pei, M.F., 2018. Effect of COD/SO 42- ratio on anaerobic digestion of penicillin bacterial residues. Huanjing Kexue/Environmental Science 39 (7), 3443–3451. Available at https://doi.org/10.13227/j.hjkx.201709199.

Quiñones, T.S., Plöchl, M., Budde, J., Heiermann, M., 2012. Results of batch anaerobic digestion test - effect of enzyme addition. Agricultural Engineering International: CIGR Journal 14 (1), 38–50.

Ramachandran, T., Singh, S., Singh, A.K., 2023. The Analysis of Latest Technologies in Producing Bio Gas. 2023 International Conference on Power Energy, Environment & Intelligent Control (PEEIC), Greater Noida, India. IEEE, pp. 1218–1222. Available at 10.1109/PEEIC59336.2023.10450178.

Rodrigues Oliveira, H., Mendes Anacleto, T., Carraro, G., Abreu, F., Enrich-Prast, A., 2024. A novel approach to estimate methanogenic pathways in biogas reactors via stable carbon isotope analysis. Biomass and Bioenergy 183, 107167. Available at https://doi.org/10.1016/j.biombioe.2024.107167.

Rojas, M.A., Netto, A.O., Zaiat, M., 2008. Specific methanogenic activity in an anaerobic-aerobic reactor applied to the treatment of domestic residual water. Interciencia 33 (4), 284–289.

Santhappan, J.S., Arumuganainar, R., Glivin, G., Jaganathan, V.M., 2024. Chapter 12 - Biogas from organic wastes. In: Sarkar, A., Lassi, U. (Eds.), Processing of Biomass Waste: Technological Upgradation and Advancement. Elsevier Science Ltd., pp. 165–183. Available at https://doi.org/10.1016/B978-0-323-95179-1.00012-8.

Schnürer, A., Nordberg, Å., 2008. Ammonia, a selective agent for methane production by syntrophic acetate oxidation at mesophilic temperature. Water Science and Technology 57 (5), 735–740. Available at https://doi.org/10.2166/wst.2008.097.

Şenol, H., 2026. Improvement of methane yield from hazelnut shells by potassium ferrate and mild ultrasonic pretreatment: kinetics study. Fuel 403, 135989. Available at https://doi.org/10.1016/j.fuel.2025.135989.

Shahbaz, M., Ammar, M., Zou, D., Korai, R.M., Li, X.J., 2019. An insight into the anaerobic co-digestion of municipal solid waste and food waste: influence of co-substrate mixture ratio and substrate to inoculum ratio on biogas production. Applied Biochemistry and

Biotechnology 187 (4), 1356–1370. Available at https://doi.org/10.1007/s12010-018-2891-3.

Shrestha, S., Pandey, R., Aryal, N., Lohani, S.P., 2023. Recent advances in co-digestion conjugates for anaerobic digestion of food waste. Journal of Environmental Management 345. Available at https://doi.org/10.1016/j.jenvman.2023.118785.

Silva, A.F.V., Santos, L.A., Valença, R.B., Porto, T.S., Da Motta Sobrinho, M.A., Gomes, G.J.C., et al., 2019. Cellulase production to obtain biogas from passion fruit (Passiflora edulis) peel waste hydrolysate. Journal of Environmental Chemical Engineering 7 (6), 103510. Available at https://doi.org/10.1016/j.jece.2019.103510.

Singh, M., 2025. 4 - Enzymatic reactions and their impact on bioenergy in anaerobic digestion. In: Sharma, P., Tong, Y.W., Mohapatra, S., Purchase, D., Khuntia, H.K., Singh, S.P. (Eds.), Waste-to-Energy: Sustainable Approaches for Emerging Economies. Elsevier, pp. 67–97. Available at https://doi.org/10.1016/B978-0-443-22356-3.00004-X.

Song, C., Zhang, Z., Wang, X., Hu, X., Chen, C., Liu, G., 2024. Machine learning-aided unveiling the relationship between chemical pretreatment and methane production of lignocellulosic waste. Waste Management 187, 235–243. Available at https://doi.org/10.1016/j.wasman.2024.07.004.

Swamy, N.K., Kotaiah, B., 1996. Digestibility of different organic industrial wastewaters by anaerobic upflow contact filter. Fresenius Environmental Bulletin 5 (7-8), 406–411.

Tabatabaei, M., Aghbashlo, M., Valijanian, E., Kazemi Shariat Panahi, H., Nizami, A.S., Ghanavati, H., et al., 2020. A comprehensive review on recent biological innovations to improve biogas production, Part 1: Upstream strategies. Renewable Energy 146, 1204–1220. Available at https://doi.org/10.1016/j.renene.2019.07.037.

Tang, F., Tian, J., Zhu, N., Lin, Y., Zheng, H., Xu, Z., et al., 2022. Dry anaerobic digestion of ammoniated straw: Performance and microbial characteristics. Bioresource Technology 351, 126952. Available at https://doi.org/10.1016/j.biortech.2022.126952.

Town, J.R., Dumonceaux, T.J., 2016. Laboratory-scale bioaugmentation relieves acetate accumulation and stimulates methane production in stalled anaerobic digesters. Applied Microbiology and Biotechnology 100 (2), 1009–1017. Available at https://doi.org/10.1007/s00253-015-7058-3.

Triolo, J.M., Sommer, S.G., Møller, H.B., Weisbjerg, M.R., Jiang, X.Y., 2011. A new algorithm to characterize biodegradability of biomass during anaerobic digestion: influence of lignin concentration on methane production potential. Bioresource Technology 102 (20), 9395–9402. Available at https://doi.org/10.1016/j.biortech.2011.07.026.

Villalobos Solis, M.I., Chirania, P., Hettich, R.L., 2022. In silico evaluation of a targeted metaproteomics strategy for broad screening of cellulolytic enzyme capacities in anaerobic microbiome bioreactors. Biotechnology for Biofuels and Bioproducts 15 (1), 32. Available at https://doi.org/10.1186/s13068-022-02125-x.

Wang, Y., Zhang, Y., Wang, J., Meng, L., 2009. Effects of volatile fatty acid concentrations on methane yield and methanogenic bacteria. Biomass and Bioenergy 33 (5), 848–853. Available at https://doi.org/10.1016/j.biombioe.2009.01.007.

Weng, N., Singh, A., Ohlsson, J.A., Dolfing, J., Westerholm, M., 2024. Catabolism and interactions of syntrophic propionate- and acetate oxidizing microorganisms under mesophilic, high-ammonia conditions. Frontiers in Microbiology 15, 1389257. Available at https://doi.org/10.3389/fmicb.2024.1389257.

Xu, R., Zhang, K., Liu, P., Khan, A., Xiong, J., Tian, F., et al., 2018. A critical review on the interaction of substrate nutrient balance and microbial community structure and function in anaerobic co-digestion. Bioresource Technology 247, 1119–1127. Available at https://doi.org/10.1016/j.biortech.2017.09.095.

Zheng, Y., Wang, P., Yang, X., Lin, P., Wang, Y., Cheng, M., et al., 2022. Process performance and microbial communities in anaerobic co-digestion of sewage sludge and food waste with a lower range of carbon/nitrogen ratio. Bioenergy Research 15 (3), 1664–1674. Available at https://doi.org/10.1007/s12155-021-10357-2.

Zhong, L., Lin, F., Wo, D., Yang, X., Sun, Y., Feng, X., et al., 2024. The dominant-substrate driven the enhanced performance in co-digestion of Pennisetum hybrid and livestock waste. Journal of Environmental Management 366, 121700. Available at https://doi.org/10.1016/j.jenvman.2024.121700.

Zhu, W., Lai, X., Liu, C., Wu, X., Bai, X., Cai, Y., et al., 2025. Enhancement of biomethane yield from spent mushroom substrate: biological pretreatment with the chlamydospores of Trichoderma viride. Fermentation 11 (3), 152. Available at https://doi.org/10.3390/fermentation11030152.

Zhu, X., Treu, L., Kougias, P.G., Campanaro, S., Angelidaki, I., 2018. Converting mesophilic upflow sludge blanket (UASB) reactors to thermophilic by applying axenic methanogenic culture bioaugmentation. Chemical Engineering Journal 332, 508–516. Available at https://doi.org/10.1016/j.cej.2017.09.113.

Digestate: from waste product to valuable resource

3

3.1 Introduction

Anaerobic digestion of organic matter is a key technology for renewable energy generation and, simultaneously, a widely applied approach to the treatment and valorisation of organic wastes (Jameel et al., 2024). Nevertheless, the process produces a considerable quantity of residual material, referred to as digestate. Digestate is the residual slurry remaining after biogas recovery; it comprises partially degraded organic fractions and water, and it contains a wide range of macro- and micronutrients as well as inorganic impurities and contaminants (Nayak and Ranade, 2025). To date, the predominant management route for digestate has been its direct application to agricultural land as an organic fertiliser (Barłóg et al., 2020; Lamolinara et al., 2022). This practice is aligned with the principles of the circular economy, insofar as nutrients embedded in waste-derived streams are returned to soils, thereby reducing reliance on mineral fertilisers and supporting the closure of biogeochemical nutrient cycles (Dahlin et al., 2015). Numerous studies have shown that digestate can promote plant growth and deliver crop yields comparable to those obtained with synthetic fertilisers, while concurrently providing stabilised organic matter that improves soil structure and enhances soil fertility (Barłóg et al., 2020; Mayerová et al., 2023) (Fig. 3.1).

Despite its evident advantages, the utilisation of digestate is associated with a range of challenges. Digestate is increasingly regarded as a valuable secondary product. However, its environmentally sound and marketable use requires compliance with appropriate quality standards and the resolution of environmental and logistical constraints (Lamolinara et al., 2022; Vilanova Plana et al., 2016). This is largely attributable to the pronounced variability of digestate composition, which is governed by feedstock characteristics and process operating conditions (Czekała et al., 2022), its typically high water content (Tymchuk et al., 2023), substantial concentrations of ammoniacal nitrogen (Drosg et al., 2015), and the potential presence of contaminants such as heavy metals, pathogenic microorganisms, and pharmaceutical residues (Golovko et al., 2022; Wolak et al., 2023).

This chapter presents the composition and key properties of digestate and discusses the principal challenges associated with its management and use. It then outlines the legal framework governing the agricultural utilisation of digestate, with particular emphasis on European Union regulations. The overarching aim is to demonstrate how this by-product of biogas production can be transformed from a problematic residue into a valuable resource in line with the principles of a circular economy.

Sustainable Biogas Production. DOI: https://doi.org/10.1016/B978-0-443-40557-0.00011-1

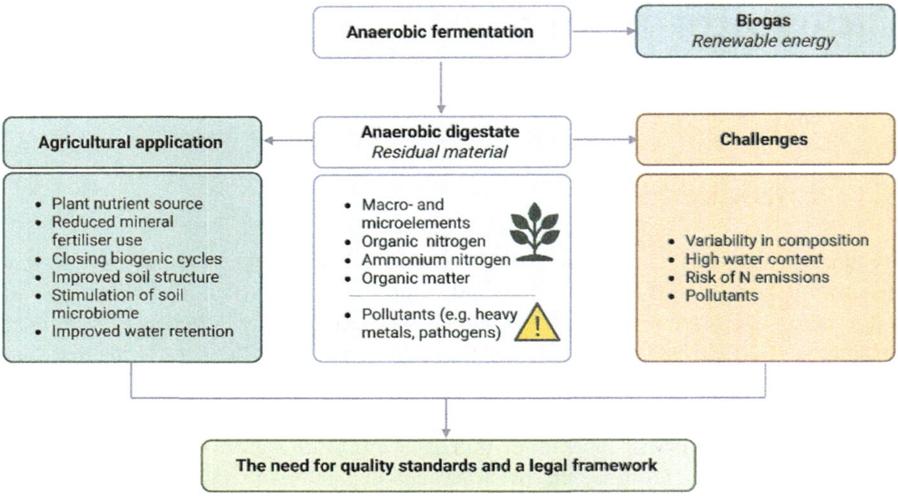

Figure 3.1 Overview of anaerobic digestate use in agriculture, linking biogas production with nutrient recycling, agronomic benefits, and environmental challenges.

3.2 Digestate properties

3.2.1 Liquid and solid fractions

Digestate leaves the anaerobic digester as a mixture of liquid and solid phases. In many cases, the digestate is separated into liquid and solid fractions immediately after the digestion process, for example by sedimentation, filtration, or filter presses (Huang et al., 2019; Martín-Sanz-Garrido et al., 2025). The primary purpose of this separation is to reduce the volume of material requiring transport and to facilitate subsequent utilisation pathways (Al Seadi et al., 2017). The division of digestate into liquid and solid fractions provides an additional advantage, as it generates two streams with distinctly different compositions, thereby enabling more targeted and efficient management strategies (Jankauskienė et al., 2024).

Experimental studies have shown that mechanical solid -liquid separation redistributes nutrients unevenly: the liquid fraction may retain most total Kjeldahl nitrogen and a substantial share of P_2O_5, whereas the solid fraction is generally richer in dry matter and particle-bound phosphorus/heavy metals. In contrast, the liquid fraction predominantly contains nitrogen in mineral form, mainly as ammonium or ammonia (Tambone et al., 2017). Phosphorus most frequently accumulates in the solid phase, which is partly attributed to the release of phosphates from organic matter during digestion (Bolzonella et al., 2018). At the same time, some studies indicate that the majority of organic nitrogen and potassium may remain in the aqueous phase of the digestate (Vázquez-Rowe et al., 2015).

This compositional differentiation suggests that digestate separation can serve as a practical tool for the recovery of valuable plant nutrients and for reducing environmental pressure by directing individual fractions to the most suitable applications. Ammoniacal nitrogen is highly mobile and largely remains in the liquid fraction of digestate (Petrova et al., 2021), which makes the liquid phase a valuable nitrogen-rich fertiliser (Nowak and Czekała, 2024). However, it also necessitates storage in sealed tanks in order to limit ammonia losses to the atmosphere (Sommer et al., 2022). The solid fraction, which is richer in slowly mineralising organic nitrogen and phosphorus, is often subjected to composting to improve its fertilising properties and hygienic quality or is applied directly as an organic fertiliser with prolonged nutrient release (Kovačić et al., 2022; Maucieri et al., 2017; Rossi et al., 2025).

3.2.2 Macronutrients

Digestate is rich in macronutrients that are essential for plant growth. It contains substantial amounts of nitrogen (N), phosphorus (P), and potassium (K) originating from the digested biomass (Dhaouefi et al., 2025). The concentrations of these nutrients depend strongly on the type of feedstock used in the biogas plant (Skrzypczak et al., 2023). For example, digestate derived from food waste typically shows a relatively low nitrogen concentration of approximately 0.06%–1.24% of d.m., yet it may be comparatively richer in phosphorus. By contrast, digestate from municipal waste may contain from 0.21% up to 7.8% N on a dry-matter basis. Digestate produced from agricultural substrates such as slurry, manure, and silage generally contains about 0.14%–2.1% N in d.m., while digestate generated from animal slurry alone may contain approximately 0.05%–0.62% N in d.m.

For comparison, organic matter can account for up to roughly 75% of the digestate d.m., indicating a high carbon content. However, this fraction largely consists of more recalcitrant compounds, including lignin, that remain after anaerobic degradation (Risberg et al., 2017; Skrzypczak et al., 2023). The phosphorus content of digestate typically reaches up to about 2.4% of d.m., potassium up to 4.0% of d.m., sulphur up to 0.4%, magnesium up to 0.5%, and calcium up to approximately 2.5% of d.m. (Risberg et al., 2017; Skrzypczak et al., 2023).

It is important to note that the N:P:K ratios in digestate do not always match crop nutrient requirements. Digestate produced from agricultural feedstocks often contains relatively higher concentrations of nitrogen and phosphorus and lower potassium contents (EBA, 2024). These differences primarily reflect the composition of the original biomass (Häfner et al., 2022). The digestion of slurry and other organic fertilisers typically increases the proportion of nitrogen and phosphorus, whereas the digestion of plant-derived residues, for example, green waste, may increase potassium levels.

Nitrogen in digestate occurs in both organic forms, meaning nitrogen bound within residual organic matter, and inorganic forms, primarily as ammonium ions. Anaerobic digestion promotes ammonification. As a result, digestate typically contains a higher proportion of ammoniacal nitrogen than the untreated feedstocks prior to digestion (Drosg et al., 2015; Jankauskienė et al., 2024). For example, in a typical digestate derived from animal slurry, up to around 50% of total nitrogen may occur as plant-available ammoniacal nitrogen. This distinguishes digestate from conventional organic fertilisers, in which a larger share of nitrogen remains in organic forms that are released more slowly (Cavalli et al., 2016). A higher proportion of mineral nitrogen is also associated with a higher pH. Digestate is generally neutral to slightly alkaline, often within the range of 7.5–8.5 (Hallat-Sanchez et al., 2023), whereas raw manures and slurries can be more acidic. Digestate typically also exhibits a lower carbon-to-nitrogen ratio (C:N) than the original feedstock. This reflects the conversion of a substantial fraction of organic carbon into biogas (CO_2 and CH_4) during digestion, while most nitrogen is retained within the remaining matrix (Drosg et al., 2015; Gao et al., 2020).

3.2.3 Micronutrients and potential contaminants

In the agricultural use of digestate, the term micronutrients refers to trace elements that perform essential nutritional functions in plants, such as Fe, Mn, Zn, Cu, B, Mo, and Ni. By contrast, inorganic contaminants primarily include potentially toxic elements and heavy metals, for example Cd, Pb, Hg, Cr, and As, as well as undesirable excesses of mineral salts. Elevated salt concentrations may lead to increased soil solution salinity and higher electrical conductivity. In agricultural practice, the boundary between a micronutrient and a contaminant is often not clear-cut, since several trace metals are essential at low concentrations but may become phytotoxic and/or accumulate in soil when present in excess.

The literature on digestate emphasises that this material represents a complex mixture of organic and inorganic components, and that its elemental profile comprises both macro- and micronutrients (Mammarella et al., 2024). At the same time, digestate composition is strongly influenced by several factors, including the type and quality of the feedstock, the operational parameters of the anaerobic digestion process such as temperature regime, hydraulic retention time, and organic loading rate, as well as post-digestion treatment methods, for example fraction separation, chemical conditioning, or concentration processes (Senevirathne and Kaparaju, 2025; Mammarella et al., 2024; Weiland, 2009).

3.2.3.1 Salinity

In addition to metals, salinity is an important quality parameter of digestate. In practice, it is most commonly assessed using electrical conductivity (EC) and the concentrations of dissolved ions. A review compiling data from full-scale biogas plants reported that digestate EC can be highly variable, ranging from 1.38 to 21.7 mS/cm (Senevirathne and Kaparaju, 2025). Excessively high EC may increase soil salinity and thereby induce osmotic stress in plants (Jin et al., 2022) showed that digestate application resulted in a two- to three-fold increase in soil EC, which implies a substantial risk of salinity stress (Jin et al., 2022).

Regarding the ions contributing to salinity, high sodium concentrations have been reported to adversely affect soil physicochemical properties and plant development. Under long-term digestate use, mitigation measures such as dilution and/or appropriate irrigation have been recommended to limit increases in salinity and EC

(Ragályi et al., 2025). Studies focusing on nutrient recovery from digestate have also highlighted that salinity may increase during processing. The use of inorganic acids introduces additional anions, including Cl^-, SO_4^{2-}, and NO_3^-, which can raise electrical conductivity and the salinity of the final product stream (Aka et al., 2024).

It is also noteworthy that specific mineral salts may occur in products derived from digestate. In a study on acidification and evaporation of liquid digestate produced from food waste, the authors identified compounds such as potassium chloride (KCl) and sodium chloride (NaCl) in the resulting solid products. The identification was supported by FTIR, XRD, and SEM-EDS analyses (Chuka-ogwude et al., 2022). This provides direct evidence that ions such as Na^+ and Cl^- can contribute to the mineral fraction of digestate-derived products and may therefore influence EC and salinity stress.

3.2.3.2 Trace elements

In the agricultural and environmental literature on digestate, trace metals are commonly discussed in two parallel roles. They are considered both as micronutrients that may enhance fertiliser value and as potential inorganic contaminants that can limit the safety of land application. A review of technologies aimed at improving the agronomic value of digestate indicated that the presence of trace amounts of elements such as Fe, Mn, Zn, Cu, B, and Mo can increase its fertiliser value, as these are micronutrients that are essential for plant growth and development (Senevirathne and Kaparaju, 2025). At the same time, a review addressing heavy-metal levels in digestates of different origins emphasised that digestate may contain elevated concentrations of heavy metals. When digestate is applied repeatedly, these metals may accumulate in soil and potentially affect soil and aquatic environments (Czatzkowska et al., 2025).

Understanding the sources of metals in digestate requires particular attention to feedstocks of animal origin. Full-scale studies of installations digesting pig slurry have shown that copper and zinc contents in the d.m. of raw slurry can span a very wide range, from 170 to 2600 mg/ kg d.m. The authors directly linked this variability to the supplementation of animal feed with these elements (Marcato et al., 2008). From a safety perspective, it is essential to interpret analytical results in relation to applicable quality limits. In a study addressing digestates derived from manure, the authors referred to the requirements of Regulation (EU) 2019/1009, according to which digestate classified as an organic fertiliser should not exceed concentrations of 300 mg Cu/kg d.m. and 800 mg Zn/kg d.m. (Al-Wahaibi et al., 2020; European Union, 2019).

In their review, Czatzkowska et al. (2025) compiled reported concentrations of heavy metals in digestates originating from different feedstocks and indicated that digestates are most commonly enriched in zinc, followed by copper, manganese, and arsenic. The authors also demonstrated that Cu and Zn concentrations are strongly dependent on substrate type. For digestate produced from a mixture of cattle manure and maize silage, data from Poland reported zinc concentrations ranging from 595 to 790 mg/kg d.m. and lead concentrations from 7.9 to 20 mg/kg d.m., with cadmium levels between 0.9 and 1.6 mg/kg d.m. In contrast, digestate derived from a mixture of sewage sludge and food-processing by-products showed substantially higher concentrations. Reported values included copper ranging from 1387 to 2059 mg/kg d.m. and zinc from 3689 to 5530 mg/kg d.m., together with cadmium levels of 7.57 to 10.79 mg/kg d.m. and lead levels of 18.6 to 63.83 mg/kg d.m. The review further noted that, following digestate separation, metal concentrations are often higher in the solid fraction than in the liquid fraction. For example, for material originating from mechanical–biological treatment of waste in Italy, the reported copper concentrations in the solid fraction ranged from 255.35 to 315.77 mg/kg d.m. and zinc from 811.5 to 920.28 mg/kg d.m. In contrast, the corresponding liquid fraction showed copper concentrations of 1.91 to 57.58 mg/kg d.m. and zinc concentrations of 14.28 to 224.63 mg/kg d.m. The authors pointed out that increases in metal concentrations after digestion are often observed relative to the input material. This effect is attributed to reductions in substrate mass and volume during methane production, and to the fact that metals are frequently associated with the insoluble solid fraction of digestate (Czatzkowska et al., 2025).

3.2.3.3 Fractionation and bioavailability

Risk assessment should not be limited to the total concentration of elements in a material. Equally important is the fraction with which elements are associated, their chemical forms, and the extent to which they can be mobilised into the soil solution and taken up by plants. In anaerobic digestion facilities processing food waste, elevated metal concentrations in the solid digestate have been shown to result primarily from metal binding to the solid fraction. Heavy metals exhibit a natural tendency to adsorb onto solid particles, which leads to their enrichment in this fraction (Wei et al., 2024). In the same case study, the authors reported specific mass balance and risk indicators. The metal mass balance, excluding Cr and Ni, ranged from 82.2% to 126.1%, while Cr and Ni concentrations in digestate samples were 4.7 and 5.5 times higher, respectively, than in the feedstock. In addition, the potential ecological risk index for solid digestate was estimated at 279.6, with Hg and Cd contributing most strongly to this value, accounting for 46.3% and 33.6%, respectively (Wei et al., 2024).

Similar observations were reported by Marcato et al. (2008) in a study of a full-scale anaerobic digestion plant treating pig slurry. The authors demonstrated that metals were mainly associated with the fine solid fraction with particle sizes between 3 and 25 μm. Screening using a sieve with a mesh size of 250 μm removed less than 2% of Cu and Zn. This finding indicates that simple mechanical separation of larger particles by sieving is unlikely to be effective in reducing Cu and Zn concentrations in digestate (Marcato et al., 2008).

From the perspective of bioavailability, results from sequential extraction procedures are particularly informative. In a study on metal speciation during mono- and co-digestion of pig manure and maize, Cr, Cu, Fe, and Pb were found to be mainly associated with the oxidisable fraction. In contrast, Zn, Mn, and Cd were predominantly assigned to potentially bioavailable forms (Luisa de Castro e Silva et al., 2024). After land application, metal mobility may change over time. In a 28-day column experiment following the application of digestate derived from non-selectively collected organic municipal waste, it was observed that trace metals did not migrate to deeper soil layers. At the same time, metal mobility increased over time in the digestate-amended layer. The mobility factor for Pb increased by 182% in the amended layer, while that for Zn increased by 85%. Zinc exhibited a higher mobility potential than Cu and Pb, whereas the mobility factor for Cu remained constant at approximately 1.75%. The authors also noted that total metal concentrations in this system were low. Consistent findings were reported in a separate 28-day column study using digestate from non-selectively collected organic municipal waste. Trace metals such as Zn, Cu, and Pb did not migrate from the digestate-enriched layer to deeper soil horizons. However, metal mobility within the amended layer increased over time, with Zn showing a higher mobility potential than Cu and Pb (Baldasso et al., 2025).

Digestate treatment technologies can reduce phosphorus and heavy metal loads in the liquid fraction, but they may simultaneously increase salinity and the risk of phytotoxicity. In a study on chemically enhanced liquid–solid separation based on coagulation and flocculation using agents such as polyaluminium chloride and polymers followed by centrifugation, near-complete removal of phosphorus and heavy metals from the liquid fraction was achieved. At the same time, a substantial increase in aluminium content was observed in the solid fraction, reaching approximately 20 g/kg d.m. Leaching tests of the solid fraction also revealed elevated concentrations in water extracts, including chlorides up to 4.6 g/L and dissolved aluminium up to 2.3 g/L. The authors indicated that the increased phytotoxicity of the extracts was most likely related to the direct toxicity of aluminium and to enhanced salinity resulting from the addition of coagulants and polymers (Beggio et al., 2021).

In summary, the environmental impact of digestate is governed less by total element concentrations than by their association with specific fractions, their chemical forms, and changes in mobility following soil application. Metals tend to accumulate in the solid and fine-grained fractions, and their bioavailability may increase over time even when total concentrations are low. Moreover, digestate treatment technologies, although effective in reducing contaminant loads in the liquid fraction, may introduce new environmental risks, including increased salinity and phytotoxicity.

3.2.4 Challenges in the management of digestate

With the expansion of the biogas sector, digestate is increasingly becoming a stream that, in some regions, exceeds the nutrient assimilation capacity of surrounding agricultural land. This situation can lead to over-application and increases the risk of eutrophication through nutrient leaching and surface runoff (Horta and Carneiro, 2021). Horta and Carneiro (2021) reported that digestate can effectively supply nitrogen and other nutrients to crops. However, in soils with low phosphorus sorption capacity and a history of excessive phosphorus fertilisation, the use of digestate as the sole nitrogen source can substantially increase phosphorus losses via surface runoff when compared with mineral fertilisation (Horta and Carneiro, 2021) (Table 3.1).

Table 3.1 Key challenges in anaerobic digestate management across environmental, technical, and system scales.

Challenge area	Nature of the challenge	Scale of occurrence	Main practical implications	Mitigation and management approaches reported in the literature
Regional digestate surpluses	Concentration of digestate production exceeding the nutrient uptake capacity of surrounding agricultural land	Regional, system-wide	Risk of over-application, nutrient accumulation in soils, eutrophication of surface and groundwater	Diversification of end uses, export of nutrients, development of non-agricultural outlets
Logistics and storage	High water content resulting in large volumes, high transport costs, and seasonal storage requirements	Operational, regional	Increased capital and operating costs, limited transport radius, reduced economic value	Solid–liquid separation, volume reduction, stabilisation prior to transport
Nutrient balance mismatch	Discrepancy between digestate N:P:K ratios and crop nutrient demand	Field-scale, long-term	Excess P or K accumulation when application rates are set to meet N demand	Integrated nutrient management, blending with mineral fertilisers
Nitrogen losses and emissions	Emissions of NH_3 and N_2O during storage and land application	Local to regional	Reduced nitrogen use efficiency, air pollution, contribution to climate change	Improved application techniques, rapid incorporation, advanced digestate treatment
Organic micropollutants	Presence of antibiotics, antimicrobial residues, and antibiotic resistance genes	Environmental	Potential dissemination of antimicrobial resistance in soils	Feedstock control, monitoring, post-treatment of digestate
Microplastics	Transfer of plastic particles from municipal waste and sewage sludge into digestate	Long-term, environmental	Accumulation in soils, unknown long-term ecological effects	Improved source separation, removal of contaminants in upstream treatment

Biological risks	Incomplete inactivation of pathogens under certain digestion conditions	Local	Restrictions on land use, sanitary concerns for food production	Hygienisation, thermal treatment, compliance with veterinary regulations
Digestate processing trade-offs	Volume reduction achieved at the expense of increased salinity or secondary pollution	Technological	Shifting of environmental burdens between digestate fractions	Technology selection based on end use, comprehensive impact assessment
Market acceptance	Perception of digestate as a waste rather than a product	Socio-economic	Limited market penetration, low added value	Quality certification, standardisation, transparency
Regulatory gaps	Insufficient regulation of emerging contaminants such as antibiotics and microplastics	Systemic	Uncertainty for operators and regulators	Development of harmonised standards and monitoring frameworks

With respect to current regulatory frameworks, they often do not include specific limit values for antibiotics and microplastics. This also constitutes a tangible environmental concern (Sobhi et al., 2024; Lessa Belone et al., 2025). In addition, a review focused on nutrient recovery from digestate reported that 80%–90% of the mass input to anaerobic digestion remains as digestate. The practical barriers to its management include high water content and the associated transport costs, as well as seasonal and regional restrictions on land application periods. These constraints can result in prolonged storage and, in some cases, a negative economic value of digestate (Lamolinara et al., 2022; Vilanova Plana et al., 2016).

Emissions represent an important hidden cost in a number of digestate management scenarios. An assessment of greenhouse gas emissions across digestate management stages showed that raw digestate has an emission potential of approximately 139 g CO_2 eq per kg. About 10% of this potential was attributed to two months of storage, while stabilisation in heaps and subsequent curing reduced the emission potential to approximately 80 g CO_2 eq per kg (Zeshan and Visvanathan, 2014). At the same time, nitrogen emissions at the point of field application remain a key issue. In a study comparing the liquid fraction of digestate with urea at an equivalent dose of 150 kg N/ha, total nitrogen emissions over 25 days amounted to 2.87 kg N/ha for digestate and 3.76 kg N/ha for urea. Digestate was associated with 23% higher N_2O emissions than urea, whereas urea emitted 66% more NH_3 than digestate. No significant differences in d.m. yield were observed (Alburquerque et al., 2025).

In practice, effective measures exist to mitigate these emissions. A full-scale field experiment in silage maize demonstrated that subsurface injection of digestate and digestate-derived products reduced NH_3 emissions by 69% and 77% relative to surface application in the 2012 and 2013 campaigns. It also significantly reduced odour nuisance (Riva et al., 2016).

A review examining the effects of digestate on soil biota indicated that organisms inhabiting the surface layer, including earthworms and springtails, may respond negatively to digestate application under conditions of toxicity associated with high ammonia concentrations. The authors also emphasised that generalising such effects is challenging because digestate properties vary substantially with feedstock type and processing technology (van Midden et al., 2023). In addition, quality-related risks arise from the presence of microcontaminants and from sanitary safety concerns, as supported by microbiological and molecular investigations of digestate. In samples collected from agricultural biogas plants, incomplete elimination of antimicrobial substances during anaerobic digestion was reported, together with the presence of antibiotic resistance genes (Wolak et al., 2023). Similarly, for digestate produced from sewage sludge, it was shown that, despite a reduction in pathogen abundance following anaerobic digestion, risks persisted. These were associated, among other factors, with spore-forming genera such as Bacillus and Clostridium and with the incomplete reduction of antibiotic resistance genes (Franchitti et al., 2024).

From an investment decision perspective, it is also important that the logistical benefits of digestate processing may need to be balanced against environmental burdens across the entire value chain. A life cycle assessment compared direct land application (S1), a scenario involving centrifugation into fractions (S2), and an advanced scenario including centrifugation, drying of the solid fraction, and polishing of the liquid fraction by ultrafiltration and reverse osmosis (S3). The functional unit was defined as the management of 1 kg of raw digestate d.m. The abstract reported a global warming potential ranging from 0.14 to 0.36 kg CO_2 eq per functional unit and a terrestrial acidification potential between 0.09 and 0.18 kg SO_2 eq. Among their conclusions, the authors noted that transporting dried digestate may be the preferred option when local utilisation is not feasible (Angouria-Tsorochidou et al., 2022).

In summary, challenges in digestate management encompass both environmental pressures associated with over-application, emissions, and contamination risks, and logistical and economic constraints resulting from its volume, compositional variability, and seasonal application windows. Effective and sustainable digestate utilisation therefore requires the integration of technologies that mitigate emissions and quality risks with processing and logistics solutions that enable its safe incorporation into nutrient cycling.

3.3 Legal framework governing the use of digestate in the European Union

Within the European Union, the use of digestate on agricultural land is regulated by several legal instruments spanning waste legislation, fertiliser law, and water protection. For a long time, digestate was treated primarily as waste. Its agricultural application was therefore governed by regulations on the spreading of waste on arable land rather than by fertiliser legislation. This situation changed with the adoption of Regulation (EU) 2019/1009 on the making available on the market of EU fertilising products. This regulation, which has been fully applicable since July 2022, replaced earlier EU rules on mineral fertilisers and, for the first time, explicitly included digestate within the catalogue of recognised fertilising products (European Union, 2019; Martín-Sanz-Garrido et al., 2025).

Under Regulation (EU) 2019/1009, digestate is defined as a component of specific Component Material Categories, namely CMC4 and CMC5. CMC4 refers to digestate derived from fresh crop materials such as silage and plant residues, excluding food and feed waste. CMC5 covers digestate produced from separately collected bio-waste and other biodegradable waste that is not classified as food or feed (European Union, 2019). In practical terms, the regulation allows digestate to be placed on the market as a CE-marked fertiliser, provided that it is produced from permitted biomass types and complies with defined quality requirements.

Regulation (EU) 2019/1009 establishes, among other provisions, contaminant limits for organic fertilisers including digestate. These include maximum allowable concentrations of heavy metals such as Cd, Pb, Hg, Ni, Cr(VI), Cu, and Zn, as well as sanitary requirements concerning the absence of *Salmonella* and low levels of *Escherichia coli*. In addition, digestate intended to qualify as an EU fertilising product must meet quality criteria comparable to those applied to compost, including requirements related to the stabilisation of organic matter and adequate nutrient content. The regulation also introduces obligations related to labelling and traceability. Producers of digestate-based fertilisers are required to ensure batch identification and compliance with applicable standards. Through these measures, the EU seeks to harmonise the digestate market as a fully-fledged commercial product, thereby supporting circular economy objectives and strengthening user confidence (Martín-Sanz-Garrido et al., 2025).

It should be noted that recognition of digestate as an EU fertilising product does not automatically imply that all digestate ceases to be classified as waste. Producers may choose either to place digestate on the market in accordance with Regulation (EU) 2019/1009 or to use it under national rules, which in many cases continue to fall under waste legislation. Despite the existence of a harmonised regulation, the legal status of digestate therefore remains complex. Depending on the country and the specific circumstances, digestate may be classified as a product, a by-product, or still as waste if recovery conditions are not met. For example, in Spain a specific decree adopted in 2022 recognises digestate meeting the CMC4 and CMC5 criteria as a recovered product that exits the waste regime, provided that national quality standards and monitoring requirements are fulfilled (Martín-Sanz-Garrido et al., 2025). Similar country-specific provisions clarifying the legal status of digestate are in place in other EU Member States (Table 3.2).

A second key legal pillar is the Nitrates Directive 91/676/EEC (European Union, 2008), which aims to reduce water pollution caused or induced by nitrates from agricultural sources and to prevent further such pollution. Under the Directive, Member States are required to identify waters affected by pollution or at risk of pollution and, on this basis, to designate vulnerable zones, defined as areas from which nitrogen runoff contributes to water contamination. The criteria set out in the Directive include, in particular, situations where groundwater contains or could contain more than 50 mg of nitrates per liter in the absence of action, as well as cases where surface waters are eutrophic or may become eutrophic without intervention.

For designated vulnerable zones, Member States must establish action programmes within specified timeframes and subject them to regular review. The Directive also allows Member States to apply an action programme across their entire territory instead of designating individual vulnerable zones. Action programs must include at least the measures listed in Annex III. These measures encompass the designation of periods during which the application of certain fertilisers is prohibited, requirements for the storage capacity of livestock manure, and rules for fertiliser application that take into account soil conditions, slope, climate, agricultural practices, and the need to maintain a balance between the foreseeable nitrogen requirements of crops and the nitrogen supplied from soil reserves and fertilisation.

Within vulnerable zones, the Directive introduces a limit whereby the amount of nitrogen from livestock manure applied to agricultural land must not exceed 170 kg nitrogen per hectare per year. This limit refers to the total annual nitrogen input from such fertilisers. The Directive provides that, during the first 4-year action programme, Member States may allow a higher limit of up to 210 kg nitrogen per hectare per year. It also allows for the establishment of other limits where justified by objective criteria

Table 3.2 Contaminant limits and hygienic requirements for organic and organo-mineral fertilisers under Regulation (EU) 2019/1009 (European Union (2019). Regulation (EU) 2019/1009 of the European Parliament and of the Council. Official Journal of the European Union, no date)

Product category	Parameter	Requirement
PFC 1(A) Organic fertiliser	Cadmium (Cd)	≤1.5 mg/kg d.m.
	Chromium (VI)	≤2 mg/kg d.m.
	Mercury (Hg)	≤1 mg/kg d.m.
	Nickel (Ni)	≤50 mg/kg d.m.
	Lead (Pb)	≤120 mg/kg d.m.
	Inorganic arsenic (As)	≤40 mg/kg d.m.
	Copper (Cu)	≤300 mg/kg d.m.
	Zinc (Zn)	≤800 mg/kg d.m.
	Biuret	Must not be present
	Salmonella spp.	Absence in 25 g or 25 ml (n=5, c=0, m=0)
	E. coli or Enterococcaceae	≤1000 CFU per g or ml (n=5, c=5, m=0)
PFC 1(B) Organo-mineral fertiliser	Cadmium (Cd)	≤3 mg/kg d.m. if P < 5% P_2O_5 eq; ≤60 mg/kg P_2O_5 for phosphate fertilisers
	Chromium (VI)	≤2 mg/kg d.m.
	Mercury (Hg)	≤1 mg/kg d.m.
	Nickel (Ni)	≤50 mg/kg d.m.
	Lead (Pb)	≤120 mg/kg d.m.
	Inorganic arsenic (As)	≤40 mg/kg d.m.
	Copper (Cu)	≤600 mg/kg d.m. (not applicable if Cu is intentionally added and declared in accordance with Annex III)
	Zinc (Zn)	≤1500 mg/kg d.m. (not applicable if Zn is intentionally added and declared in accordance with Annex III)
	Biuret	≤12 g/kg d.m.
	Salmonella spp.	Absence in 25 g or 25 mL (n = 5, c = 0, m = 0)
	E. coli or Enterococcaceae	≤1000 CFU per g or mL (n = 5, c = 5, m = 0)

specified in the Directive and where such limits do not compromise its objectives, provided that the Member State informs the European Commission accordingly.

In the context of materials such as digestate, it is important that the Directive defines livestock manure as including the excreta and urine of farmed animals, including in processed form. The term fertiliser is defined broadly as any substance containing a nitrogen compound applied to land for the purpose of promoting plant growth. The Directive also obliges Member States to establish a code of good agricultural practice. This code includes recommendations on situations in which fertiliser application is inappropriate and measures to limit nitrogen losses. As an optional element, it identifies the maintenance of a minimum vegetative cover during periods when there is a risk of nitrate leaching. The Directive further establishes monitoring requirements, including the sampling of surface waters at least once per month, and provides for periodic reporting by Member States to the Commission.

A third important regulatory area within the EU concerns waste legislation and the concept of end of waste. Under the Waste Framework Directive 2008/98/EC (European Union, 2025), certain waste materials may cease to be classified as waste after undergoing a recovery operation, including recycling, provided that the conditions set out in Article 6(1) are met. These conditions require that the material is commonly used for specific purposes, that a market or demand exists, that it fulfils relevant technical requirements and complies with applicable product legislation, and that its use does not lead to overall adverse impacts on the environment or human health. The Directive further provides that, where detailed end-of-waste criteria have not been established at Union level, Member States may decide on a case-by-case basis at national level in accordance with Article 6(4). In practice, this provision underpins national approaches to determining when compost or digestate, following appropriate treatment and quality control, may cease to be considered waste.

At EU level, Regulation (EU) 2019/1009 on fertilising products is also of particular relevance, as it establishes a mechanism whereby a material that is classified as waste under Directive 2008/98/EC may cease to be waste if it forms part of a compliant EU fertilising product. This transition is linked to fulfilment of the requirements set out in the Regulation and its annexes, and to the moment at which the EU declaration of conformity is drawn up (European Union, 2019). With regard to contaminants, it can be cautiously noted that neither the main text of Regulation (EU) 2019/1009 nor its annexes explicitly define standalone limit values specifically for microplastics or antibiotics. This implies that any requirements in this respect must derive from other legislative instruments or from future amendments to the regulatory framework (European Union, 2019).

Beyond the EU-level framework, Member States maintain their own national regulations that further specify the conditions for placing fertilising materials on the market and for their agricultural use, including digestate, as well as the associated administrative and quality requirements. In Poland, the legal basis is provided by the Act of 10 July 2007 on fertilisers and fertilisation, as consolidated and published in 2024. This act regulates the rules for placing fertilisers and plant growth enhancers on the market. Detailed implementing provisions were updated in the Regulation of the Minister of Agriculture and Rural Development of August 9, 2024 on the implementation of certain provisions of the Act on fertilisers and fertilisation. This regulation is currently in force, and the Ministry indicated that the update specifically addressed digestate-derived products (MARD, 2024).

In Germany, the rules governing the placing on the market of fertilisers and related products are set out in the *Düngemittelverordnung*. In parallel, the *Bioabfallverordnung* applies to the utilisation of biowaste. At the federal level, additional requirements are also communicated regarding the reduction of foreign contaminants, including plastics, in biowaste streams (DüMV, 2012). In Italy, a key legal instrument is the decree of February 25, 2016, which establishes general criteria and technical standards for the production and agricultural use of digestate. These provisions are subsequently operationalised through regional regulations, an example being the regulation of the Emilia-Romagna region of 19 March 2024 governing the agricultural use of digestate (MIPA, 2016).

In summary, despite a common EU framework, the practical rules governing the placing on the market and agricultural use of digestate are largely defined by national and regional regulations. These include specific quality requirements, administrative procedures, monitoring obligations, and restrictions linked to water protection and nutrient management. As a result, the legal status and utilisation pathways of digestate may differ substantially between Member States.

3.4 Conclusion

Digestate generated through anaerobic digestion constitutes a valuable by-product that can play an important role in the circular economy by supplying plants with essential macro- and micronutrients and by contributing to the improvement of soil properties. Its agronomic value derives both from the presence of readily available nitrogen forms and from a complex mixture of mineral components and organic matter that exerts soil-stabilising effects. At the same time, digestate is characterised by considerable variability in chemical composition, high water content, and the potential presence of contaminants, which poses challenges related to storage, transport, and safe agricultural application.

Effective digestate management therefore requires consideration of both agronomic benefits and environmental risks, including greenhouse gas emissions, microbiological contamination, heavy metals, and excessive soil salinity. Appropriate treatment technologies, such as fraction separation, composting, or chemical stabilisation, can enhance the safety of digestate use and enable the recovery of valuable nutrients. However, these processes may also introduce new environmental burdens. The legal frameworks established at the European Union and Member State levels provide a basis for placing digestate on the market as a fertilising product, while defining contaminant limits, quality requirements, and monitoring obligations. These measures increase the safety of digestate use and support its integration into sustainable agricultural systems. Nevertheless, regulatory complexity and the absence of harmonised limits for certain contaminants, such as microplastics and antibiotics, highlight the need for further development of quality standards and management practices.

In summary, digestate can be transformed from a waste material into a valuable agricultural resource. However, realising its full potential requires an integrated approach that combines quality control, processing technologies, appropriate application strategies, and compliance with regulatory frameworks. Such an approach enables the safe and efficient incorporation of digestate into nutrient cycles in line with the principles of the circular economy.

AI disclosure

During the preparation of this work the author(s) used DEEPL, ChatGPT, Grammarly, Mendeley in order to Translation, language correction, bibliography management. After using this tool/service, the author(s) reviewed and edited the content as needed and take(s) full responsibility for the content of the publication.

References

Aka, R.J.N., Hossain, M.M., Nasir, A., Zhan, Y., Zhang, X., Zhu, J., et al., 2024. Enhanced nutrient recovery from anaerobically digested poultry wastewater through struvite precipitation by organic acid pre-treatment and seeding in a bubble column electrolytic reactor. Water Res.earch 252. Available at https://doi.org/10.1016/j.watres.2024.121239.

Al Seadi, T., Drosg, B., Fuchs, W., Rutz, D., Janssen, R., 2017. Comprehensive characterization of the liquid fraction of digestates from full-scale anaerobic co-digestion. Waste Management 59, 118–128. Available at https://doi.org/10.1533/9780857097415.2.267.

Alburquerque, J.A., de la Fuente, C., Bernal, M.P., 2025. Application of digestate from low-tech digesters for degraded soil restoration: Effects on soil fertility and carbon sequestration. Science of The Total Environment 967, 178854. Available at https://doi.org/10.1016/j.agee.2011.03.007.

Al-Wahaibi, A., Osman, A.I., Al-Muhtaseb, A.H., Alqaisi, O., Baawain, M., Fawzy, S., et al., 2020. Techno-economic evaluation of biogas production from food waste via anaerobic digestion. Scientific Reports 10 (1). Available at https://doi.org/10.1038/s41598-020-72897-5.

Angouria-Tsorochidou, E., Seghetta, M., Trémier, A., Thomsen, M., 2022. Life cycle assessment of digestate post-treatment and utilization. Science of the Total Environment 815. Available at https://doi.org/10.1016/j.scitotenv.2021.152764.

Baldasso, V., Tomasino, M.P., Sayen, S., Guillon, E., Frunzo, L., Gomes, C.A.R., et al., 2025. Effects of digestate soil amendment on the fate of trace metals and on the soil microbial community. Environmental Pollution 371. Available at https://doi.org/10.1016/j.envpol.2025.125961.

Barłóg, P., Hlisnikovský, L., Kunzová, E., 2020. Effect of digestate on soil organic carbon and plant-available nutrient content compared to cattle slurry and mineral fertilization. Agronomy 10 (3). Available at https://doi.org/10.3390/agronomy10030379.

Beggio, G., Peng, W., Lü, F., Cerasaro, A., Bonato, T., Pivato, A., 2021. Chemically Enhanced Solid–Liquid Separation of Digestate: Suspended Solids Removal and Effects on Environmental Quality of Separated Fractions. Waste and Biomass Valorization 13 (2), 1029–1041. Available at https://doi.org/10.1007/s12649-021-01591-y.

Bolzonella, D., Fatone, F., Gottardo, M., Frison, N., 2018. Nutrients recovery from anaerobic digestate of agro-waste: Techno-economic assessment of full scale applications. Journal of Environmental Management 216, 111–119. Available at https://doi.org/10.1016/j.jenvman.2017.08.026.

Calabi-Floody, M., Medina, J., Rumpel, C., Condron, L.M., Hernandez, M., Dumont, M., Mora, M.de la L., 2018. Smart fertilizers as a strategy for sustainable agriculture. Advances in Agronomy 147, 119–157. Available at https://doi.org/10.1016/bs.agron.2017.10.003.

Cavalli, D., Cabassi, G., Borrelli, L., Geromel, G., Bechini, L., Degano, L., et al., 2016. Nitrogen fertilizer replacement value of undigested liquid cattle manure and digestates. European Journal of Agronomy 73, 34–41. Available at https://doi.org/10.1016/j.eja.2015.10.007.

Cheong, J.C., Lee, J.T.E., Lim, J.W., Song, S., Tan, J.K.N., Chiam, Z.Y., et al., 2020. Closing the food waste loop: Food waste anaerobic digestate as fertilizer for the cultivation of the leafy vegetable, xiao bai cai (Brassica rapa). Science of the Total Environment 715. Available at https://doi.org/10.1016/j.scitotenv.2020.136789.

Chuka-ogwude, D., Nafisi, M., Taher, H., Ogbonna, J.C., Moheimani, N.R., 2022. Food waste digestate as a source of nitrogen for the cultivation of *Dunaliella salina*: influence on growth and carotenogenesis under hyper osmotic stress. Journal of Applied Phycology 34 (1), 101–112. Available at https://doi.org/10.1007/s10811-021-02663-1.

Czatzkowska, M., Rolbiecki, D., Korzeniewska, E., Harnisz, M., 2025. Heavy metal and antimicrobial residue levels in various types of digestate from biogas plants—a review. Sustainability 17 (2), 416. Available at https://doi.org/10.3390/su17020416.

Czekała, W., Jasiński, T., Grzelak, M., Witaszek, K., Dach, J., 2022. Biogas plant operation: digestate as the valuable product. Energies 15 (21), 8275. Available at https://doi.org/10.3390/en15218275.

Dahlin, J., Herbes, C., Nelles, M., 2015. Biogas digestate marketing: qualitative insights into the supply side. Resources, Conservation and Recycling 104, 152–161. Available at https://doi.org/10.1016/j.resconrec.2015.08.013.

Dhaouefi, Z., Taktek, S., Bélanger, F., Fortin, P., Charbonneau, J., Lange, S., et al., 2025. Optimized biogas yield and safe digestate valorization through intensified anaerobic digestion of invasive plant biomass. Energies 18 (19). Available at https://doi.org/10.3390/en18195151.

Drosg, B., Fuchs, W., Al, T., Madsen, S.M., Linke, B., 2015. Nutrient recovery by biogas digestate processing. IEA Bioenergy. http://www.iea-biogas.net/files/daten-redaktion/download/Technical%20Brochures/NUTRIENT_RECOVERY_RZ_web1.pdf

DüMV. (2012). DüMV - Verordnung über das Inverkehrbringen von Düngemitteln, Bodenhilfsstoffen, Kultursubstraten und Pflanzenhilfsmitteln 1. https://www.gesetze-im-internet.de/d_mv_2012/BJNR248200012.html?utm_source=chatgpt.com.

EBA. (2024). Exploring digestate's contribution to healthy soils. www.europeanbiogas.eu.

European Union. (2008). *Consolidated text: Council Directive of 12 December 1991 concerning the protection of waters against pollution caused by nitrates from agricultural sources (91/676/EEC).* https://eur-lex.europa.eu/legal-content/EN/TXT/?uriCELEX%3A01991L0676-20081211

European Union. (2019). Regulation (EU) 2019/1009 of the European Parliament and of the Council. Official Journal of the European Union.

European Union. (2025). Consolidated text: Directive 2008/98/EC of the European Parliament and of the Council of 19 November 2008 on waste and repealing certain Directives (Text with EEA relevance). https://eur-lex.europa.eu/legal-content/EN/TXT/?uri=CELEX%3A02008L0098-20251016.

Franchitti, E., Pedullà, M., Madsen, A.M., Traversi, D., 2024. Effect of anaerobic digestion on pathogens and antimicrobial resistance in the sewage sludge. Environment International 191. Available at https://doi.org/10.1016/j.envint.2024.108998.

Gao, J., Li, J., Wachemo, A.C., Yuan, H., Zuo, X., Li, X., 2020. Mass conversion pathway during anaerobic digestion of wheat straw. RSC Advances 10 (46), 27720–27727. Available at https://doi.org/10.1039/d0ra02441d.

Golovko, O., Ahrens, L., Schelin, J., Sörengård, M., Bergstrand, K.-J., Asp, H., et al., 2022. Organic micropollutants, heavy metals and pathogens in anaerobic digestate based on food waste. Journal of Environmental Management 313, 114997. Available at https://doi.org/10.1016/j.jenvman.2022.114997.

Häfner, F., Hartung, J., Möller, K., 2022. Digestate Composition Affecting N Fertiliser Value and C Mineralisation. Waste and Biomass Valorization 13 (8), 3445–3462. Available at https://doi.org/10.1007/s12649-022-01723-y.

Hallat-Sanchez, J., Smith, J., Norton, G.J., 2023. Impact of adding anaerobic digestate to soil and consequences on crop performance. Agronomy 13 (12). Available at https://doi.org/10.3390/agronomy13122889.

Horta, C., Carneiro, J.P., 2021. Phosphorus Losses to Surface Runoff Waters After Application of Digestate to a Soil Over Fertilised with Phosphorus. Water, Air, & Soil Pollution 232 (10), 439. Available at https://doi.org/10.1007/s11270-021-05382-y.

Huang, J., Han, L., Huang, G., 2019. Characterization of digestate composting stability using fluorescence EEM spectroscopy combining with PARAFAC. Waste Management and Research 37 (5), 486–494. Available at https://doi.org/10.1177/0734242X19828181.

Jameel, M.K., Mustafa, M.A., Ahmed, H.S., Mohammed, A.J., Ghazy, H., Shakir, M.N., et al., 2024. Biogas: Production, properties, applications, economic and challenges: a review. Results in Chemistry 7. Available at https://doi.org/10.1016/j.rechem.2024.101549.

Jankauskienė, J., Laužikė, K., Kaupaitė, S., 2024. The use of anaerobic digestate for greenhouse horticulture. Agronomy 14 (10). Available at https://doi.org/10.3390/agronomy14102437.

Jin, K., Ran, Y., Alengebawy, A., Yang, G., Jia, S., Ai, P., 2022. Agro-environmental sustainability of using digestate fertilizer for solanaceous and leafy vegetables cultivation: Insights on fertilizer efficiency and risk assessment. Journal of Environmental Management 320, 115895. Available at https://doi.org/10.1016/j.jenvman.2022.115895.

Kovačić, Đ., Lončarić, Z., Jović, J., Samac, D., Popović, B., Tišma, M., 2022. Digestate management and processing practices: a review. Applied Sciences 12 (18), 9216. Available at https://doi.org/10.3390/app12189216.

Lamolinara, B., Pérez-Martínez, A., Guardado-Yordi, E., Guillén Fiallos, C., Diéguez-Santana, K., Ruiz-Mercado, G.J., 2022. Anaerobic digestate management, environmental impacts, and techno-economic challenges. Waste Management 140, 14–30. Available at https://doi.org/10.1016/j.wasman.2021.12.035.

Lessa Belone, M.C., Yli-Rantala, E., Sarlin, E., Kokko, M., 2025. Microplastics in an anaerobic digester treating sewage sludge: Occurrence and factors affecting their identification with Raman spectroscopy. Journal of Hazardous Materials 491, 138015. Available at https://doi.org/10.1016/j.jhazmat.2025.138015.

Luisa de Castro e Silva, H., Robles-Aguilar, A.A., Akyol, Ç., Willems, B., Meers, E., 2024. Potential availability of metals in anaerobic mono- and co-digestion of pig manure and maize. Bioresource Technology 402, 130818. Available at https://doi.org/10.1016/j.biortech.2024.130818.

Mammarella, D., Di Giuliano, A., Gallucci, K., 2024. Reuse and Valorization of Solid Digestate Ashes from Biogas Production. Energies 17 (3). Available at https://doi.org/10.3390/EN17030751.

Marcato, C.E., Pinelli, E., Pouech, P., Winterton, P., Guiresse, M., 2008. Particle size and metal distributions in anaerobically digested pig slurry. Bioresource Technology 99 (7), 2340–2348. Available at https://doi.org/10.1016/j.biortech.2007.05.013.

MARD. (2024). Rozporządzenie Ministra Rolnictwa i Rozwoju Wsi z dnia 9 sierpnia 2024 r. w sprawie wykonania niektórych przepisów ustawy o nawozach i nawożeniu. https://eli.gov.pl/eli/DU/2024/1261/ogl?utm_source=chatgpt.com.

Martín-Sanz-Garrido, C., Revuelta-Aramburu, M., Santos-Montes, A.M., Morales-Polo, C., 2025. A review on anaerobic digestate as a biofertilizer: characteristics, production, and environmental impacts from a life cycle assessment perspective. Applied Sciences 15 (15), 8635. Available at https://doi.org/10.3390/app15158635.

Maucieri, C., Nicoletto, C., Caruso, C., Sambo, P., Borin, M., 2017. Effects of digestate solid fraction fertilisation on yield and soil carbon dioxide emission in a horticulture succession. Italian Journal of Agronomy 12 (2), 800. Available at https://doi.org/10.4081/ija.2017.800.

Mayerová, M., Šimon, T., Stehlík, M., Madaras, M., Koubová, M., Smatanová, M., 2023. Long-term application of biogas digestate improves soil physical properties. Soil and Tillage Research 231, 105715. Available at https://doi.org/10.1016/j.still.2023.105715.

van Midden, C., Harris, J., Shaw, L., Sizmur, T., Pawlett, M., 2023. The impact of anaerobic digestate on soil life: A review. Applied Soil Ecology 191. Available at https://doi.org/10.1016/j.apsoil.2023.105066.

MIPA. (2016). Criteri e norme tecniche generali per la disciplina regionale dell'utilizzazione agronomica degli effluenti di allevamento e delle acque reflue, nonche' per la produzione e l'utilizzazione agronomica del digestato. (16A02762). https://www.gazzettaufficiale.it/eli/id/2016/04/18/16A02762/sg?utm_source=chatgpt.com.

Moheimani, N.R., Vadiveloo, A., Ayre, J.M., Pluske, J.R., 2018. Nutritional profile and in vitro digestibility of microalgae grown in anaerobically digested piggery effluent. Algal Research 35, 362–369. Available at https://doi.org/10.1016/j.algal.2018.09.007.

Nayak, J.K., Ranade, V.V., 2025. Valorisation of digestate: characteristics, products, processes and potential. Chemical Engineering Journal Advances 24. Available at https://doi.org/10.1016/j.ceja.2025.100887.

Nowak, M., Czekała, W., 2024. Sustainable use of digestate from biogas plants: separation of raw digestate and liquid fraction processing. Sustainability 16 (13), 5461. Available at https://doi.org/10.3390/su16135461.

Petrova, I.P., Pekrun, C., Möller, K., 2021. Organic matter composition of digestates has a stronger influence on n2o emissions than the supply of ammoniacal nitrogen. Agronomy 11 (11). Available at https://doi.org/10.3390/agronomy11112215.

Ragályi, P., Szécsy, O., Uzinger, N., Magyar, M., Szabó, A., Rékási, M., 2025. Factors influencing the impact of anaerobic digestates on soil properties. Soil Systems 9 (3). Available at https://doi.org/10.3390/soilsystems9030078.

Risberg, K., Cederlund, H., Pell, M., Arthurson, V., Schnürer, A., 2017. Comparative characterization of digestate versus pig slurry and cow manure – chemical composition and effects on soil microbial activity. Waste Management 61, 529–538. Available at https://doi.org/10.1016/j.wasman.2016.12.016.

Riva, C., Orzi, V., Carozzi, M., Acutis, M., Boccasile, G., Lonati, S., Tambone, F., D'Imporzano, G., Adani, F., 2016. Short-term experiments in using digestate products as substitutes for mineral (N) fertilizer: Agronomic performance, odours, and ammonia emission impacts. Science of The Total Environment 547, 206–214. Available at https://doi.org/10.1016/J.SCITOTENV.2015.12.156.

Rossi, G., Beni, C., Socciarelli, S., Neri, U., 2025. Zootechnical and municipal solid waste digestates: effects on soil nitrogen mineralization and kinetics. Biomass 5 (1), 5. Available at https://doi.org/10.3390/biomass5010005.

Senevirathne, N., Kaparaju, P., 2025. Enhancing the agronomic value of anaerobic digestate: a review of current vs. emerging technologies, challenges and future directions. Agriculture (Switzerland) 15 (20). Available at https://doi.org/10.3390/agriculture15202108.

Skrzypczak, D., Krzysztof, T., Mironiuk, M., Katarzyna, M., Grzegorz, I., Polomska, X., Wiśniewski, J., Mielko, K., Moustakas, K., Katarzyna, C., 2023. Recent innovations in fertilization with treated digestate from food waste to recover nutrients for arid agricultural fields. Environmental Science and Pollution Research 31 (29), 41563–41585. Available at https://doi.org/10.1007/S11356-023-31211-2.

Sobhi, M., Elsamahy, T., Zakaria, E., Gaballah, M.S., Zhu, F., Hu, X., et al., 2024. Characteristics, limitations and global regulations in the use of biogas digestate as fertilizer: a comprehensive overview. Science of The Total Environment 957, 177855. Available at https://doi.org/10.1016/j.scitotenv.2024.177855.

Sommer, S.G., Hafner, S.D., Laubach, J., van der Weerden, T.J., Leytem, A.B., Pacholski, A., 2022. Model for calculating ammonia emission from stored animal liquid manure. Biosystems Engineering 223, 41–55. Available at https://doi.org/10.1016/j.biosystemseng.2022.08.007.

Tambone, F., Orzi, V., D'Imporzano, G., Adani, F., 2017. Solid and liquid fractionation of digestate: Mass balance, chemical characterization, and agronomic and environmental value. Bioresource Technology 243, 1251–1256. Available at https://doi.org/10.1016/j.biortech.2017.07.130.

Tymchuk, I., Malovanyy, M., Zhuk, V., Kochubei, V., Yatsukh, K., Luchyt, L., 2023. Towards Increasing the utilization of anaerobic digestate from biogas production in agrotechnologies. Ecological Questions 34 (4), 1–18. Available at https://doi.org/10.12775/EQ.2023.054.

Vázquez-Rowe, I., Golkowska, K., Lebuf, V., Vaneeckhaute, C., Michels, E., Meers, E., et al., 2015. Environmental assessment of digestate treatment technologies using LCA methodology. Waste Management 43, 442–459. Available at https://doi.org/10.1016/j.wasman.2015.05.007.

Vilanova Plana, P., Noche, B., 2016. A review of the current digestate distribution models: storage and transport. WIT Transactions on Ecology and The Environment 202. Available at https://doi.org/10.2495/WM160311.

Wei, J., Zhang, J., Li, H., Liu, J., Deng, Z., Lian, H., et al., 2024. Mass flow and ecological risk of heavy metals in anaerobic digestion of food waste. Environmental Technology & Innovation 36, 103863. Available at https://doi.org/10.1016/j.eti.2024.103863.

Weiland, P., 2009. Biogas production: current state and perspectives. Applied Microbiology and Biotechnology 85 (4), 849–860. Available at https://doi.org/10.1007/s00253-009-2246-7.

Wolak, I., Bajkacz, S., Harnisz, M., Stando, K., Męcik, M., Korzeniewska, E., 2023. Digestate from agricultural biogas plants as a reservoir of antimicrobials and antibiotic resistance genes—implications for the environment. International Journal of Environmental Research and Public Health 20 (3). Available at https://doi.org/10.3390/ijerph20032672.

Zeshan, Visvanathan, C., 2014. Evaluation of anaerobic digestate for greenhouse gas emissions at various stages of its management. International Biodeterioration and Biodegradation 95 (PA), 167–175. Available at https://doi.org/10.1016/j.ibiod.2014.06.020.

Advanced digestate treatment technologies

4

4.1 Introduction

Anaerobic methane fermentation is a widely used technology for processing organic waste and biomass, enabling energy recovery in the form of biogas. This process also generates a by-product called digestate, which is rich in stable organic compounds and mineral nutrients (Abdelfatah-Aldayyat and Gómez, 2025; Mikula et al., 2023). Digestate is characterised by a high content of macro- and microelements, including nitrogen, phosphorus, and potassium, which makes it a potentially valuable organic fertiliser (Sobhi et al., 2024). It is estimated that within the European Union alone, annual digestate production reaches approximately 180 million tons, with nearly half generated in Germany (Wood Environment & Infrastructure Solutions UK Limited, 2019). Globally, Europe accounts for about 50% of world biogas production, while Asia contributes around 35%, according to recent literature summaries (Sobhi et al., 2024). Without proper management, digestate surpluses may exceed the nutrient absorption capacity of surrounding agricultural land, leading to overfertilization and environmental risks.

Excessive or improper application of digestate to fields results in the leaching of nitrogen and phosphorus into groundwater and surface waters, causing eutrophication and contamination of water bodies (Proskynitopoulou et al., 2024). It has been estimated that without control, greenhouse gas emissions associated with traditional slurry or digestate spreading may reach approximately 139 g CO_2eq per kg of digestate (Sobhi et al., 2024). On the other hand, the use of anaerobic digestion contributes to the reduction of methane emissions from raw wastes such as manure by capturing this gas in biogas and preventing its release into the atmosphere (Burg et al., 2018). However, the fermentation process does not remove biogenic elements. Most of the nitrogen and phosphorus present in the feedstocks remain in the digestate, while nitrogen initially bound in organic forms is mineralised into ammonium nitrogen NH_4 (Bolzonella et al., 2018; Möller and Müller, 2012). As a result, digestate is rich in ammonium nitrogen that is readily available to plants, which on one hand increases its fertiliser value and on the other creates a risk of nitrogen losses during storage and application (Drapanauskaite et al., 2021).

A key challenge is therefore the sustainable management of digestate in a way that allows the recovery of its resources while minimising negative environmental impacts. Raw digestate is characterised by a very high water content, often exceeding 90%, and by significant variability in composition depending on the feedstock and process conditions (Van Puffelen et al., 2022). The material may also contain contaminants such as heavy metals, pathogens, antibiotic residues, microplastic particles, or other organic pollutants, especially when municipal waste or sewage sludge is used as feedstock (Barampouti et al., 2020; Cristina et al., 2020). Studies indicate that excessive accumulation of digestate in soil leads to the buildup of heavy metals, causing toxicity to crops and posing risks to human health (Luisa de Castro e Silva et al., 2023; Zhao et al., 2022).

Sustainable Biogas Production. DOI: https://doi.org/10.1016/B978-0-443-40557-0.00001-9

The presence of pathogens may further limit the direct use of digestate if it has not undergone sufficient sanitisation (Nag et al., 2020). Many countries have introduced regulations defining maximum allowable levels of heavy metals and sanitary requirements for digestate based fertilisers, but standards for antibiotics or microplastics are still lacking (Sobhi et al., 2024). These regulatory gaps highlight the need for additional digestate treatment or stricter quality standards to ensure safe application.

In response to these challenges, advanced digestate processing technologies are being developed with the aim of increasing its fertiliser value while reducing its environmental burden. In line with the principles of the circular economy, digestate is viewed as a source of valuable resources that should be recovered and returned to the soil (Abdelfatah Aldayyat and Gómez, 2025). The European Union promotes such solutions, among others, through new regulations. Regulation EU 2019/1009 on fertilising products allows digestate based fertilisers to be placed on the market with CE marking, provided that they meet defined quality requirements. Digestates derived from sewage sludge are not eligible as CMC 4 or CMC 5 materials under Regulation (EU) 2019/1009; however, other recovered materials derived from sludge treatment may be eligible only if they meet the specific criteria of another applicable CMC. (European Union, 2019).

This chapter presents a review of advanced digestate management technologies, with particular emphasis on the most recent developments and trends. Methods for converting digestate into high quality fertilisers that meet stringent quality standards are discussed, and selected case studies of innovative solution implementations are presented.

4.2 Latest advances in digestate processing

4.2.1 Phase separation

Most installations begin digestate processing with separation into a solid and a liquid fraction. This is a key step because the solid fraction contains a large share of phosphorus and dry matter, while the liquid fraction contains most of the ammonium nitrogen and potassium (Drosg et al., 2015; Tambone et al., 2017). Separation of the solid fraction facilitates further treatment. The solid material can be subjected to thermal processes or composting, whereas the liquid fraction can be treated using physicochemical methods (Tambone et al., 2017).

Numerous improvements have been introduced to increase separation efficiency. For example, sorptive additives such as biochar impregnated with magnesium or calcium salts can significantly enhance the transfer of nutrients into the solid phase. In the study by Morozova and Lemmer (2022), the addition of biochar impregnated with $MgCl_2$ and synthesised at 600°C for 20 hours increased the transfer of ammonium nitrogen and phosphorus to the solid fraction by approximately 76% and 47%, respectively (Morozova and Lemmer, 2022). This approach improves separation performance and results in a more concentrated solid fertiliser.

At the same time, the liquid fraction is increasingly treated using advanced membrane filtration systems based on microfiltration, ultrafiltration, and reverse osmosis (Zielińska and Bułkowska, 2025). Cascaded membrane configurations allow the production of clean water and fertiliser concentrates. In membrane installations, the permeate can account for up to 50% of the total volume under favourable configurations (Bolzonella et al., 2018). At the same time, a concentrated stream containing most of the nutrients is produced, which significantly reduces fertiliser transport and storage costs. For example, in the GENIUS technology implemented in the Netherlands, two stage separation followed by reverse osmosis produces purified water, while nitrogen and potassium are concentrated in the remaining 31% of the volume as a liquid fertiliser in the form of a reverse osmosis concentrate (Van Puffelen et al., 2022).

Modern membranes, especially ceramic membranes used in microfiltration and ultrafiltration, are characterised by high durability and are becoming increasingly cost effective due to improvements in cleaning procedures and resistance to fouling. In addition, membrane technologies have already reached a high level of technological maturity at TRL 9 and are successfully operating in many biogas plants, significantly reducing the volume of digestate applied directly to land (EBA, 2024).

4.2.2 *Chemical hydrolysis*

Alongside the concentration of mineral nutrients, technologies that also utilise the organic fraction of digestate are gaining increasing importance. Digestate processing makes use of acid, alkaline, or thermal hydrolysis processes (Samoraj et al., 2022). These processes lead to protein degradation, resulting in the formation of free amino acids and short peptides. These compounds are readily taken up by plants and therefore stimulate plant growth and development . Acid and alkaline hydrolysis mainly employ strong acids such as H_2SO_4 and HCl, as well as bases such as KOH and NaOH (Mikula et al., 2023). Both significant decreases and increases in pH result in a sanitising effect, reducing the microbial/pathogen load in the treated digestate. In the valorisation of renewable raw materials, acid hydrolysis is often more efficient, although it requires an additional neutralisation step (Kurbanoglu, 2004). In contrast, alkaline hydrolysis proceeds more slowly and often does not lead to complete protein breakdown (Coward- Kelly et al., 2006).

Mikula et al. (2023) described digestate sanitisation carried out through acid or alkaline hydrolysis, aimed at both material disinfection and protein conversion into free amino acids and short peptides. It was shown that acid hydrolysis resulted in a much higher pool of amino acids than alkaline hydrolysis, with total concentrations of 1981 mg per liter and 186 mg/L, respectively. Subsequently, Skrzypczak et al. presented a method for producing a solid fertiliser product through the neutralisation and granulation of an acid digestate based hydrolysate using ash from sewage sludge incineration. It was demonstrated that slow dosing of ash into the hydrolysate reduced nitrogen losses sixfold compared to the reverse order of addition (Skrzypczak et al., 2023). In the upscaled variant, the composition was adjusted with nitrogen in the form of urea with a urease inhibitor, and the resulting product met the requirements of Regulation EU 2019/1009 for a solid inorganic macronutrient fertiliser.

This approach enables the simultaneous utilisation of the organic fraction of digestate and an increase in its fertiliser value through the production of compounds that are more readily available to plants. At the same time, sanitisation combined with the possibility of stabilisation and product shaping, for example through neutralisation and granulation, supports safer agricultural use and facilitates the implementation of solutions compliant with regulatory requirements.

4.2.3 Ammonia recovery

Nitrogen is a key fertiliser component in digestate, but in its ammonium form it is prone to losses. The most widely used approach is ammonia stripping combined with its capture in an acidic solution, commonly referred to as ammonia stripping with acid absorption. The process involves converting ammonium ions present in digestate into free ammonia NH_3, which is achieved by increasing pH, for example through the addition of alkaline agents, and by raising the temperature, typically to values from 70°C to 80°C. Subsequently, a gas stream such as air or steam is passed through the liquid, forcing the desorption of ammonia into the gas phase (EBA, 2024).

The carrier gas containing ammonia is then directed to an absorption column containing a strong acid, most commonly sulphuric acid, where ammonia is bound to form stable ammonium sulphate (EBA, 2024). The resulting product is an ammonium sulphate solution with a typical concentration of 30%–40%, which is a fully valuable nitrogen fertiliser used in agriculture. The efficiency of nitrogen removal using this method is very high, and in practice a reduction of ammonium nitrogen of 50%–96% is achieved depending on process conditions (Ochs et al., 2022). Under optimal conditions, up to approximately 95% of total ammonium nitrogen can be recovered (Limoli et al., 2016; Rahman et al., 2026).

Ammonia stripping technology is also mature, corresponding to TRL 9, and is commercially applied. However, it requires prior removal of suspended solids from digestate in order to avoid scaling and fouling of the equipment (Bolzonella et al., 2018; EBA, 2024). One of the remaining challenges is the relatively high demand for thermal energy required to heat the liquid, as well as the cost of the chemical reagent, namely the acid. Despite these drawbacks, many biogas plants have implemented stripping based nitrogen removal systems in order to comply with local regulatory constraints.

An alternative to conventional air stripping is offered by membrane based ammonia capture systems, known as gas permeable membranes, which use hydrophobic membranes to extract NH_3 and absorb it on the opposite side of the membrane in an acidic solution. Pilot scale studies of such systems, for example using polytetrafluoroethylene tubing, have shown that they can recover ammonia with efficiencies comparable to air stripping while requiring lower energy input, as the process can operate at lower temperatures (Palakodeti et al., 2022; Riaño et al., 2021). Although these technologies are still at the demonstration stage, they represent a promising pathway for increasing the efficiency of nitrogen recovery from digestate.

4.2.4 Struvite precipitation

Phosphorus present in digestate is largely associated with the solid fraction, mainly in the form of calcium and magnesium phosphates or organic bound particles (Melgaço et al., 2021; Tuszynska et al., 2021). Its direct availability to plants is often limited, while excessive phosphorus application to agricultural land increases the risk of water eutrophication. For this reason, processes aimed at precipitating phosphorus into an isolated mineral salt that is easy to transport and apply are being increasingly developed. The most well-known approach is struvite precipitation.

Struvite forms spontaneously under conditions of elevated pH and in the presence of magnesium, ammonium, and phosphate ions (Doyle and Parsons, 2002; Le Corre et al., 2009). While this process is commonly associated with problematic scale formation in pipelines, controlled precipitation allows it to be converted into a valuable fertiliser product. In modern installations, a source of magnesium ions such as magnesium oxide or magnesium chloride is added to digestate, and phosphate may also be supplemented if molar ratios are unbalanced. The pH is then adjusted to alkaline values of approximately 8–9 to initiate struvite crystallisation.

Recent studies demonstrate very high efficiency of this method. Nagarajan et al. (2024) achieved at least 95% recovery of total ammoniacal nitrogen in the form of struvite from various types of digestate, while simultaneously reducing volatile fatty acid concentrations by approximately 61.5%. The precipitated struvite occurs as a crystalline powder or fine granules that are easy to transport. The nitrogen and phosphorus contained in struvite are released slowly, which makes it a slow release fertiliser (Valle et al., 2022; Wang et al., 2025). Application of struvite does not result in sudden nutrient release into water bodies, while still providing effective long term plant nutrition (Valle et al., 2022).

In one study, precipitation yields ranging from 0.33 to 1.0 g of struvite per 10 mL of digestate were reported, depending on digestate type (Nagarajan et al., 2024). This corresponds to a potential production of approximately 30–100 kg of struvite per cubic metre of digestate. Struvite precipitation technology is being intensively studied and implemented primarily in wastewater treatment plants, but it is also increasingly adopted in agricultural biogas plants as a method for phosphorus recovery (Trotta et al., 2023).

Importantly, struvite has a growing market value. It is marketed as an environmentally friendly phosphorus fertiliser, and increasing interest from farmers has been observed due to the depletion of global phosphate rock resources (Scholz et al., 2013). Other forms of phosphorus precipitation from digestate include calcium phosphates, for example those formed through lime addition resulting in hydroxyapatite, as well as multicomponent NPK type precipitates.

4.2.5 Thermal treatment processes

The solid fraction of digestate contains a substantial share of dry matter and organic compounds that were not degraded during anaerobic digestion, such as lignin. Instead of applying this material directly as a soil amendment, there is growing interest in thermal processing aimed at recovering energy, carbon, and mineral nutrients in more stable forms (Abdelfatah -Aldayyat et al., 2024). Several technologies are applied for this purpose, including combustion (Dziedzic et al., 2022), gasification, pyrolysis (Wiśniewski et al., 2015), and hydrothermal carbonisation (Mikusińska et al., 2023).

Combustion of digestate, most often after drying, enables complete utilisation of organic matter and generates ash rich in phosphorus and potassium. However, it is associated with nitrogen losses and requires compliance with emission standards similar to those applied to biomass combustion (Mammarella et al., 2024; Mellmann et al., 2024). More advanced options include gasification, which involves partial oxidation (Gao et al., 2023), and pyrolysis, defined as thermal decomposition under oxygen free conditions (V and Panwar, 2024). Digestate pyrolysis is typically conducted at temperatures of 400°C–600°C in the absence of oxygen, resulting in the formation of bio oil as a condensable liquid fraction, synthesis gas, and solid biochar (EBA, 2024; Gao et al., 2023; V and Panwar, 2024). Bio oil and gas can be used for energy generation, for example by combustion in boilers or engines, thereby increasing the overall energy output of biogas plants. Biochar represents a carbon rich residue containing stable organic carbon and mineral nutrients, primarily phosphorus (Patel and Ahammed, 2025).

Biochar derived from digestate exhibits valuable properties. Its high porosity and high content of stable carbon make it an effective sorbent for contaminants, for instance in the removal of heavy metals from water (Inyang et al., 2012), and a means of carbon sequestration in soil (Bekchanova et al., 2024). The incorporation of biochar into soil can improve soil structure and water holding capacity, and it can also act as a buffer for nutrients, reducing nitrogen leaching (Edeh et al., 2020).

Another category of technologies includes hydrothermal processes, in particular hydrothermal carbonisation and hydrothermal liquefaction (Tekin et al., 2014; Toor et al., 2011). Hydrothermal carbonisation involves the treatment of wet material, such as digestate without prior drying, in water under elevated pressure at temperatures of approximately 180°C–250°C for periods ranging from several tens of minutes to several hours (EBA, 2022; Hämäläinen et al., 2021). Under these conditions, partial biomass decomposition and polymerisation of organic compounds occur, leading to the formation of hydrochar. This solid product exhibits properties similar to lignite in terms of carbonisation degree and calorific value (Cao et al., 2021; Libra et al., 2011).

Hydrochar produced from digestate retains most of the phosphorus, often in forms more available than in raw digestate, as well as a portion of nitrogen. In contrast, a significant share of nitrogen and potassium remains in the aqueous post reaction liquor (Abdelfatah Aldayyat and Gómez, 2025). An important advantage of hydrothermal carbonisation is the elimination of the need for feedstock drying, as the process occurs directly in water. The resulting hydrochar can be used as a solid soil amendment, similarly to biochar by increasing the pool of stable carbon in soil (Hagner et al., 2025), or as a solid fuel, as its calorific value is usually higher than that of raw biomass (Xiao et al., 2012). Key challenges associated with hydrothermal carbonisation include management of the liquid byproduct and the costs related to energy input and pressurised equipment (Abdelfatah Aldayyat and Gómez, 2025).

Similarly, hydrothermal liquefaction, conducted at higher temperatures of approximately 250 to 370 degrees Celsius, enables the conversion of wet digestate without drying (Fan et al., 2022). In this case, the primary objective is the production of bio crude oil from the organic fraction of digestate (Hegdahl et al., 2022). This oil can subsequently be refined into liquid biofuels. Hydrothermal liquefaction remains at the research stage, but preliminary results suggest that it is promising for digestate management. Nevertheless, challenges related to effluent treatment and process stability still need to be addressed (Abdelfatah Aldayyat and Gómez, 2025).

In summary, thermal processes allow a drastic reduction in the volume of digestate residues and enable the production of stable products such as ash, biochar, and hydrochar that are rich in mineral nutrients and may find agricultural or industrial applications. Moreover, the concept of integrating anaerobic digestion with thermal processes increases the overall energy efficiency of the system and supports the idea of a closed loop biorefinery (Monlau et al., 2015). However, full scale implementation of such solutions still requires further refinement of economic aspects, including high capital and operational costs that are only partially offset by energy recovery, as well as environmental considerations such as emissions from combustion and the need for proper management of process waters.

4.2.6 Biological conversion

The technologies discussed above are complemented by solutions based on biological processes. These include, among others, constructed wetland systems, the cultivation of algae and aquatic plants on digestate, and microbiological biotechnologies (Chuka-ogwude et al., 2020; Yan et al., 2024; Comino et al., 2013). Constructed wetlands rely on passing digestate, especially the liquid fraction after preliminary treatment, through gravel and vegetation beds where microorganisms and plant roots remove contaminants and uptake nutrients (Sengupta and Dalwani, 2008; Stottmeister et al., 2003). This technology has proven effective in wastewater treatment and is also applied to digestate, particularly in small scale farms due to its simplicity and low operating costs (Comino et al., 2013).

Another approach involves the use of microalgae for digestate bioremediation (Chuka- ogwude et al., 2020; Fernandes et al., 2022). Algal cultures such as Chlorella and Scenedesmus grown in digestate based media can efficiently assimilate nitrogen and phosphorus while simultaneously producing biomass rich in proteins or lipids (Chuka ogwude et al., 2020; Judd et al., 2015). Studies show that microalgae are capable of growing on appropriately diluted digestate effluents and removing significant amounts of ammonium nitrogen (Salbitani and Carfagna, 2021). A key requirement is the adjustment of cultivation conditions, including pH, dilution rate, and the availability of carbon dioxide and light, to algal needs. High ammonia concentrations or the dark colour of digestate may inhibit algal growth, but pretreatment measures such as digestate filtration or pH control can mitigate these effects (Fernandes et al., 2022). An important advantage of algal processes is the simultaneous utilisation of nutrients and reduction of pollutant loads. Algae are also capable of capturing heavy metals and pharmaceutical residues (Morillas España et al., 2025; Sobolewska et al., 2024).

In parallel, phytoremediation approaches using aquatic plants are being developed. Species such as duckweed of the genus Lemna and water lettuce Pistia stratiotes can be cultivated on diluted digestate or post digestion effluents, where their rapid growth enables intensive uptake of nitrogen and phosphorus from water (Chau et al., 2023; Pena et al., 2017). The harvested plant biomass can then be valorised, for example as a feed component or as a substrate for further anaerobic digestion, thereby supporting nutrient cycle closure.

Advanced solutions often combine these approaches into integrated systems. For example, after separation of the solid fraction and partial nitrogen recovery through ammonia stripping, the remaining liquid effluent can be directed to algal ponds or constructed wetlands to remove residual nitrogen, phosphorus, and trace contaminants. Such multistage concepts are referred to as integrated digestate management. Analyses emphasise that life cycle assessment and techno economic analysis are essential tools in the design of such systems (Mancuso et al., 2024). These methods allow the identification of optimal technology combinations, for example balancing energy costs with environmental benefits. Life cycle assessment results indicate that intensive digestate processing can significantly reduce eutrophication potential and global warming potential compared with simple land application scenarios, despite additional energy inputs (Drapanauskaite et al., 2021).

For instance, Drapanauskaite et al. (2021) compared a baseline scenario involving liquid digestate application to fields with two digestate processing variants leading to the production of solid ammonium bicarbonate fertiliser via distillation and via conventional ammonia stripping to ammonium sulphate. The authors demonstrated that the advanced approaches reduced the overall carbon footprint and eutrophication potential (Drapanauskaite et al., 2021). Similar assessments confirm that advanced digestate processing methods enable nutrient cycle closure and limit negative emissions, in line with the principles of sustainable agriculture and the circular economy (Mancuso et al., 2024).

It should be noted, however, that these benefits come at the cost of additional technological inputs. Full scale intensive digestate treatment can be expensive, and estimates indicate that advanced treatment of the liquid fraction may reach costs on the order of ten to fifteen euros per cubic metre of digestate in the analysed scenarios (Mancuso et al., 2024). Consequently, ongoing research focuses on improving the energy efficiency of these processes, for example through heat recovery and the use of renewable energy sources, as well as on identifying lower cost reagents and materials, such as natural sorbents or ashes used as magnesium sources for struvite precipitation.

4.3 Methods for converting digestate into high quality fertilisers

One of the main objectives of the advanced technologies discussed earlier is the production of digestate derived products with the characteristics of high-quality fertilisers that could compete with or complement conventional mineral fertilisers. High quality in this context refers to stability and predictability of composition, ease of transport and application, sanitary and environmental safety, and compliance with legal standards for agricultural inputs.

4.3.1 Nitrogen fertilisers

Thanks to nitrogen recovery technologies, it is possible to transform diluted ammonium nitrogen from digestate into concentrated ammonium salt solutions

(Ghyselbrecht et al., 2018). The most commonly obtained product is an ammonium sulphate solution, which is widely used for crops such as rapeseed, maize, and grasses (Menkveld and Broeders, 2018). As a result, digestate derived fertilisers can become fully functional substitutes for synthetic fertilisers. Their broader adoption, however, depends on meeting legal requirements. In the European Union, a digestate-derived nitrogen fertiliser may be placed on the market as an EU fertilising product only if the final product meets the requirements of an applicable PFC and contains only component materials allowed under the relevant CMCs of Regulation (EU) 2019/1009. Otherwise, it remains subject to national rules. (European Union, 2019).

Another nitrogen compound that can be produced is ammonium bicarbonate NH_4HCO_3. Although it is not a widely used fertiliser due to its thermal instability, research has shown that it can be obtained from digestate in the form of solid crystals. Drapanauskaitė et al. (2020) developed a process for converting the liquid fraction of digestate into a solid nitrogen fertiliser in the form of ammonium bicarbonate. The process is based on digestate distillation under elevated pressure, condensation of a stream containing components of the NH_3–CO_2–H_2O system, and low temperature crystallisation of the product, which allows the production of highly pure ammonium bicarbonate. This product, considered a potentially marketable fertiliser, contains approximately 18% nitrogen.

The study also compared the distillation based pathway producing ammonium bicarbonate with an alternative nitrogen recovery route involving ammonium sulphate production via air stripping. Lower capital costs and reduced demand for utilities were demonstrated for the ammonium bicarbonate process. Life cycle assessment indicated a significant reduction in eutrophication potential compared with direct land application of digestate, while showing a similar global warming potential. At the same time, the analysis highlighted a substantial ecotoxicity burden associated with the required use of antifoaming agents (Drapanauskaite et al., 2021). Although this solution remains at the development stage, it represents an example of an innovative direction focused on producing nonstandard nitrogen fertilisers from digestate, provided that they prove advantageous in terms of efficiency or environmental impact.

4.3.2 Phosphorus fertilisers

Struvite precipitated in crystallisation processes is a valuable phosphorus fertiliser. It is a well-defined magnesium ammonium phosphate compound, which means it can be produced as a high purity product when crystalliszation and separation conditions are properly optimised (Zhang et al., 2019). At the same time, it is a concentrated source of phosphorus with limited solubility, which translates into gradual nutrient release in soil and a lower risk of losses compared with readily soluble phosphorus fertilisers (Wang et al., 2023). Unlike raw digestate or compost, struvite has a strictly defined composition, which facilitates standardisation and dosing (González- Morales et al., 2021).

Another phosphorus fertiliser form that can be obtained from digestate is phosphorus containing ash produced through combustion of the solid digestate fraction or of digestate that has undergone prior treatment (Mammarella et al., 2024).

In ash, phosphorus occurs in mineral forms that are often highly concentrated, creating potential for further use either directly as a fertiliser material or as a feedstock for manufacturing mineral phosphorus fertilisers (Herzel et al., 2016). Regulation EU 2019/1009 also covers ashes, which may enable their use as a raw material for phosphorus fertilisers. However, this depends on ash quality, including low heavy metal content and the absence of contamination originating from fossil fuels. Digestate derived from clean agricultural feedstocks typically produces ash with low contaminant levels, making fertiliser use feasible in many cases.

In summary, advanced methods make it possible to isolate phosphorus from digestate in forms that are comparable in quality to traditional phosphorus fertilisers, which is increasingly important given the shrinking global reserves of phosphate rock.

4.3.3 Multinutrient fertilisers

Digestate itself is a multinutrient fertiliser because it contains nitrogen, phosphorus, and potassium, as well as micronutrients and organic matter (Martín- Sanz- Garrido et al., 2025). The challenge, however, has been its diluted form and variable composition. With the technologies described above, it is possible to produce products with more predictable characteristics.

A good example of this approach is the full scale Dutch GENIUS system described by Van Puffelen et al. (2022), in which the separation cascade includes two decanters, microfiltration, double reverse osmosis, and an ion exchange unit. In this configuration, phosphorus is largely captured before the reverse osmosis stage. Specifically, 66% of phosphorus was recovered in the first solid fraction, and in total 98% of phosphorus was removed prior to reverse osmosis. The reverse osmosis concentrate contained 34% of the total nitrogen originally present in the digestate, and the authors emphasised that nitrogen occurred exclusively in mineral form, supporting the interpretation of this stream as an alternative to synthetic nitrogen fertiliser. This means that two separate fertiliser products were obtained: a liquid NK concentrate and a solid organic phosphorus fertiliser. Such a separation is advantageous because it enables more precise fertilisation with nitrogen and phosphorus independently, according to agronomic needs.

In parallel with stream splitting into an NK concentrate and a solid fraction, the literature also describes approaches based on the deliberate blending of multiple feedstocks and additives in order to obtain multinutrient fertilisers with a targeted composition. In the study by Mikula et al. (2023), digestate underwent sanitisation and hydrolysis using strong inorganic acids. After centrifugation, the liquid phase was used to formulate a foliar fertiliser, while the solid fraction together with the hydrolysate, ash from sewage sludge incineration, and mineral additives were used to produce several variants of granular fertilisers. As a result, formulations with declared macronutrient content were obtained, including a granulated product classified as a multinutrient fertiliser, in which the presence of at least two primary nutrients as well as secondary nutrients at levels allowing declaration was confirmed (Mikula et al., 2023). Skrzypczak et al. (2023) presented a method for producing a digestate based granular fertiliser in which, following acid sanitisation, neutralisation and granulation were carried out using ash from sewage sludge incineration. At the upscaled stage, the composition was further adjusted with nitrogen in the form of urea combined with a urease inhibitor. The authors showed that the final product met the requirements for a solid organo-mineral fertiliser (Skrzypczak et al., 2023).

Advanced processing technologies can overcome the key limitations of digestate, namely its dilution and compositional variability, by converting it into standardised fertiliser products with predictable properties. The reported studies show that, through stream separation as well as targeted blending and granulation, it is possible to obtain both mineral fertilisers suitable for precision application and solid organo-mineral fertilisers that meet quality and regulatory requirements.

4.4 Conclusion

The analysis of modern digestate management methods clearly shows that the future development of biogas plants in Europe no longer depends solely on energy production, but increasingly on the ability to manage the material streams generated after digestion effectively. Digestate is no longer just a problematic waste stream and is becoming a key link in the cycling of nutrients and carbon. The available evidence indicates that, without additional processing, digestate cannot be managed safely and sustainably at large scale. Its high water content, variable composition, and the presence of contaminants mean that simple land application can overload local agroecosystems, drive nutrient losses, and increase environmental risk. In this sense, untreated digestate becomes a bottleneck for the entire biogas sector.

At the same time, progress in separation technologies, nitrogen and phosphorus recovery, thermal processing, and biological methods demonstrates that technical solutions already exist. A key conclusion is that the constituents of digestate can be converted into products with quality comparable to mineral fertilisers or soil amendments, rather than only into low value organic fertilisers. This creates a realistic opportunity to substitute synthetic fertilisers and reduce dependence on imported resources such as phosphate rock and natural gas. From a systems perspective, it is particularly important that digestate processing technologies make it possible to separate nitrogen cycling from phosphorus cycling. In practice, this enables more precise fertilisation aligned with crop and soil needs, which significantly reduces losses to the environment. Digestate no longer needs to be applied as a single heterogeneous material, but can instead serve as a source of specialised fertiliser streams. The results indicate that intensive digestate processing delivers clear environmental benefits compared with direct land application. Reduced eutrophication potential and lower greenhouse gas emissions mean that advanced digestate management systems should be viewed as an integral part of environmental protection infrastructure, not merely as an add on to biogas plants. At the same time, economics remains a major barrier. Capital and operating costs mean that full digestate processing is not always justified for small installations. This points to the need for regional models, technology clusters, and support instruments that can enable economies of scale.

In summary, digestate should be seen not as a secondary problem of anaerobic digestion, but as its most important carrier of material value. Whether biogas plants become a pillar of sustainable agriculture and resource security will depend primarily on the extent to which digestate is transformed into standardised, safe, and marketable fertiliser products.

AI disclosure

During the preparation of this work the author(s) used DEEPL, ChatGPT, Grammarly, Mendeley in order to Translation, language correction, bibliography management. After using this tool/service, the author(s) reviewed and edited the content as needed and take(s) full responsibility for the content of the publication.

References

Abdelfatah-Aldayyat, E., Gómez, X., 2025. Hydrothermal Treatment of Digestates: Challenges and Perspectives. Environments 12 (10), 347. Available at https://doi.org/10.3390/environments12100347.

Abdelfatah-Aldayyat, E., González-Rojo, S., Gómez, X., 2024. Reviewing Digestate Thermal Valorization: Focusing on the Energy Demand and the Treatment of Process Water. Environments 11 (11), 239. Available at https://doi.org/10.3390/environments11110239.

Barampouti, E.M., Mai, S., Malamis, D., Moustakas, K., Loizidou, M., 2020. Exploring technological alternatives of nutrient recovery from digestate as a secondary resource. Renewable and Sustainable Energy Reviews 134. Available at https://doi.org/10.1016/j.rser.2020.110379 110379.

Bekchanova, M., Kuppens, T., Cuypers, A., Jozefczak, M., Malina, R., 2024. Biochar's effect on the soil carbon cycle: a rapid review and meta-analysis. Biochar 6 (1). Available at https://doi.org/10.1007/s42773-024-00381-8.

Bolzonella, D., Fatone, F., Gottardo, M., Frison, N., 2018. Nutrients recovery from anaerobic digestate of agro-waste: Techno-economic assessment of full scale applications. Journal of Environmental Management 216, 111–119. Available at https://doi.org/10.1016/j.jenvman.2017.08.026.

Cao, Y., He, M., Dutta, S., Luo, G., Zhang, S., Tsang, D.C.W., 2021. Hydrothermal carbonization and liquefaction for sustainable production of hydrochar and aromatics. Renewable and Sustainable Energy Reviews 152, 111722. Available at https://doi.org/10.1016/j.rser.2021.111722.

Burg, V., Bowman, G., Haubensak, M., Baier, U., Thees, O., 2018. Valorization of an untapped resource: energy and greenhouse gas emissions benefits of converting manure to biogas through anaerobic digestion. Resources, Conservation and Recycling 136, 53–62. Available at https://doi.org/10.1016/j.resconrec.2018.04.004.

Chau, N.N.V., Van, T.H., Cong, T.N., Kim, L., Van Pham, D., 2023. Water lettuce (*Pistia stratiotes* L.) increases biogas effluent pollutant removal efficacy and proves a positive substrate for renewable energy production. PeerJ 11. Available at https://doi.org/10.7717/peerj.15879.

Chuka-ogwude, D., Ogbonna, J., Moheimani, N.R., 2020. A review on microalgal culture to treat anaerobic digestate food waste effluent. Algal Research 47, 101841. Available at https://doi.org/10.1016/j.algal.2020.101841.

Comino, E., Riggio, V.A., Rosso, Maurizio, 2013. Constructed wetland treatment of agricultural effluent from an anaerobic digester. Ecological Engineering 54, 165–172. Available at https://doi.org/10.1016/j.ecoleng.2013.01.027.

Coward-Kelly, G., Agbogbo, F.K., Holtzapple, M.T., 2006. Lime treatment of keratinous materials for the generation of highly digestible animal feed: 2. Animal hair. Bioresource Technology 97 (11), 1344–1352. Available at https://doi.org/10.1016/j.biortech.2005.05.017.

Cristina, G., Camelin, E., Tommasi, T., Fino, D., Pugliese, M., 2020. Anaerobic digestates from sewage sludge used as fertilizer on a poor alkaline sandy soil and on a peat substrate: effects on tomato plants growth and on soil properties. Journal of Environmental Management 269. Available at https://doi.org/10.1016/j.jenvman.2020.110767 110767.

Doyle, J.D., Parsons, S.A., 2002. Struvite formation, control and recovery. Water Research 36 (16), 3925–3940. Available at https://doi.org/10.1016/S0043-1354(02)00126-4.

Drapanauskaite, D., Handler, R.M., Fox, N., Baltrusaitis, J., 2021. Transformation of liquid digestate from the solid-separated biogas digestion reactor effluent into a solid NH_4HCO_3 fertilizer: sustainable process engineering and life cycle assessment. ACS Sustainable Chemistry and Engineering 9 (1), 580–588. Available at https://doi.org/10.1021/acssuschemeng.0c08374.

Drosg, B., Fuchs, W., Al, T., Madsen, S.M., Linke, B., 2015. Nutrient Recovery by Biogas Digestate Processing. IEA Bioenergy. Available at https://www.ieabioenergy.com/wp-content/uploads/2015/08/NUTRIENT_RECOVERY_RZ_web2.pdf.

Dziedzic, K., Łapczyńska-Kordon, B., Jurczyk, M., Wróbel, M., Jewiarz, M., Mudryk, K., et al., 2022. Solid digestate—mathematical modeling of combustion process. Energies 15 (12). Available at https://doi.org/10.3390/en15124402.

EBA (2022). A mature business opportunity to enhance sustainable carbon cycles and untap the circularity and climate benefits of biogas production. https://www.europeanbiogas.eu/wp-content/uploads/2025/06/Biogenic-CO2-from-the-biogas-industry_Sept2022-1.pdf

Edeh, I.G., Mašek, O., Buss, W., 2020. A meta-analysis on biochar's effects on soil water properties – New insights and future research challenges. Science of The Total Environment 714, 136857. Available at https://doi.org/10.1016/j.scitotenv.2020.136857.

EBA (2024). Exploring digestate's contribution to healthy soils. www.europeanbiogas.eu.

European Union, 2019. Regulation (EU) 2019/1009 of the European Parliament and of the Council. Official Journal of the European Union.

Fan, Y., Prestigiacomo, C., Gong, M., Tietz, T., Hornung, U., Dahmen, N., 2022. Comparative investigation on the value-added products obtained from continuous and batch hydrothermal liquefaction of sewage sludge. Frontiers in Environmental Science 10. Available at https://doi.org/10.3389/fenvs.2022.996353.

Fernandes, F., Silkina, A., Gayo-Peláez, J.I., Kapoore, R.V., de la Broise, D., Llewellyn, C.A., 2022. Microalgae cultivation on nutrient rich digestate: the importance of strain and digestate tailoring under PH control. Applied Sciences (Switzerland) 12 (11). Available at https://doi.org/10.3390/app12115429.

Gao, Y., Wang, M., Raheem, A., Wang, F., Wei, J., Xu, D., et al., 2023. Syngas production from biomass gasification: influences of feedstock properties, reactor type, and reaction parameters. ACS Omega 8 (35), 31620–31631. Available at https://doi.org/10.1021/acsomega.3c03050.

Ghyselbrecht, K., Monballiu, A., Somers, M.H., Sigurnjak, I., Meers, E., Appels, L., et al., 2018. Stripping and scrubbing of ammonium using common fractionating columns to prove ammonium inhibition during anaerobic digestion. International Journal of Energy and Environmental Engineering 9 (4), 447–455. Available at https://doi.org/10.1007/s40095-018-0283-7.

González-Morales, C., Fernández, B., Molina, F.J., Naranjo-Fernández, D., Matamoros-Veloza, A., Camargo-Valero, M.A., 2021. Influence of pH and Temperature on Struvite Purity and Recovery from Anaerobic Digestate. Sustainability 13 (19), 10730. Available at https://doi.org/10.3390/su131910730.

Hagner, M., Salmela, M.J., Ahmadi, S., Yaah, V.K., Ojala, S., Laitinen, T., et al., 2025. Biochar and hydrochar from organic side-streams induce species-specific responses in plants. Journal of Soil Science and Plant Nutrition 25 (3), 6262–6280. Available at https://doi.org/10.1007/s42729-025-02529-2.

Hämäläinen, A., Kokko, M., Kinnunen, V., Hilli, T., Rintala, J., 2021. Hydrothermal carbonisation of mechanically dewatered digested sewage sludge—Energy and nutrient recovery in centralised biogas plant. Water Research 201. Available at https://doi.org/10.1016/j.watres.2021.117284 117284.

Hegdahl, S.H., Løhre, C., Barth, T., 2022. Hydrothermal liquefaction of sewage sludge anaerobic digestate for bio-oil production: Screening the effects of temperature, residence time and KOH catalyst. Waste Management & Research: The Journal for a Sustainable Circular Economy 41 (5), 977–986. Available at https://doi.org/10.1177/0734242x221138497.

Herzel, H., Krüger, O., Hermann, L., Adam, C., 2016. Sewage sludge ash—a promising secondary phosphorus source for fertilizer production. Science of The Total Environment 542 (Pt B), 1136–1143. Available at https://doi.org/10.1016/j.scitotenv.2015.08.059.

Inyang, M., Gao, B., Yao, Y., Xue, Y., Zimmerman, A.R., Pullammanappallil, P., et al., 2012. Removal of heavy metals from aqueous solution by biochars derived from anaerobically digested biomass. Bioresource Technology 110, 50–56. Available at https://doi.org/10.1016/j.biortech.2012.01.072.

Judd, S., van den Broeke, L.J.P., Shurair, M., Kuti, Y., Znad, H., 2015. Algal remediation of CO2 and nutrient discharges: a review. Water Research 87, 356–366. Available at https://doi.org/10.1016/j.watres.2015.08.021.

Kurbanoglu, E.B., 2004. Enhancement of citric acid production with ram horn hydrolysate by Aspergillus niger. Bioresource Technology 92 (1), 97–101. Available at https://doi.org/10.1016/j.biortech.2003.08.002.

Le Corre, K.S., Valsami-Jones, E., Hobbs, P., Parsons, S.A., 2009. Phosphorus recovery from wastewater by struvite crystallization: a review. Critical Reviews in Environmental Science and Technology 39 (6), 433–477. Available at https://doi.org/10.1080/10643380701640573.

Limoli, A., Langone, M., Andreottola, G., 2016. Ammonia removal from raw manure digestate by means of a turbulent mixing stripping process. Journal of Environmental Management 176, 1–10. Available at https://doi.org/10.1016/j.jenvman.2016.03.007.

Libra, J.A., Ro, K.S., Kammann, C., Funke, A., Berge, N.D., Neubauer, Y., et al., 2011. Hydrothermal carbonization of biomass residuals: a comparative review of the chemistry, processes and applications of wet and dry pyrolysis. Biofuels 2 (1), 71–106. Available at https://doi.org/10.4155/bfs.10.81.

Luisa de Castro e Silva, H., Sigurnjak, I., Robles-Aguilar, A., Adriaens, A., Meers, E., 2023. Development of conversion factors to estimate the concentrations of heavy metals in manure-derived digestates. Waste Management 168, 334–343. Available at https://doi.org/10.1016/j.wasman.2023.06.008.

Mancuso, G., Habchi, S., Maraldi, M., Valenti, F., El Bari, H., 2024. Comprehensive review of technologies for separate digestate treatment and agricultural valorisation within circular and green economy. Bioresource Technology 409. Available at https://doi.org/10.1016/j.biortech.2024.131252 131252.

Martín-Sanz-Garrido, C., Revuelta-Aramburu, M., Santos-Montes, A.M., Morales-Polo, C., 2025. A Review on Anaerobic Digestate as a Biofertilizer: Characteristics, Production, and Environmental Impacts from a Life Cycle Assessment Perspective. Applied Sciences 15 (15), 8635. Available at https://doi.org/10.3390/app15158635.

Mellmann, J., Salamat, R., Kharaghani, A., 2024. Drying behavior of solid digestate and reaction kinetics of ammonium degradation during laboratory-scale drying. Waste Management 173, 75–86. Available at https://doi.org/10.1016/j.wasman.2023.11.007.

Mammarella, D., Di Giuliano, A., Gallucci, K., 2024. Reuse and Valorization of Solid Digestate Ashes from Biogas Production. Energies 17 (3), 751. Available at https://doi.org/10.3390/en17030751.

Melgaço, L., Robles-Aguilar, A., Meers, E., Mota, C., 2021. Phosphorus recovery from liquid digestate by chemical precipitation using low-cost ion sources. Journal of Chemical Technology and Biotechnology 96 (10), 2891–2900. Available at https://doi.org/10.1002/jctb.6842.

Menkveld, H.W.H., Broeders, E., 2018. Recovery of ammonia from digestate as fertilizer. Water Practice and Technology 13 (2), 382–387. Available at https://doi.org/10.2166/wpt.2018.049.

Mikula, K., Trzaska, K., Skrzypczak, D., Izydorczyk, G., Mironiuk, M., Gil, F., et al., 2023. Efficient anaerobic digestate valorization: nutrient recovery strategy for enhancing soil fertility in arid agricultural regions. Journal of Environmental Chemical Engineering 11 (5). Available at https://doi.org/10.1016/j.jece.2023.110522 110522.

Mikusińska, J., Kuźnia, M., Czerwińska, K., Wilk, M., 2023. Hydrothermal carbonization of digestate produced in the biogas production process. Energies 16 (14). Available at https://doi.org/10.3390/en16145458.

Möller, K., Müller, T., 2012. Effects of anaerobic digestion on digestate nutrient availability and crop growth: a review. Engineering in Life Sciences 12 (3), 242–257. Available at https://doi.org/10.1002/elsc.201100085.

Monlau, F., Sambusiti, C., Antoniou, N., Barakat, A., Zabaniotou, A., 2015. A new concept for enhancing energy recovery from agricultural residues by coupling anaerobic digestion and pyrolysis process. Applied Energy 148, 32–38. Available at https://doi.org/10.1016/j.apenergy.2015.03.024.

Morillas-España, A., Ruiz-Nieto, Á., Lafarga, T., Acién, G., Arbib, Z., González-López, C.V., 2022. Biostimulant Capacity of Chlorella and Chlamydopodium Species Produced Using Wastewater and Centrate. Biology 11 (7), 1086. Available at https://doi.org/10.3390/biology11071086.

Morozova, I., Lemmer, A., 2022. Nutrient recovery from digestate of agricultural biogas plants: a comparative study of innovative biocoal-based additives in laboratory and full-scale experiments. Molecules 27 (16). Available at https://doi.org/10.3390/molecules27165289.

Nagarajan, A., Chen, Y., Raghavan, V., Goyette, B., Rajagopal, R., 2024. Sustainable nutrient recovery through struvite precipitation from poultry and multi-substrate agricultural waste digestates. Bioresource Technology Reports 27. Available at https://doi.org/10.1016/j.biteb.2024.101924 101924.

Nag, R., Whyte, P., Markey, B.K., O'Flaherty, V., Bolton, D., Fenton, O., et al., 2020. Ranking hazards pertaining to human health concerns from land application of anaerobic digestate. Science of the Total Environment 710. Available at https://doi.org/10.1016/j.scitotenv.2019.136297.

Ochs, P., Martin, B., Germain-Cripps, E., Stephenson, T., van Loosdrecht, M., Soares, A., 2023. Techno-economic analysis of sidestream ammonia removal technologies: Biological options versus thermal stripping. Environmental Science and Ecotechnology 13, 100220. Available at https://doi.org/10.1016/j.ese.2022.100220.

Palakodeti, A., Azman, S., Dewil, R., Appels, L., 2022. Ammonia recovery from organic waste digestate via gas–liquid stripping: application of the factorial design of experiments and comparison of the influence of the stripping gas. Sustainability (Switzerland) 14 (24). Available at https://doi.org/10.3390/su142417000.

Patel, N.P., Ahammed, M.M., 2025. Biochar from anaerobic digestates: Production, characteristics and applications in wastewater remediation. Bioresource Technology Reports 31. Available at https://doi.org/10.1016/j.biteb.2025.102258.

Pena, L., Oliveira, M., Fragoso, R., Duarte, E., 2017. Potential of duckweed for swine wastewater nutrient removal and biomass valorisation through anaerobic co-digestion. Journal of Sustainable Development of Energy, Water and Environment Systems 5 (2), 127–138. Available at https://doi.org/10.13044/j.sdewes.d5.0137.

Proskynitopoulou, V., Lorentzou, S., Yaman, R., Blanch, G., Herbert, B., Rubio Rincon, F.J., et al., 2024. Mobile solution for digestate transformation to high added-value products. Journal of Cleaner Production 477. Available at https://doi.org/10.1016/j.jclepro.2024.143915.

Rahman, M.S., Shapiro Ellis, T., Freeburn, I.J.R., Rose, A., Thornton, A.W., Erler, D., 2026. Removal and recovery of ammonium nitrogen from dairy processing wastewater using air stripping technology: a pilot-scale study. Water (Switzerland) 18 (2). Available at https://doi.org/10.3390/w18020196.

Riaño, B., Molinuevo-Salces, B., Vanotti, M.B., García-González, M.C., 2021. Ammonia recovery from digestate using gas-permeable membranes: A pilot-scale study. Environments 8 (12), 133. Available at https://doi.org/10.3390/environments8120133.

Salbitani, G., Carfagna, S., 2021. Ammonium utilization in microalgae: a sustainable method for wastewater treatment. Sustainability 13 (2), 956. Available at https://doi.org/10.3390/su13020956.

Scholz, R.W., Ulrich, A.E., Eilittä, M., Roy, A., 2013. Sustainable use of phosphorus: a finite resource. Science of the Total Environment 461-462, 799–803. Available at https://doi.org/10.1016/j.scitotenv.2013.05.043.

Skrzypczak, D., Trzaska, K., Mikula, K., Gil, F., Izydorczyk, G., Mironiuk, M., et al., 2023. Conversion of anaerobic digestates from biogas plants: Laboratory fertilizer formulation, scale-up and demonstration of applicative properties on plants. Renewable Energy 203, 506–517. Available at https://doi.org/10.1016/j.renene.2022.12.080.

Samoraj, M., Mironiuk, M., Izydorczyk, G., Witek-Krowiak, A., Szopa, D., Moustakas, K., et al., 2022. The challenges and perspectives for anaerobic digestion of animal waste and fertilizer application of the digestate. Chemosphere 295, 133799. Available at https://doi.org/10.1016/j.chemosphere.2022.133799.

Sobhi, M., Elsamahy, T., Zakaria, E., Gaballah, M.S., Zhu, F., Hu, X., et al., 2024. Characteristics, limitations and global regulations in the use of biogas digestate as fertilizer: A comprehensive overview. Science of The Total Environment 957. Available at https://doi.org/10.1016/j.scitotenv.2024.177855 177855.

Sobolewska, E., Borowski, S., Nowicka-Krawczyk, P., 2024. Cultivation of Microalgae in Liquid Digestate to Remove Nutrients and Organic Contaminants. BioEnergy Research 17 (3), 1843–1855. Available at https://doi.org/10.1007/s12155-024-10753-4.

Stottmeister, U., Wießner, A., Kuschk, P., Kappelmeyer, U., Kästner, M., Bederski, O., et al., 2003. Effects of plants and microorganisms in constructed wetlands for wastewater treatment. Biotechnology advances 22 (1-2), 93–117. Available at https://doi.org/10.1016/j.biotechadv.2003.08.010.

Tambone, F., Orzi, V., D'Imporzano, G., Adani, F., 2017. Solid and liquid fractionation of digestate: mass balance, chemical characterization, and agronomic and environmental value. Bioresource Technology 243, 1251–1256. Available at https://doi.org/10.1016/j.biortech.2017.07.130.

Tekin, K., Karagöz, S., Bektaş, S., 2014. A review of hydrothermal biomass processing. Renewable and Sustainable Energy Reviews. 40, 673–687. Available at https://doi.org/10.1016/j.rser.2014.07.216.

Toor, S.S., Rosendahl, L., Rudolf, A., 2011. Hydrothermal liquefaction of biomass: A review of subcritical water technologies. Energy 36 (5), 2328–2342. Available at https://doi.org/10.1016/J.ENERGY.2011.03.01.

Trotta, S., Adani, F., Fedele, M., Salvatori, M., 2023. Nitrogen and phosphorus recovery from cow digestate by struvite precipitation: Process optimization to maximize phosphorus recovery. Results in Engineering 20. Available at https://doi.org/10.1016/j.rineng.2023.101478 101478.

Tuszynska, A., Czerwionka, K., Obarska-Pempkowiak, H., 2021. Phosphorus concentration and availability in raw organic waste and post fermentation products. Journal of Environmental Management 278, 111468. Available at https://doi.org/10.1016/j.jenvman.2020.111468.

V, K.K., Panwar, N.L., 2024. Pyrolysis technologies for biochar production in waste management: a review. Clean Energy 8 (4), 61–78. Available at https://doi.org/10.1093/ce/zkae036.

Valle, S.F., Giroto, A.S., Dombinov, V., Robles-Aguilar, A.A., Jablonowski, N.D., Ribeiro, C., 2022. Struvite-based composites for slow-release fertilization: a case study in sand. Scientific Reports 12 (1). Available at https://doi.org/10.1038/s41598-022-18214-8.

Van Puffelen, J.L., Brienza, C., Regelink, I.C., Sigurnjak, I., Adani, F., Meers, E., et al., 2022. Performance of a full-scale processing cascade that separates agricultural digestate and its nutrients for agronomic reuse. Separation and Purification Technology 297. Available at https://doi.org/10.1016/j.seppur.2022.121501 121501.

Vipat, V., Singh, U.R., Billore, S., 2008. Efficacy of Rootzone Technology for Treatment of Domestic Wastewater: Field Scale Study of a Pilot Project in Bhopal (MP), India. In: Sengupta, M., Dalwani, R. (Eds.), Proceedings of Taal2007: The 12th World Lake Conference, pp. 995–1003. Available at https://www.researchgate.net/publication/228484323_Efficacy_of_Rootzone_Technology_for_Treatment_of_Domestic_Wastewater_Field_Scale_Study_of_a_Pilot_Project_in_BhopalMP_India.

Wang, Y., Feng, X., Wen, Z., Chu, C., Luo, Y., Chen, H., et al., 2025. Expanding application of struvite as fertilizer in hydroponic cultivation by providing accumulative nutrients supply. Journal of Cleaner Production 532. Available at https://doi.org/10.1016/j.jclepro.2025.146976 146976.

Wang, J., Xue, L., Hou, P., Hao, T., Xue, L., Zhang, X., et al., 2023. Struvite as P fertilizer on yield, nutrient uptake and soil nutrient status in the rice–wheat rotation system: A two-year field observation. Agronomy 13 (12), 2948. Available at https://doi.org/10.3390/agronomy13122948.

Wiśniewski, D., Gołaszewski, J., Białowiec, A., 2015. The pyrolysis and gasification of digestate from agricultural biogas plant / Piroliza i gazyfikacja pofermentu z biogazowni rolniczych. Archives of Environmental Protection 41 (3), 70–75. Available at https://doi.org/10.1515/aep-2015-0032.

Wood Environment & Infrastructure Solutions UK Limited, 2019. Digestate and compost as fertilisers: risk assessment and risk management options. Report for the European Commission, DG Environment.

Xiao, L.P., Shi, Z.J., Xu, F., Sun, R.C., 2012. Hydrothermal carbonization of lignocellulosic biomass. Bioresource Technology 118, 619–623. Available at https://doi.org/10.1016/j.biortech.2012.05.060.

Yan, J., Wu, L., Ye, W., Zhou, J., Ji, Q., Alberto Gomez, M., et al., 2024. Ferric and sulfate coupled ammonium oxidation enhanced nitrogen removal in two-stage partial nitrification - anammox/denitrification process for food waste liquid digestate treatment. Bioresource Technology 398. Available at https://doi.org/10.1016/j.biortech.2024.130533 130533.

Zhang, H., Gong, W., Luo, X., Xie, B., Li, G., Liang, H., 2019. Obtaining high-purity struvite from anaerobically digested wastewater: effects of pH, Mg/P, and Ca 2+ Interactions. Environmental Engineering Science 36 (1), 102–113. Available at https://doi.org/10.1089/ees.2018.0065.

Zhao, H., Wu, Y., Lan, X., Yang, Y., Wu, X., Du, L., 2022. Comprehensive assessment of harmful heavy metals in contaminated soil in order to score pollution level. Scientific Reports 12 (1). Available at https://doi.org/10.1038/s41598-022-07602-9.

Zielińska, M., Bułkowska, K., 2025. Use of Membrane Techniques for Removal and Recovery of Nutrients from Liquid Fraction of Anaerobic Digestate. Membranes 15 (2). Available at https://doi.org/10.3390/membranes15020045.

Optimising biogas production: emerging trends and technologies

5

Biogas technology lies at the interface of renewable energy, waste management, and sustainable agriculture, embodying circular-economy principles. Converting organic wastes into methane-rich biogas and nutrient-rich digestate enables anaerobic digestion (AD) systems to generate renewable energy while producing bio-fertilisers that can displace synthetic alternatives (Adnane et al., 2025). Optimising biogas production is not only about improving energy yields but also about maximising environmental and economic benefits across the waste-to-energy-to-fertiliser continuum. Enhancements that increase methane output per unit of feedstock improve the overall energy balance and lower the unit cost of biogas, thereby supporting climate-change mitigation and resource recovery, outcomes widely recognised for AD. Optimisation strategies must also ensure system-level sustainability; any additional inputs (energy, chemicals) or process modifications should be evaluated through life-cycle assessment (LCA) and techno-economic analysis (TEA) to confirm net gains in greenhouse-gas (GHG) reduction and cost-effectiveness (Tian et al., 2021). Biogas process optimisation directly supports this book's circular-economy and sustainability goals by maximising energy recovery from organic waste, enhancing digestate use as bio-fertiliser, and maintaining favourable LCA/TEA profiles.

Research to improve biogas yield and process efficiency has intensified as AD has matured from a basic waste treatment method into a sophisticated bioresource technology (Náthia-Neves et al., 2018). Upstream interventions address feedstock limitations, including recalcitrant lignocellulosic structures and nutrient imbalances, alongside process innovations designed to enhance microbial performance (Mirmohamadsadeghi et al., 2021; Rahmani et al., 2022; Tabatabaei et al., 2020). A recent book chapter details practical methods to enrich feedstock characteristics and fuel properties for improved biogas generation (Chozhavendhan et al., 2020), complementing these research advances. Simultaneously, downstream upgrades, such as gas purification and power-to-gas (PtG) systems, are being integrated into biogas plants to maximise energy recovery and product quality (Assunção et al., 2021; Jensen et al., 2021; Miltner et al., 2016). Such technologies align with broader sustainability goals; for example, upgrading biogas to biomethane enables grid injection or use as vehicle fuel, displacing fossil fuels (Baena-Moreno et al., 2020; Gustafsson and Anderberg, 2022). Improvements in digestate management, including nutrient-recovery processes, ensure that the residual solids and liquids from AD can be safely and effectively used as fertilisers, closing the resource loop (Adnane et al., 2025). This chapter examines recent advances in three themes: enhancing biogas yield (Section 5.1) through pretreatment, co-digestion and additives; integrating novel technologies in biogas plants (Section 5.2), including advanced monitoring, biogas upgrading and PtG systems; and future directions (Section 5.3), which

Sustainable Biogas Production. DOI: https://doi.org/10.1016/B978-0-443-40557-0.00007-X

consider upcoming research frontiers and the evolving role of biogas in sustainable energy systems.

5.1 Latest research in enhancing biogas yield

Maximising the methane yield from available substrates is a primary objective in AD optimisation. A higher biogas yield delivers more energy from the same quantity of organic waste, improving overall efficiency and project economics (Yousuf et al., 2016). Contemporary research converges on three levers to enhance yield, broadly categorised as feedstock pretreatments, codigestion approaches and additive supplementation techniques (Tabatabaei et al., 2020). Recent reviews synthesise these strategies and document consistent methane gains across substrates when pretreatment, co-digestion and additive dosing are applied judiciously (Tabatabaei et al., 2020; Chuenchart et al., 2024; Pilarska et al., 2024). The strategies address distinct limiting factors within the AD process.

Many organic wastes contain lignocellulosic fibers that are only partially digestible by anaerobic microbiota; appropriate pretreatment increases biodegradability and reduces lignocellulose recalcitrance (Mirmohamadsadeghi et al., 2021). Among biological options, controlled enzymatic or microbial pretreatments can reduce lignocellulose recalcitrance and improve hydrolysis efficiency (Saggi et al., 2017).

A single substrate may also lack key nutrients or exhibit inhibitory imbalances, such as excess nitrogen or sulphur; codigestion with complementary substrates can deliver a more balanced feedstock mix (Chuenchart et al., 2024). The use of additives, ranging from trace-element supplements to novel materials, can relieve specific biochemical bottlenecks or toxicity issues and thereby boost methane production (Pilarska et al., 2024). A bibliometric review outlines nanomaterial-enhanced AD opportunities and challenges (Parra-Orobio et al., 2025). Table 5.1 summarises the major pretreatment families, their mechanisms for yield enhancement, typical improvements in methane output, energy and cost trade-offs, and best-fit substrate types (Mosleh Uddin et al., 2022; Tian et al., 2021; Sinsuw et al., 2024).

Pretreatment has long been recognised as crucial to anaerobic digestion (AD) performance (Mudhoo, 2012), and current work is moving toward more efficient, integrated schemes. Reviews point to strategic coupling of methods to capture synergy while limiting reagent use and severity (Li et al., 2022).

Combinations of thermochemical or biochemical steps can outperform single treatments. A mild alkali soak followed by brief thermal conditioning often delivers higher solubilisation at lower chemical dose, as reviewed by Li et al. (2022) and demonstrated for lignocellulosic residues such as wheat straw (Novakovic et al., 2021). Feedstock traits at the crop level (e.g., cultivar, fertilisation and harvest timing) can markedly affect biogas yield and process stability (Jankovičová et al., 2022). Researchers have also explored translating pretreatment advances from bioethanol production to biogas systems; for example, strategies for deconstructing lignocellulose developed in bioethanol research could inform AD feedstock processing (Kululo et al., 2025).

Table 5.1 Pretreatment families versus mechanisms, typical gains, energy/cost trade-offs, and best-fit substrates.

Pretreatment family (example methods)	Mechanism of action in AD context	Typical methane yield gains (vs. no pretreat)	Energy/cost trade-offs	Best-fit substrates (examples)
Mechanical (milling, grinding, high-shear mixing)	Physical size reduction increases surface area and disrupts fibrous structure, accelerating hydrolysis (Mirmohamadsadeghi et al., 2021).	Often 5%–20% increase in CH_4 yield for lignocellulosics; faster hydrolysis kinetics (Mirmohamadsadeghi et al., 2021).	High electricity demand for intensive milling; diminishing returns at very small particle sizes; no reagent costs, but equipment wear is non-trivial.	Fibrous biomass with large particle size, for example, crop stalks, yard waste, straw; manure fibers to improve pumping/handling.
Thermal (steam explosion, autoclave, thermal hydrolysis at 160°C–180°C)	Heat and pressure disrupt lignin structures and hydrolyse hemicellulose, increasing cell-wall permeability; also provides hygienisation/pathogen reduction in sludges (Mirmohamadsadeghi et al., 2021; Li et al., 2022).	Typically 10%–30% higher CH_4 yields, occasionally >30% for waste-activated sludge or straw; improved digestion rate and dewaterability in sludges (Mirmohamadsadeghi et al., 2021; Li et al., 2022).	Substantial heat/steam requirement and pressure-rated equipment; excessive severity can generate inhibitors; strongest cases with waste-heat integration and favourable LCA/TEA (Tian et al., 2021; Sinsuw et al., 2024).	Waste-activated sludge, organic fraction of municipal solid waste (OFMSW), recalcitrant crop residues; sites requiring hygienisation (sewage sludge).
Chemical (alkali: NaOH, Ca $(OH)_2$; acid: H_2SO_4; ammonia-based, e.g., AFEX; oxidative: H_2O_2)	Alkali swells fibers, saponifies lignin and removes acetyl groups; acid hydrolyses hemicellulose; ammonia-based routes promote deacetylation and lignin disruption; oxidative routes cleave aromatic rings in lignin (Li et al., 2022).	High potential for lignocellulosics. Example: NaOH-pretreated wheat straw reached ~338 mL CH_4 g^{-1} VS and about 1.5–2× higher methane than untreated controls (Novakovic et al., 2021).	Reagent costs and handling; alkali/acid require neutralisation with possible salt build-up; corrosion and inhibitor risks (e.g., furfural/HMF) at high severity; effective, but downstream impacts must be managed.	Straw, corn stover, grasses (high-lignin agro-residues) for alkali/ammonia; dilute-acid for selected food/crop residues targeting hemicellulose solubilisation; oxidative routes mainly at research stage for AD.

(*Continued*)

Table 5.1 (Continued)

Pretreatment family (example methods)	Mechanism of action in AD context	Typical methane yield gains (vs. no pretreat)	Energy/cost trade-offs	Best-fit substrates (examples)
Biological (fungal pretreatment, ensiling, microbial consortia, enzyme addition)	Microorganisms or enzymes selectively degrade lignin and complex carbohydrates; ensiling via lactic fermentation preserves biomass and partially pre-digests it; direct enzymatic hydrolysis of polymers (Ferdeş et al., 2020). Comprehensive reviews detail biological pretreatment methods to enhance lignocellulose digestibility (Saggi et al., 2017).	Variable but often meaningful under controlled conditions: fungal pretreatments ~20%–40% higher CH_4; enzyme addition often >20% with adequate dosing/ contact time; larger boosts reported for tailored consortia in specific cases (Zhao et al., 2019).	Mild operating conditions and low energy use, but longer residence times (days-weeks) and additional volume; enzymes can be costly; some volatile solids may be consumed during pretreatment.	Agricultural residues (straw, corn stover, bagasse) where time/space are available; ensiling for energy crops or high-sugar residues; enzyme cocktails for high-solids or short-HRT systems.

VS, volatile solids; *HRT*, hydraulic retention time; *OFMSW*, organic fraction of municipal solid waste; *LCA/TEA*, life-cycle assessment/techno-economic analysis.

A comprehensive survey of biological–chemical hybrids indicates that pairing gentle chemical steps with targeted enzymes or microbial consortia can yield substantially higher methane outputs (Sunar et al., 2025). However, the performance of hybrid configurations remains highly sensitive to operating conditions such as mixing intensity, hydraulic retention time and temperature set-points (Wahid and Horn, 2021). Nanomaterial-assisted pretreatments have also emerged, aiming to improve substrate accessibility and enzyme-substrate contact while curbing the use of harsh reagents (Ingle et al., 2019). Targeted use of nanomaterials has been reported to accelerate hydrolysis and methanogenesis while enabling lower reagent dosages (Tian et al., 2017). Recent bibliometric analyses highlight opportunities and challenges for nanomaterial-enhanced anaerobic digestion (Parra-Orobio et al., 2025). A specific case study reports higher CH_4 with SnO_2 nanoparticle-doped mica (Kaskun et al., 2023).

Selecting the right pretreatment intensity remains pivotal. Over-pretreatment introduces diminishing returns or inhibition through by-product formation and excess energy input. Optimal severity is increasingly set by detailed feedstock characterisation and biochemical methane potential (BMP) tests (Mirmohamadsadeghi et al., 2021; Ohemeng-Ntiamoah and Datta, 2018).

Pretreatment can reshape digester microbiology, with implications for pathway selection and stability. Changes in substrate architecture and solubilised intermediates may shift the balance between acetoclastic and hydrogenotrophic methanogenesis, and biological strategies can enrich specific functional guilds; metataxonomic work at field scale links these shifts to performance, supporting coordinated pretreatment-microbial management (Kim et al., 2023; Zhao et al., 2019).

Specialised techniques are also being developed for unconventional feedstocks. Marine macroalgae require tailored approaches to address high salt and unique polysaccharides, and suitable pretreatments have been shown to raise biogas yields from algal biomass (Thompson et al., 2019).

Beyond pretreatment, anaerobic codigestion is a widely adopted and effective route to higher biogas yields and improved process stability (Chuenchart et al., 2024). Codigestion blends substrates so their complementary properties produce a more balanced microbial diet and dilute inhibitors that may be present in any one waste. Additional strategies include adjusting feedstock composition to improve fuel properties and biodegradability (Chozhavendhan et al., 2020). The approach often delivers synergistic gains, where the combined output exceeds the sum of separate digestions under identical conditions.

A recent meta-analysis quantified these effects. Codigesting lignocellulosic biomass with nutrient-rich manures can nearly double methane yield relative to separate digestion; for example, adding pig manure to lignocellulosic residues gave a relative synergistic index (RSI) of ~1.91, while chicken manure with crop residues yielded ~1.71. Pairing easily degradable food waste with fibrous or high-nitrogen wastes also improved performance, with RSI values of ~1.41 for food waste plus straw and ~1.28 for food waste plus manure (Chuenchart et al., 2024). The meta-analysis attributes these improvements to better C/N balance, dilution of toxic compounds and complementary microbial functions that underpin a more robust degradation network.

Not all substrate combinations yield synergy; poorly selected blends can be antagonistic if they introduce inhibition or imbalance (Kemka et al., 2024; Chuenchart et al., 2024). Substrate characterisation is therefore central, with the most effective pairs chosen so that one waste compensates for the limitations of another, rather than compounding them.

The meta-analysis by Chuenchart et al. (2024) shows that sewage sludge, with moderate carbon and high nitrogen plus an active inoculum, co-digested with food waste, which is carbon-rich and acidifies readily, can curb volatile fatty acid accumulation and raise methane yields relative to mono-digestion. The same synthesis warns that combining two protein-rich wastes risks free-ammonia inhibition unless a high-carbon cosubstrate is included (Chuenchart et al., 2024; Rahmani et al., 2022).

Mixing ratios are a key lever. Operators typically screen options using BMP tests and, where available, simple predictive tools to locate proportions that maximise gas output while maintaining stability (Chuenchart et al., 2024; Kunatsa et al., 2022). Data-driven methods are increasingly applied at scale, with machine-learning (ML) models predicting codigestion performance from substrate descriptors and operating conditions (Yalcinkaya and Yucel, 2025; Wang et al., 2020). In a Malaysian case study, a pilot-scale biogas system treating industrial bio-waste was successfully scaled up, demonstrating the real-world applicability of optimised codigestion and upgrading techniques (Park and Sun, 2021).

When multiple substrates are fed, the dominant component often governs performance, underscoring the need to tune the blend to achieve the intended synergy (Zhong et al., 2024). Operating conditions interact with substrate effects: thermophilic regimes can accelerate lignocellulose breakdown and improve handling of fats and oils, raising yields for mixes such as energy crops with food waste (Bi et al., 2025; Lv et al., 2019).

Commonly successful pairings include manure with crop residues or energy crops, sewage sludge with organic solid waste, and food waste with yard waste, combinations that leverage nutrient balancing and moisture adjustment. Manures contribute buffering capacity and trace metals essential for methanogenesis (e.g., Co, Ni), while fibrous residues supply structural carbon and can absorb excess moisture (Chuenchart et al., 2024; Rahmani et al., 2022). Implementation should also account for secure feedstock supply and the rheology of the blend (Tabatabaei et al., 2020). Lipid-rich streams such as fats, oils, and grease codigested with waste-activated sludge can yield high specific methane but require attention to foaming and long-chain fatty-acid management (Wang et al., 2018).

Across farm-scale and municipal facilities, codigestion remains a cost-effective route to increase biogas output and stabilise operation, for instance, by reducing foaming episodes or acid spikes. Current reviews position codigestion as a mainstream strategy to boost yields, valorise multiple waste streams and advance circular-economy goals (Tabatabaei et al., 2020; Adnan et al., 2025).

Alongside codigestion, the use of additives and supplements in anaerobic digesters offers a means to fine-tune yields and process resilience. Additives are commonly grouped into nutrient supplements, inhibitor mitigants and bioactivity enhancers, with some materials spanning categories (Pilarska et al., 2024; Tabatabaei et al., 2020).

Trace-metal dosing illustrates the nutrient class. Many high-rate systems treating carbon-rich or mono-substrates lack micronutrients such as cobalt, nickel, selenium, and iron that serve as cofactors for methanogenic enzymes. Methanogens require Ni in coenzyme F_{430} and Co in vitamin B_{12}-dependent pathways; balanced supplementation can raise CH_4 yields, often by 10%–30%, and prevent drift toward instability in long-running reactors (Pilarska et al., 2024). In an anaerobic sludge digester, small additions of $NiCl_2$ and $CoCl_2$ increased activity and stability, although gains in methane yield were modest relative to other enhancers (Xie et al., 2020).

Inhibitor-mitigating agents target ammonia, hydrogen sulphide, and volatile-acid accumulation. Iron salts or zero-valent iron precipitate sulphide as FeS, simultaneously lowering H_2S in biogas and supplying bioavailable iron; controlled microaeration can also metabolise inhibitory intermediates and increase methane production under stress (Pilarska et al., 2024; Ruan et al., 2019).

Adsorptive and conductive additives such as activated carbon, biochar, and natural zeolite both adsorb inhibitors and, when conductive, promote direct interspecies electron transfer (DIET). For cornstalk digestion, powdered activated carbon enhanced biofilm development and electron transport, accelerating kinetics and increasing methane yield by ~12% (Pilarska et al., 2024; Xie et al., 2022).

Low-cost, waste-derived materials are gaining attention. In chicken-manure digestion, coal slime improved buffering, shortened the lag phase and enriched *Methanosarcina*, delivering significantly higher CH_4 than controls; integrating such carbonaceous residues into plant-scale TEA/LCA can improve both performance and economics (Guo et al., 2022; Mosleh Uddin et al., 2022).

Bioaugmentation, adding specialised microbial cultures or consortia, is best positioned as an operational intervention for start-up and upset recovery rather than a routine yield booster. Targeted inocula (e.g., propionate-oxidising or cellulolytic cultures) can shorten commissioning and clear volatile fatty acid accumulations after overload, restoring methane production more rapidly than passive recovery. Durable steady-state gains are less consistent because introduced strains often wash out or are outcompeted unless their ecological niche is maintained through feed composition and attachment sites; repeated dosing may be required (Tabatabaei et al., 2020; Braz et al., 2018).

Co-digestion and targeted additives act on complementary constraints, C/N balance, inhibitor control, micronutrient supply and electron transfer (DIET), and, when selected to match substrate characteristics and operating goals, can materially raise CH_4 output and resilience. Table 5.2 summarises representative substrate pairings and additive strategies, outlining their rationale, reported effects on CH_4 yield and key implementation considerations.

Decentralised AD systems that pair household cooking-fuel production with nutrient recycling (e.g., pelletised biofertilisers or struvite) deliver high social value through cleaner kitchens, reduced fuel-collection burdens and improved soil fertility (Wang, 2014; Adnan et al., 2025). At the meso-scale, regional biomethane clusters allow small producers to pool raw biogas for central upgrading and certification and inject biomethane into gas grids, improving project economics and market access while keeping feedstock logistics local (Sesini et al., 2024; Baena-Moreno et al., 2020).

Table 5.2 Codigestion and additive strategies, synergistic combinations, mechanisms, effects, and references.

Codigestion combination or additive strategy	Mechanism/rationale	Reported effect on biogas yield (illustrative)	Other Benefits/considerations	References
Lignocellulosic biomass + manure (e.g., crop straw + pig manure)	C/N balancing; manure supplies alkalinity, moisture and active inoculum; straw adds structural carbon.	Strong synergy in meta-analysis: straw with pig manure RSI ~ 1.91; chicken manure with crop residues RSI ~ 1.71.	Faster fiber degradation and buffering; chop straw to aid mixing and avoid floatation.	Chuenchart et al. (2024); Rahmani et al. (2022)
Food waste + sewage sludge (or food waste + manure)	Food waste is readily fermentable and carbon-rich; sludge/manure buffers VFAs and contributes enzymes and micronutrients.	Food waste + straw RSI ~ 1.41; food waste + manure RSI ~ 1.28.	Mitigates souring under high food-waste loads; ensure hygienisation and plan for dewatering where sludge is used.	Chuenchart et al. (2024)
High-N waste + high-C waste (e.g., chicken litter + straw; fish waste + paper waste)	Carbon-rich co-substrate moderates ammonia release and toxicity from protein-rich wastes.	Substantial CH_4 increase when ratios keep free ammonia below inhibitory range; chicken manure + crop residues RSI ~ 1.71.	Reduces need for dilution or stripping; consider trace-metal dosing if both streams are micronutrient-poor.	Chuenchart et al. (2024); Rahmani et al. (2022)
Multi-feedstock blends (e.g., manure + crop residues + food waste + yard waste)	Diversifies nutrients and microbial niches; dilutes single inhibitors and contaminants.	Frequently steadier CH_4 profiles than single/dual feeds; quantitative gains depend on blend composition.	Improves supply flexibility; requires pre-processing and contaminant removal for municipal streams.	Kemka et al. (2024)
Ratio optimisation (BMP-guided or model-based)	Identifies proportions that maximise synergy and stability; accounts for dominant-substrate effects.	Optimal-ratio feeding improved biogas potential in tri-substrate systems; dominant substrate often governs performance.	Screen with BMP tests before scale-up; revise with seasonal shifts.	Kunatsa et al. (2022); Zhong et al. (2024)

Trace-metal supplementation (Ni, Co, Fe, Se)	Supplies essential cofactors for methanogenesis, Ni in coenzyme F_{430}; Co in cobalamin-dependent steps.	Typically ~10%–30 % higher CH_4 in deficient systems; Ni/Co salts improved activity and stability in sludge digestion.	Low doses effective; avoid toxicity; metals partition to digestate, check agronomic limits if land-applying.	Pilarska et al. (2024); Xie et al. (2020)
Adsorptive/alkaline additives (biochar, powdered activated carbon, zeolite, $CaCO_3$)	Adsorb VFAs/NH_3/H_2S and improve buffering; conductive carbons can support DIET.	Powdered activated carbon increased CH_4 by ~12% with faster kinetics in cornstalk systems; biochar often shows measurable CH_4 gains and H_2S reduction.	Solid dosing raises TS, ensure mixing; digestate-derived biochar can close the loop and align with plant-scale TEA/LCA.	Xie et al. (2022); Mosleh Uddin et al. (2022)
Conductive materials for DIET (e.g., magnetite Fe_3O_4, graphite, PAC, carbon fibres)	Provide electron-transfer conduits that ease H_2/formate limitations and accelerate syntrophic oxidation.	Research trials report ~ 5–15 % CH_4 increases with faster degradation; PAC improved yield and kinetics under optimised dosing.	Prefer recoverable or inert media for scale-up; nanoparticle fate in digestate warrants evaluation.	Xie et al. (2022); Tian et al. (2017)
Iron-based dosing for sulphide control ($FeCl_2$, zero-valent iron)	Precipitates sulphide as FeS, lowering H_2S and alleviating toxicity; also supplies bioavailable Fe.	Documented stability gains and improved biogas quality in high-sulphur feeds under appropriate dosing.	Integrates gas-clean-up with process stability; adjust dosing to avoid excess iron carry-over.	Pilarska et al. (2024)
Microaeration (controlled O_2 dosing)	Small O_2 inputs stimulate oxidation of inhibitory intermediates and support syntrophic activity.	Semi-continuous trials showed higher methane production and increased enzyme activity versus anoxic controls.	Useful for overload recovery; requires tight control to prevent aerobic losses.	Ruan et al. (2019)
Waste-derived mineral/carbon additive (coal slime in chicken-manure digestion)	Porous, ion-exchange matrix adsorbs ammonia and metals; provides attachment sites and micronutrients.	Significantly higher CH_4 than controls; shortened lag phase and enrichment of Methanosarcina.	Low-cost where locally available; characterise contaminants prior to use.	Guo et al. (2022)

(*Continued*)

Table 5.2 (Continued)

Codigestion combination or additive strategy	**Mechanism/rationale**	**Reported effect on biogas yield (illustrative)**	**Other Benefits/considerations**	**References**
Thermophilic regime during co-digestion	Higher temperature accelerates hydrolysis and alters community structure; improves handling of fats/oils.	Increased methane productivity across OLRs in food-waste AD; temperature shifts modified bacteria-methanogen linkages.	Enables shorter HRT and improved sanitisation; monitor ammonia inhibition risk and rheology.	Bi et al. (2025); Lv et al. (2019)
Inoculum/bioaugmentation additives (specialised cultures)	Introduces propionate-oxidising or cellulolytic guilds to aid start-up and upset recovery.	Primarily faster recovery and commissioning rather than sustained yield uplift at steady state.	Persistence can be limited without a maintained niche; may require re-dosing.	Tabatabaei et al. (2020); Braz et al. (2018)

Abbreviations: *BMP*, biochemical methane potential; CH_4, methane; *DIET*, direct interspecies electron transfer; H_2, hydrogen; H_2S, hydrogen sulphide; *HRT*, hydraulic retention time; NH_3, ammonia; *OLR*, organic loading rate; *RSI*, relative synergistic index; *TEA/LCA*, techno-economic analysis/life-cycle assessment; *VFAs*, volatile fatty acids.

Enhancing biogas yield typically requires a multi-faceted optimisation strategy. Single measures can raise performance, but larger gains arise when a mild pretreatment to increase substrate accessibility is paired with codigestion of complementary feedstocks to balance nutrients and dilute inhibitors, operating parameters are tuned (organic loading rate, OLR; hydraulic retention time, HRT; temperature; mixing) and, where appropriate, micronutrients or conductive media are added to promote DIET. The meta-analysis by Chuenchart et al. (2024) and targeted pretreatment trials with alkaline-treated straw (Novakovic et al., 2021) indicate that integrated bundles can approach a doubling of CH_4 yields relative to unoptimised baselines. Translation to full-scale practice depends on verified energy and cost trade-offs, with LCA/TEA showing when additional inputs deliver net GHG and economic gains (Tian et al., 2021; Sinsuw et al., 2024). Tighter operation under intensified conditions is increasingly supported by advanced monitoring and control, including soft-sensing and ML-assisted optimisation that stabilise loading and anticipate upsets (Rutland et al., 2025).

5.2 Integration of novel technologies in biogas plants

Biogas production is evolving from an empirical, farm-based practice to an advanced biotechnology. Modern plants integrate digital monitoring and automation, innovative reactor configurations (e.g., two-stage digestion, membrane-based systems), biogas upgrading to biomethane and emerging PtG routes for biological methanation, while treating heat, power, biomethane and fertilisers as coproducts within a unified design (Assunção et al., 2021; Elhaus et al., 2024). Demonstrated scale-up with membrane purification is reported in Malaysia (Park and Sun, 2021).

5.2.1 Real-time monitoring and intelligent control

As operators push digesters to higher organic loading rates and manage more variable feedstocks, real-time monitoring and automated control have become central to stable operation. Modern systems track gas composition (CH_4, CO_2, H_2S), intermediate metabolites such as volatile fatty acids (VFAs), alkalinity and ammonia, alongside proxy signals for microbial activity, with data streams integrated into supervisory control to support rapid set-point adjustment (Premalatha et al., 2024). Metagenomic profiling further informs performance diagnostics in food-waste digesters (da Silva Martins et al., 2017).

Continuous, high-resolution data also strengthen life-cycle and techno-economic performance by reducing flaring, avoiding off-spec gas and minimising corrective dosing, effects that are particularly evident in multi-stage configurations evaluated through LCA/TEA (Raketh et al., 2024; Sinsuw et al., 2024).

Artificial intelligence (AI) and ML tools are increasingly used to convert routine signals into actionable diagnostics and forecasts. Soft sensors can estimate the FOS/TAC ratio, volatile acids to alkalinity, from readily available measurements, providing early warning of souring without frequent laboratory VFA assays, while data-driven models predict stability and methane productivity under changing feeds and loads (Rutland et al., 2025; He et al., 2017; Wang et al., 2020).

Intelligent control closes the loop. Historical and real-time data inform temperature management and feed modulation, enabling proactive interventions such as transient feed holds or alkali addition on detection of VFA excursions, and demand-responsive gas production via model-based control (Cinar et al., 2022; Peters et al., 2019).

Operationally, tighter digital control allows plants to run closer to performance limits, supporting higher throughputs for a given reactor volume, accommodating challenging codigestion blends and enabling remote supervision, thereby lowering manual oversight and widening the scope for decentralised deployment (Cheon et al., 2022; Meola et al., 2023). Despite technical advances, persistent scale-up barriers—including feedstock logistics, process stability and integration with existing infrastructure—continue to limit deployment (Patinvoh and Taherzadeh, 2019).

5.2.2 Innovative reactor designs and configurations

Reactor design is moving beyond the single-stage continuous stirred-tank reactor to better match operational conditions to microbial functions and improve contact between biomass and substrates. Two-stage digestion separates hydrolysis/acidogenesis from acetogenesis/methanogenesis so that pH, retention time and loading can be tuned for each phase, which improves stability and methane productivity for complex, slowly hydrolysing feeds (Heiker et al., 2021; Sinsuw et al., 2024).

Comparative assessments report that two-stage systems can deliver higher CH_4 production from the same feedstock than single-stage baselines, with LCA showing that added infrastructure may raise some per-unit impacts, yet gains in yield and gas quality can offset these penalties at scale (Raketh et al., 2024; Sinsuw et al., 2024). AnMBRs offer a different route by retaining biomass through separation, effectively decoupling hydraulic and solids retention times. High cell densities and full retention of slow-growing methanogens are achievable, although membrane fouling and associated energy demands remain design targets for low-shear, low-energy operation (Heiker et al., 2021). Temperature-phased anaerobic digestion places a thermophilic stage ahead of a mesophilic stage to accelerate hydrolysis and pathogen reduction while preserving methanogenic stability; gradual acclimation and stepwise temperature control are critical for robust operation (Westerholm et al., 2017; Lv et al., 2019).

5.2.3 Biogas upgrading and power-to-gas systems

Upgrading raw biogas (typically a CH_4/CO_2 mixture with trace impurities) to pipeline-grade biomethane expands use options and is increasingly favored in European policy frameworks for gas-grid decarbonisation and market access (Sesini et al., 2024; Baena-Moreno et al., 2020). Commercial technologies include pressure-swing adsorption, membrane separation, water scrubbing and amine scrubbing, with cryogenic and biological routes emerging; selection is dictated by scale, target specification, and site economics (Assunção et al., 2021). Membrane purification systems have been successfully scaled in practice, as demonstrated by a Malaysian biogas case study (Park and Sun, 2021). Multicriteria decision models can support technology pathway selection (Luengo et al., 2025).

Biological methanation within broader PtG schemes upgrades biogenic CO_2 using renewable H_2 to produce additional CH_4. Performance hinges on reactor configuration and H_2 gas-liquid mass transfer, and recent techno-economic studies outline conditions under which coupling AD with biological methanation is cost-effective (Jensen et al., 2021; Elhaus et al., 2024). Table 5.3 summarises major upgrading and PtG options, indicating achievable biomethane purity, energy requirements, and salient sustainability considerations.

Integrating PtG can convert nearly all biogenic CO_2 into CH_4, raising methane content from ~60% to ~95%; this corresponds to roughly a 66% increase in CH_4 production; pilot and demonstration systems report high conversions and robust operation with impure CO_2 streams (Jensen et al., 2021; Jønson et al., 2022). The net benefit depends on the hydrogen source. When H_2 derives from surplus renewable electricity, PtG markedly improves the GHG balance by converting CO_2 that would otherwise be released and by curbing methane slip, while coupling electricity and gas systems for seasonal energy storage. Reported step efficiencies are around 80%, and the additional methane displaces fossil natural gas (Elhaus et al., 2024).

Foresight analyzes identify carbon capture and utilisation, including biological methanation, as a strategic driver of biogas-system evolution toward 2050 (Assunção et al., 2021; Rojas Arboleda et al., 2021).

Biogas plants are increasingly configured as multioutput energy hubs: upgraded biomethane for grid injection is paired with cascaded heat from upgrading and combined heat and power to supply local heating, cooling via absorption chilling and, in some designs, desalinated water, maximising resource utilisation (Horgen et al., 2023; Wu, 2025). Farm-scale cogeneration analyses indicate overall energy efficiencies on the order of 86%–90%, reinforcing the value of heat-cascade integration (Horgen et al., 2023).

Digestate processing is integral to plant integration. Although higher biogas yield increases energy output, overall environmental and economic benefits require verification by life-cycle and techno-economic analyses to ensure net sustainability gains (Tian et al., 2021). New installations commonly couple solid–liquid separation with fiber drying or composting and nutrient-recovery units, for example, struvite precipitation for phosphorus or ammonia stripping to ammonium sulphate, producing concentrated fertilisers that generate revenue and lower land-application volumes in nutrient-surplus regions (Adnan et al., 2025). In some concepts, digestate fibres are pyrolysed to biochar that is recycled as a digester additive to moderate inhibitors and enhance stability, then applied to soils for carbon sequestration; technoeconomic and environmental assessments indicate feasible pathways when carbonaceous residues are integrated into manure-to-biomethane systems (Mosleh Uddin et al., 2022). At farm scale, techno-economic analysis reported that incorporating corn-stover biochar can reduce the selling price of renewable natural gas by approximately 15%–22% while preserving GHG benefits (Mosleh Uddin et al., 2022). Nutrient-rich liquor can also be valorised via algal cultivation, broadening the product slate of biogas facilities (Adnan et al., 2025).

Small-scale and modular technologies constitute a parallel integration trend. Clusters of farm- or community-scale units equipped with smart control and, where

Table 5.3 Upgrading and PtG options, performance, energy use, sustainability, and references.

Biogas upgrading/ PtG option	Biomethane purity and performance	Energy requirements/ operating cost	LCA/sustainability notes	References
Pressure swing adsorption (PSA), cyclic adsorption of CO_2 on solids with tail-gas regeneration	Product typically 96%–98% CH_4; ≥98 % with multibed designs and tail-gas recycle; CH_4 recovery commonly ≥96%–99% with off-gas management.	Electricity for compression (feed ~4–10 bar); no process heat; ~0.1–0.2 kWh/Nm³ raw biogas in modern units.	Robust, solvent-free; methane slip must be oxidised or recycled to protect GHG performance; low overall footprint with low-carbon power.	Assunção et al. (2021); Miltner et al. (2016)
Membrane separation, staged polymer membranes (permeate recycle/ polishing)	Grid-grade biomethane ≥97%–99% CH_4 with 2–3 stages; CH_4 recovery ~95%–99 % with permeate recycle.	Electricity for compression (typically 8–20 bar); ~0.15–0.3 kWh/Nm³; periodic module replacement.	Compact and modular; very low fugitive CH_4 when permeate is recycled; impacts dominated by electricity carbon intensity; no solvent waste.	Assunção et al. (2021); Miltner et al. (2016)
Water scrubbing, high-pressure water absorbs CO_2 and H_2S	Typical product 95%–98% CH_4; CH_4 loss ~2%–5% via dissolution unless stripped; drying required.	Electricity for pumps/compressors (~8–10 bar); ~0.1–0.2 kWh/Nm³ with recirculation and off-gas stripping; make-up water and biofouling control.	Simultaneous H_2S removal; sustainability hinges on CH_4 recovery from wash water and renewable electricity supply.	Assunção et al. (2021); Miltner et al. (2016)
Chemical solvent (amine) scrubbing, MEA/DEA or advanced amines	Pipeline-grade biomethane (≥99% CH_4); negligible CH_4 slip; high-purity CO_2 co-product on regeneration.	Heat dominates: solvent regeneration typically ~0.5–0.8 MJ/Nm³ raw biogas (process-dependent), plus minor electricity; solvent make-up for degradation/losses.	Very high product quality; LCA driven by heat source and solvent management; waste-heat or low-carbon steam yields strong GHG performance.	Assunção et al. (2021); Miltner et al. (2016)

Cryogenic separation, deep cooling to condense CO_2; optional LBM	97%–99% CH_4 feasible; can co-produce liquid CO_2 and liquefied biomethane (LBM); near-zero CH_4 loss with full condensate capture.	Electricity-intensive refrigeration/compression; ~0.3–0.6 kWh/Nm³ with economies of scale at ≥1,000 Nm³/h.	Chemical-free and yields saleable co-products; electricity demand dominates impacts, favoring renewable power and larger plants.	Assunção et al. (2021); Miltner et al. (2016)
Biological methanation (ex situ PtG), separate reactor converts biogenic $CO_2 + 4\ H_2 \rightarrow CH_4 + 2\ H_2O$	~98%–100% CH_4 achievable with sufficient H_2 transfer; tolerant to some impurities; essentially zero CH_4 slip.	Energy dominated by H_2 production via electrolysis; reactor power modest; exothermic heat recoverable; biological systems handle intermittent H_2.	Converts biogenic CO_2 to additional CH_4 and enables sector coupling; GHG benefit depends on low-carbon H_2; TEA points to niches with cheap renewable power and grid access.	Jensen et al. (2021); Elhaus et al. (2024)
In-situ biological methanation, H_2 sparged directly into the AD digester	Raises raw-gas CH_4 from ~60% to ~70%–80% in favourable cases; conversion limited by H_2 gas-liquid mass transfer and community shifts.	Same stoichiometric H_2 need as ex situ; no extra reactor, but additional mixing/aeration energy and careful dosing control.	Minimal retrofit; control of H_2 partial pressure is critical to avoid inhibition; sustainability again hinges on renewable H_2; generally less complete conversion than ex situ.	Wahid and Horn (2021); Jensen et al. (2021)

Abbreviations: *AD*, anaerobic digestion; *CH_4*, methane; *CO_2*, carbon dioxide; *GHG*, greenhouse gas; *H_2*, hydrogen; *H_2S*, hydrogen sulphide; *LBM*, liquefied biomethane; *LCA*, life-cycle assessment; *PtG*, power-to-gas; *TEA*, techno-economic analysis; *Nm³*, normal cubic metre.

viable, on-site upgrading can inject biomethane at distributed grid nodes, improving market access while keeping feedstock logistics local. Comparative policy analyses associate supportive grid-injection frameworks and stable incentives with greater adoption of innovative plant configurations and wider end uses, including vehicle fuel and grid balancing (Sesini et al., 2024; Baena-Moreno et al., 2020).

Forward-looking studies highlight flexible small-scale innovations and coupling to carbon capture and utilisation routes such as biological methanation (PtG) as influential for a robust biogas sector toward 2050, with demonstration-scale work outlining conditions under which AD combined with biomethanation is techno-economically attractive (Rojas Arboleda et al., 2021; Elhaus et al., 2024). Integration here is both technical and systemic: distributed units interface with energy markets, storing surplus renewable electricity as grid-compatible CH_4, and with environmental services, including waste treatment and nutrient recycling, enabled by the monitoring and control tools outlined earlier.

5.3 Future directions in biogas production optimisation

5.3.1 Microbial community engineering and biotechnology

Future biogas systems will increasingly leverage a granular understanding of the AD microbiome to raise efficiency. Meta-omics now resolves the bacterial and archaeal consortia that drive anaerobic digestion and links community structure to stability and methane productivity, opening the way to targeted, mechanism-based control (Ferguson et al., 2018; Kim et al., 2023). For example, a metagenomic study of a high-loading food waste digester showed how shifts in microbial populations correlated with performance across different feed rates (da Silva Martins et al., 2017). Building on this, predictive modeling from community data is being used to anticipate responsiveness under changing feeds, suggesting the feasibility of real-time microbial monitoring that feeds into control logic, an online biofeedback layer for digesters (Kim et al., 2023).

Operational bottlenecks can be addressed by steering the microbiome with measured interventions. Nutrient dosing, selective bioaugmentation and set-point adjustments that favor desired guilds, for example under propionate stress or high ammonia, are converging into deliberate microbial resource management in full-scale practice (Tabatabaei et al., 2020). Looking further ahead, genetic and metabolic engineering of key organisms has been proposed, for instance, more robust hydrolytics or methanogens with improved tolerance, yet any deployment will require careful oversight because digestate is land-applied (Tabatabaei et al., 2020).

Selective enrichment and adaptive evolution offer nearer-term routes. Cultivating consortia with traits such as elevated ammonia tolerance or enhanced degradation of recalcitrant fractions, then using these as inocula, aligns with evidence that higher functional diversity supports operating stability. Upstream measures also shape downstream ecology: pretreatment alters substrate architecture and intermediates, which can enrich functional groups that accelerate subsequent methanogenesis. Reviews of pretreatment report such community shifts, including gains in hydrolytic and methanogenic activity following biological or mild chemical conditioning (Mirmohamadsadeghi et al., 2021; Zhao et al., 2019).

5.3.2 Advanced process control and AI integration

Future AD plants will increasingly operate under autonomous or semi-autonomous control guided by AI. Digital twins, virtual digester replicas synchronised with plant data, can forecast upsets and optimise set-points for feeding, substrate mix, temperature, and alkalinity. Dynamic simulations already support controller design and stress-testing before deployment, complementing data-driven approaches (He et al., 2025; Ganeshan and Rajendran, 2022).

Data-centric operation depends on reliable soft sensing. Recent studies show ML models can infer stability indicators such as the FOS/TAC ratio from routine signals, enable early anomaly detection and sustain higher loading with steadier performance (Rutland et al., 2025; Cheon et al., 2022).

Near-term innovation is converging on low-cost biosensors, for example real-time VFA spectra or dissolved H_2 probes, wired directly to AI controllers. Coupling inexpensive sensing with predictive control allows operators to push organic loading rates while maintaining safety margins through automatic feed modulation or alkalinity dosing (Cheon et al., 2022).

Temperature control illustrates the value of predictive actuation. ML-assisted controllers trained on historical data have maintained thermophilic regimes with minimal energy use, while microbiome studies emphasise that gradual acclimation to new temperature set-points underpins long-term stability (Cinar et al., 2022; Westerholm et al., 2017).

Demand-responsive operation is another frontier. Model-based control has been used to ramp biogas output to follow energy demand peaks without compromising stability, and reviews of full-scale practice catalogue the modeling approaches that enable such strategies (Peters et al., 2019; Heiker et al., 2021).

Foundational work on dynamic digester models continues to inform optimisation roadmaps. Early integrated energy-AD modeling (DARIES) and process simulations in Aspen Plus have demonstrated how scenario testing and sensitivity analysis de-risk configuration choices before scale-up (Brouwer et al., 2012; Rajendran et al., 2014). As Internet of Things (IoT) devices proliferate, even farm-scale units can adopt smart controllers that learn from local data and fleet-level experience, tightening feedback loops and reducing failure rates.

5.3.3 Further integration with energy grids and sector coupling

The role of biogas is shifting from steady base-load generation toward a flexible, multi-sector balancing resource. Future plants are being configured for rapid modulation with gas holders, multiple generator sets, and grid-responsive dispatch, so operators can store gas when electricity is abundant and generate when intermittent renewables dip (Peters et al., 2019). Foresight work positions biogas within adaptable, 2050-ready energy systems where operational flexibility and sector coupling are central (Rojas Arboleda et al., 2021).

Policy design is a major lever. Strong grid-injection frameworks and stable incentives correlate with more innovative configurations, including vehicle-fuel production, grid-support services and PtG integration, relative to electricity-only regimes (Sesini et al., 2024). Comparative analyses across Europe likewise link supportive policies to broader end uses and greater market access for biomethane (Gustafsson and Anderberg, 2022). Region-specific studies underscore similar themes–for example, in Southern Africa, non-technical barriers and past programme failures highlight the need for robust policy, financing, and training frameworks to realise biogas benefits (Tembo et al., 2023). National policy also drives industry transformation: in China, for instance, biogas-specific industrial policies have measurably accelerated sectoral innovation and growth (Luo et al., 2024).

Coupling AD with biological methanation is an important vector for sector integration. Ex-situ PtG units convert biogenic CO_2 with renewable H_2 to additional CH_4, while in-situ approaches inject H_2 directly into digesters. Performance depends on reactor configuration and H_2 gas–liquid mass transfer, with careful control of retention and mixing regimes required to sustain high conversions and stable communities (Jensen et al., 2021). Techno-economic studies outline operating windows where PtG complementing AD is advantageous, particularly when low-carbon electricity and grid access align (Elhaus et al., 2024).

A broader carbon-capture-and-utilisation (CCU) perspective extends these linkages, positioning biomethane as a flexible energy vector while directing captured CO_2 into other products where policy and economics favour diversification (Assunção et al., 2021). Frameworks for lignocellulosic biorefineries emphasise AD as a core platform in integrated valorisation chains that maximise energy and material recovery (Vu et al., 2020). Recent reviews provide broad overviews of technological advances and process integration in anaerobic digestion (Náthia-Neves et al., 2018).

5.3.4 Emphasis on sustainability metrics and circularity

Future optimisation will place greater weight on rigorous life-cycle and techno-economic outcomes. Yield gains must remain net-positive once full-system impacts are counted; otherwise, electricity intensity or costly reagents can erode climate and cost benefits. Multicriteria evaluations highlight the need to assess improvements holistically, balancing energy return, GHG reduction, nutrient recycling and costs within decision frameworks for technology adoption (Budzianowski, 2016; Assunção et al., 2021).

Comparative assessments already illustrate these trade-offs. Two-stage digestion pilots show higher methane productivity and improved gas quality, yet additional infrastructure can raise per-kWh impacts; at scale, optimised operation can offset these penalties (Raketh et al., 2024; Sinsuw et al., 2024). Policy context continues to shape outcomes by determining which integrations, grid-quality biomethane, PtG units, nutrient-recovery modules, deliver the strongest system benefits (Sesini et al., 2024).

Major plant modifications, new reactor types, pretreatment steps or additive regimes, are increasingly screened through simulation or pilot-scale LCA/TEA before deployment to ensure verifiable gains in climate and economics. Integrated AD-PtG pathways are a case in point, where favourable results depend on low-carbon power and well-matched operating conditions (Tian et al., 2021; Elhaus et al., 2024).

5.3.5 Decentralisation and developing-world applications

Alongside high-tech pathways, a crucial frontier is delivering appropriate, affordable biogas for small-scale and developing-country contexts. Barriers are often logistical and socioeconomic, maintenance capacity, financing, supply chains and social acceptance, rather than purely technical (Mutahi et al., 2025).

Evidence from Pakistan highlights cost, awareness, and institutional hurdles as dominant impediments to adoption, closely tied to persistent energy poverty (Batool et al., 2023). Reviews of past initiatives in Southern Africa likewise point to non-technical shortfalls, including inadequate user training and weak supply chains, as primary causes of programme underperformance (Mabecua et al., 2024). Similar challenges are reported in East Africa—for example, inadequate maintenance support and social acceptance have constrained biogas programmes in rural Ethiopia (Sime and Jin, 2020).

Optimisation in these settings prioritises simplicity and robustness: low-cost biological pretreatments, such as controlled composting or ensiling of crop residues to improve digestibility, basic solar thermal heating for mild temperature control and modular digesters sized for household or farm needs (Wang, 2014; Sime and Jin, 2020). Financial sustainability can be strengthened through community ownership models, revenues from tipping fees for accepting organic wastes and micro-finance schemes that align with identified adoption drivers (Ali et al., 2023; Yousuf et al., 2016).

Designs must also be tailored to local constraints. Model-based analyses for sub-Saharan Africa show optimal configurations depend on site-specific resources, demand profiles and logistics, underscoring the value of structured decision tools for sustainable deployment (Rupf et al., 2016).

Digitalisation can trickle down via low-cost sensors and data-driven advice. Soft-sensing and AI-based tools that estimate stability indicators from routine signals can help small operators run at higher organic loading rates while maintaining safety margins, without full-time expert oversight (Cheon et al., 2022; Rutland et al., 2025).

Decentralised AD that couples household cooking-fuel production with nutrient recycling, for example, pelletised biofertilisers or struvite, delivers social and environmental value by improving indoor air quality, reducing fuel-collection burdens and returning nutrients to soils (Wang, 2014; Adnan et al., 2025). At the meso-scale, regional biomethane clusters can aggregate outputs from many small digesters for upgrading and grid injection or local transport fuel, where policy frameworks and market access allow (Sesini et al., 2024; Baena-Moreno et al., 2020).

5.3.6 Upscaling and industrial symbiosis

Translating laboratory advances into reliable field performance is a prerequisite for commercial success. Pilot work has shown that process gains demonstrated under controlled conditions can be replicated at industrial scale, while exposing practical constraints that must be addressed during scale-up, feed handling, mixing, foaming control and off-gas management among them (Wynn et al., 2016; Fiore et al., 2016).

Regional biomethane clusters are emerging as an effective bridge between small, distributed plants and grid-scale markets. In hub-and-spoke architectures, multiple farm- or municipal-scale digesters deliver raw gas to a central upgrading and injection hub via short pipelines or virtual-pipeline logistics. Centralising gas cleaning, certification and dispatch preserves local waste logistics yet captures economies of scale in upgrading assets, analytics and maintenance; hub integration also facilitates add-ons such as CO_2 utilisation and heat cascading that improve thermoeconomic performance (Skovsgaard and Jacobsen, 2017; Wu, 2025).

Industrial symbiosis drives sector-specific optimisation. Tailoring inocula and operating parameters to waste chemistry, integrating sulphur management where needed and deploying high-rate or phase-separating configurations—granular sludge systems, staged or baffled reactors, anaerobic membrane bioreactors (AnMBRs)—helps stabilise variable industrial streams and decouple hydraulic from solids retention for higher cell densities (Koniuszewska et al., 2020; Wang et al., 2018). Likewise, co-treating industrial wastewater streams in biogas plants can valorise pollutants while generating energy (Saikat et al., 2023). In one study, a compartmentalised baffled reactor achieved spatial separation of AD phases, improving performance and microbial stability under high loading (Ziganshin et al., 2019).

Evidence from palm-oil operations illustrates this trajectory. Physics-informed models for palm oil mill effluent (POME) are informing control strategies, while two-stage, LCA-guided codigestion schemes that pair POME with other industrial effluents have demonstrated feasibility and clarified trade-offs at pilot and pre-commercial scales (Shaw et al., 2024; Raketh et al., 2024). Beyond POME, integrated biorefinery approaches position AD alongside other green processes—for instance, coupling AD with algal treatment or composting in a unified system can maximise energy and resource recovery (Nkuna et al., 2024) and can create value in industrial wastewater contexts (Saikat et al., 2023).

Policy roadmaps increasingly couple biogas with industrial processes, aligning waste treatment, on-site renewable energy and biomethane production under supportive renewable natural gas frameworks that prioritise grid injection and diversified end uses such as vehicle fuel and grid support (Sesini et al., 2024; Gustafsson and Anderberg, 2022).

At farm level, integrated projects underscore the co-benefits of energy and nutrient recovery. A livestock installation in Japan demonstrated how on-site anaerobic digestion can simultaneously supply renewable energy and improve manure management, exemplifying practical industrial symbiosis at small scale (Richards and Yabar, 2023).

Biogas production optimisation is becoming smart, holistic, and context-specific. Facilities will operate as data-driven, biologically tuned systems that extract maximal value from wastes while responding in real time to grid conditions and environmental constraints (Cheon et al., 2022; Kim et al., 2023). Biogas will play a strategic, connective role in a low-carbon economy, linking waste management with renewable energy, electricity grids with gas grids via PtG integration, and food production with soil health through improved digestate valorisation (Adnan et al., 2025; Sesini et al., 2024). By advancing microbiome-aware operation, AI-enabled control, integrated upgrading/PtG and nutrient recovery, and by demonstrating verified LCA/TEA benefits, the sector can raise yields and resilience and consolidate biogas as a cornerstone of the circular economy (Assunção et al., 2021; Rojas Arboleda et al., 2021).

AI disclosure

During the preparation of this work the author(s) used DEEPL, ChatGPT, Grammarly, Mendeley in order to Translation, language correction, bibliography management. After using this tool/service, the author(s) reviewed and edited the content as needed and take(s) full responsibility for the content of the publication.

References

Adnan, M., Ahmed, A., Rafiq, M., Ullah, N., Ahmad, A., Sarfraz, M., et al., 2025. Green approaches for modifying lignocellulosic features: Recent advances in reducing recalcitrance for sustainable bioethanol production. Biomass and Bioenergy 203. Available at https://doi.org/10.1016/j.biombioe.2025.108304, 108304.

Adnane, I., Taoumi, H., Lahrech, K., Fertahi, S.E.-D., Ghodbane, M., 2025. From waste to resource: biogas and digestate valorisation strategies for sustainable energy and agriculture. Biomass and Bioenergy 200. Available at https://doi.org/10.1016/j.biombioe.2025.108006, 108006.

Ali, S., Yan, Q., Razzaq, A., Khan, I., Irfan, M., 2023. Modeling factors of biogas technology adoption: a roadmap towards environmental sustainability and green revolution. Environmental Science and Pollution Research 30 (5), 11838–11860. Available at https://doi.org/10.1007/s11356-022-22894-0.

Assunção, L.R.C., Mendes, P.A.S., Matos, S., Borschiver, S., 2021. Technology roadmap of renewable natural gas: identifying trends for research and development to improve biogas upgrading technology management. Applied Energy 292. Available at https://doi.org/10.1016/j.apenergy.2021.116849, 116849.

Baena-Moreno, F.M., Malico, I., Rodríguez-Galán, M., Serrano, A., Fermoso, F.G., Navarrete, B., 2020. The importance of governmental incentives for small biomethane plants in South Spain. Energy 206. Available at https://doi.org/10.1016/j.energy.2020.118158.

Batool, K., Zhao, Z.Y., Nureen, N., Irfan, M., 2023. Assessing and prioritizing biogas barriers to alleviate energy poverty in Pakistan: an integrated AHP and G-TOPSIS model. Environmental Science and Pollution Research 30 (41), 94669–94693. Available at https://doi.org/10.1007/s11356-023-28767-4.

Bi, S., Guo, L., Wang, H., Yu, X., Wei, F., Lei, L., et al., 2025. Performance of mesophilic and thermophilic anaerobic digestion of food waste at varying organic loading rate: methane production, pathogens reduction, and dominant microbial community dynamics. Renewable Energy 239. Available at https://doi.org/10.1016/j.renene.2024.122094, 122094.

Braz, G.H.R., Fernandez-Gonzalez, N., Lema, J.M., Carballa, M., 2018. The time response of anaerobic digestion microbiome during an organic loading rate shock. Applied

Microbiology and Biotechnology 102 (23), 10285–10297. Available at https://doi.org/10.1007/s00253-018-9383-9.

Brouwer, A.F., Grimberg, S.J., Powers, S.E., 2012. The dynamic anaerobic reactor & integrated energy system (DARIES) model: model development, validation, and sensitivity analysis. Water Environment Research 84 (12), 2090–2098. Available at https://doi.org/10.2175/106143012x13415215907059.

Budzianowski, W.M., 2016. A review of potential innovations for production, conditioning and utilization of biogas with multiple-criteria assessment. Renewable and Sustainable Energy Reviews 54, 1148–1171. Available at https://doi.org/10.1016/j.rser.2015.10.054.

Cheon, A., Sung, J., Jun, H., Jang, H., Kim, M., Park, J., 2022. Application of various machine learning models for process stability of bio-electrochemical anaerobic digestion. Processes 10 (1), 158. Available at https://doi.org/10.3390/pr10010158.

Chozhavendhan, S., Gnanavel, G., Karthiga Devi, G., Subbaiya, R., Praveen Kumar, R., Bharathiraja, B., 2020. Enhancement of feedstock composition and fuel properties for biogas production. In: Praveen Kumar, R., Bharathiraja, B., Kataki, R., Moholkar, V. (Eds.), Biomass Valorization to Bioenergy. Energy, Environment, and Sustainability. Springer Nature, Singapore, pp. 113–131. Available at https://doi.org/10.1007/978-981-15-0410-5_9.

Chuenchart, W., Surendra, K.C., Khanal, S.K., 2024. Understanding anaerobic co-digestion of organic wastes through meta-analysis. ACS ES and T Engineering 4 (5), 1177–1192. Available at https://doi.org/10.1021/acsestengg.3c00598.

Cinar, S.Ö., Cinar, S., Kuchta, K., 2022. Machine learning algorithms for temperature management in the anaerobic digestion process. Fermentation 8 (2). Available at https://doi.org/10.3390/fermentation8020065.

da Silva Martins, A., Ornelas Ferreira, B., Ribeiro, N.C., Martins, R., Rabelo Leite, L., Oliveira, G., et al., 2017. Metagenomic analysis and performance of a mesophilic anaerobic reactor treating food waste at various load rates. Environmental Technology (United Kingdom) 38 (17), 2153–2163. Available at https://doi.org/10.1080/09593330.2016.1247197.

Elhaus, N., Volkmann, M., Kolb, S., Schindhelm, L., Herkendell, K., Karl, J., 2024. Techno-economic evaluation of anaerobic digestion and biological methanation in power-to-methane-systems. Energy Conversion and Management 315. Available at https://doi.org/10.1016/j.enconman.2024.118787, 118787.

Ferdeș, M., Dincă, M.N., Moiceanu, G., Zăbavă, B.Ștefania, Paraschiv, G., 2020. Microorganisms and enzymes used in the biological pretreatment of the substrate to enhance biogas production: a review. Sustainability 12 (17), 7205. Available at https://doi.org/10.3390/su12177205.

Ferguson, R.M.W., Coulon, F., Villa, R., 2018. Understanding microbial ecology can help improve biogas production in AD. Science of the Total Environment 642, 754–763. Available at https://doi.org/10.1016/j.scitotenv.2018.06.007.

Fiore, S., Ruffino, B., Campo, G., Roati, C., Zanetti, M.C., 2016. Scale-up evaluation of the anaerobic digestion of food-processing industrial wastes. Renewable Energy 96, 949–959. Available at https://doi.org/10.1016/j.renene.2016.05.049.

Ganeshan, P., Rajendran, K., 2022. Dynamic simulation and optimization of anaerobic digestion processes using MATLAB. Bioresource Technology 351. Available at https://doi.org/10.1016/j.biortech.2022.126970, 126970.

Guo, H., Zhao, S., Xia, D., Zhao, W., Li, Q., Liu, X.L., et al., 2022. The biochemical mechanism of enhancing the conversion of chicken manure to biogenic methane using coal slime as additive. Bioresource Technology 344 (Pt B). Available at https://doi.org/10.1016/j.biortech.2021.126226, 126226.

Gustafsson, M., Anderberg, S., 2022. Biogas policies and production development in Europe: a comparative analysis of eight countries. Biofuels 13 (8), 931–944. Available at https://doi.org/10.1080/17597269.2022.2034380.

He, Q., Li, L., Peng, X., 2017. Early warning indicators and microbial mechanisms for process failure due to organic overloading in food waste digesters. Journal of Environmental Engineering 143 (12). Available at https://doi.org/10.1061/(ASCE)EE.1943-7870.0001280, 04017077.

He, X., Guo, J., Shen, D., Lin, R., 2025. Application of machine learning in anaerobic digestion of organic solid waste: recent advances and prospects. Journal of Southeast University (Natural Science Edition) 55 (3), 743–750. Available at https://doi.org/10.3969/j.issn.1001-0505.2025.03.014.

Heiker, M., Kraume, M., Mertins, A., Wawer, T., Rosenberger, S., 2021. Biogas plants in renewable energy systems—a systematic review of modeling approaches of biogas production. Applied Sciences 11 (8), 3361. Available at https://doi.org/10.3390/app11083361.

Horgen, C.A., Wang, Y., Huang, Y., 2023. Analysis of a cogeneration system in a norwegian farm. Energy Proceedings 32. Available at https://doi.org/10.46855/energy-proceedings-10507.

Ingle, A.P., Chandel, A.K., Antunes, F.A.F., Rai, M., da Silva, S.S., 2019. New trends in application of nanotechnology for the pretreatment of lignocellulosic biomass. Biofuels, Bioproducts and Biorefining 13 (3), 776–788. Available at https://doi.org/10.1002/bbb.1965.

Jankovičová, B., Hutňan, M., Czölderová, M.N., Hencelová, K., Imreová, Z., 2022. Comparison of acid and alkaline pre-treatment of lignocellulosic materials for biogas production. Plant, Soil and Environment 68 (4), 195–204. Available at https://doi.org/10.17221/421/2021-PSE.

Jensen, M.B., Ottosen, L.D.M., Kofoed, M.V.W., 2021. H_2 gas-liquid mass transfer: a key element in biological power-to-gas methanation. Renewable and Sustainable Energy Reviews 147. Available at https://doi.org/10.1016/j.rser.2021.111209, 111209.

Jønson, B.D., Tsapekos, P., Tahir Ashraf, M., Jeppesen, M., Ejbye Schmidt, J., Bastidas-Oyanedel, J.R., 2022. Pilot-scale study of biomethanation in biological trickle bed reactors converting impure CO_2 from a Full-scale biogas plant. Bioresource Technology 365. Available at https://doi.org/10.1016/j.biortech.2022.128160.

Kaskun, S., Çalhan, R., Akinay, Y., 2023. Enhancement of biogas production using SnO_2 nanoparticle-doped mica catalyst. Biomass Conversion and Biorefinery 13 (8), 7239–7246. Available at https://doi.org/10.1007/s13399-021-01983-z.

Kemka, U.N., Ogbulie, T.E., Oguzie, K., Akalezi, C.O., Oguzie, E.E., 2024. Impact of substrate heterogeneity on anaerobic co-digestion process: a review. EQA 62, 18–34. Available at https://doi.org/10.6092/issn.2281-4485/18498.

Kim, H.H., Basak, B., Lee, D.Y., Chung, W.J., Chang, S.W., Kwak, M.J., et al., 2023. Insights into prokaryotic metataxonomics and predictive metabolic function in field-scale anaerobic digesters treating various organic wastes. Renewable and Sustainable Energy Reviews 187. Available at https://doi.org/10.1016/j.rser.2023.113685.

Koniuszewska, I., Korzeniewska, E., Harnisz, M., Czatzkowska, M., 2020. Intensification of biogas production using various technologies: a review. International Journal of Energy Research 44 (8), 6240–6258. Available at https://doi.org/10.1002/er.5338.

Kululo, W.W., Habtu, N.G., Abera, M.K., Sendekie, Z.B., Fanta, S.W., Yemata, T.A., 2025. Advances in various pretreatment strategies of lignocellulosic substrates for the production of bioethanol: a comprehensive review. Discover Applied Sciences 7 (5). Available at https://doi.org/10.1007/s42452-025-06748-1.

Kunatsa, T., Zhang, L., Xia, X., 2022. Biogas potential determination and production optimisation through optimal substrate ratio feeding in co-digestion of water hyacinth, municipal solid waste and cow dung. Biofuels 13 (5), 631–641. Available at https://doi.org/10.1080/17597269.2020.1835452.

Li, X., Shi, Y., Kong, W., Wei, J., Song, W., Wang, S., 2022. Improving enzymatic hydrolysis of lignocellulosic biomass by bio-coordinated physicochemical pretreatment—a review. Energy Reports 8, 696–709. Available at https://doi.org/10.1016/j.egyr.2021.12.015.

Luengo, C.A.G., Amâncio-Vieira, S.F., Fidelis, R., Contani, E.Ad.R., 2025. Application of a multicriteria decision model for the selection of conversion pathways for biofuel production and management in a medium-sized municipality in the state of Paraná Energies 18 (9). Available at https://doi.org/10.3390/en18092367.

Luo, E., Yan, R., He, Y., Han, Z., Feng, Y., Qian, W., et al., 2024. Does biogas industrial policy promote the industrial transformation? Resources Policy 88. Available at https://doi.org/10.1016/j.resourpol.2023.104502, 104502.

Lv, Z., Wu, X., Zhou, B., Wang, Y., Sun, Y., Wang, Y., et al., 2019. Effect of one step temperature increment from mesophilic to thermophilic anaerobic digestion on the linked pattern between bacterial and methanogenic communities. Bioresource Technology 292. Available at https://doi.org/10.1016/j.biortech.2019.121968, 121968.

Mabecua, F., Dimande, N., Condo, A., Klintenberg, P., Lucas, C., Schwede, S., 2024. Barriers to successful implementation of small-scale biogas technology in Southern Africa: what can be learned from past initiatives in Mozambique? Energy Proceedings. 43. Available at https://doi.org/10.46855/energy-proceedings-11039.

Meola, A., Winkler, M., Weinrich, S., 2023. Metaheuristic optimization of data preparation and machine learning hyperparameters for prediction of dynamic methane production. Bioresource Technology 372. Available at https://doi.org/10.1016/j.biortech.2023.128604, 128604.

Miltner, M., Makaruk, A., Harasek, M., 2016. Selected methods of advanced biogas upgrading. Chemical Engineering Transactions 52, 463–468. Available at https://doi.org/10.3303/CET1652078.

Mirmohamadsadeghi, S., Karimi, K., Azarbaijani, R., Parsa Yeganeh, L., Angelidaki, I., Nizami, A.S., et al., 2021. Pretreatment of lignocelluloses for enhanced biogas production: a review on influencing mechanisms and the importance of microbial diversity. Renewable and Sustainable Energy Reviews 135. Available at https://doi.org/10.1016/j.rser.2020.110173.

Mosleh Uddin, M., Wen, Z., Mba Wright, M., 2022. Techno-economic and environmental impact assessment of using corn stover biochar for manure derived renewable natural gas production. Applied Energy 321. Available at https://doi.org/10.1016/j.apenergy.2022.119376, 119376.

Mudhoo, A., 2012. Biogas Production: Pretreatment Methods in Anaerobic Digestion. Wiley. Available at https://doi.org/10.1002/9781118404089.

Mutahi, P.M., Kimutai, S.K., Adaramola, M.S., 2025. Barriers to the growth and implementation of biogas technology: the Kenyan experience. Energy Science and Engineering. Available at https://doi.org/10.1002/ese3.70237.

Náthia-Neves, G., Berni, M., Dragone, G., Mussatto, S.I., Forster-Carneiro, T., 2018. Anaerobic digestion process: technological aspects and recent developments.

International Journal of Environmental Science and Technology 15 (9), 2033–2046. Available at https://doi.org/10.1007/s13762-018-1682-2.

Nkuna, R., Mohlomi, N., Mukhuba, M., Thathana, M.G., Puri, A.K., 2024. 11 - Green aspects of integrating different processes with anaerobic digestion for biogas production. In: Matambo, T.S., Christian, R. (Eds.), Innovations in the Global Biogas Industry: Applications of Green Principles. Woodhead Publishing, pp. 299–340. Available at https://doi.org/10.1016/B978-0-443-22372-3.00012-1.

Novakovic, J., Kontogianni, N., Barampouti, E.M., Mai, S., Moustakas, K., Malamis, D., et al., 2021. Towards upscaling the valorization of wheat straw residues: alkaline pretreatment using sodium hydroxide, enzymatic hydrolysis and biogas production. Environmental Science and Pollution Research 28 (19), 24486–24498. Available at https://doi.org/10.1007/s11356-020-08905-y.

Ohemeng-Ntiamoah, J., Datta, T., 2018. Evaluating analytical methods for the characterization of lipids, proteins and carbohydrates in organic substrates for anaerobic co-digestion. Bioresource Technology 247, 697–704. Available at https://doi.org/10.1016/j.biortech.2017.09.154.

Park, Y.G., Sun, Y.G., 2021. Study of bio-waste anaerobic digestion and its membrane purification technologies in oversea: I. Malaysian biogas scale-up case. Journal of Korea Society of Waste Management 38 (5), 458–472. Available at https://doi.org/10.9786/kswm.2021.38.5.458.

Parra-Orobio, B.A., Soto-Paz, J., Lara-Franco, S.J., Castañeda-Restrepo, M.F., Oviedo-Ocaña, E.R., Wang, Z., et al., 2025. Nanomaterial-enhanced anaerobic digestion for sustainable bioenergy production: opportunities, challenges and territorial issues. A systematic bibliometric review. Biomass and Bioenergy 203. Available at https://doi.org/10.1016/j.biombioe.2025.108268, 108268.

Patinvoh, R.J., Taherzadeh, M.J., 2019. Challenges of biogas implementation in developing countries. Opinion in Environmental Science and Health 12, 30–37. Available at https://doi.org/10.1016/j.coesh.2019.09.006.

Peters, L., Biernacki, P., Quaghebeur, W., Uhlenhut, F., Steinigeweg, S., 2019. Model-based control for a demand-driven biogas production to cover residual load rises. Bioprocess and Biosystems Engineering 42 (11), 1829–1841. Available at https://doi.org/10.1007/s00449-019-02179-6.

Pilarska, A.A., Pilarski, K., Kulupa, T., Kubiak, A., Wolna-Maruwka, A., Niewiadomska, A., et al., 2024. Additives improving the efficiency of biogas production as an alternative energy source—a review. Energies 17 (17), 4506. Available at https://doi.org/10.3390/en17174506.

Premalatha, S., Rajapandian, B., Swathi, K., Ahisri, M., Rajeswari, N., Rupashri, K.R., 2024. Biogas monitoring system. 2024 International Conference on Power, Energy, Control and Transmission Systems (ICPECTS), Chennai, India. IEEE, pp. 1–4. Available at https://doi.org/10.1109/ICPECTS62210.2024.10780028.

Rahmani, A.M., Tyagi, V.K., Ahmed, B., Kazmi, A.A., Ojha, C.S.P., Singh, R., 2022. Critical insights into anaerobic co-digestion of wheat straw with food waste and cattle manure: synergistic effects on biogas yield and kinetic modeling. Environmental Research 212. Available at https://doi.org/10.1016/j.envres.2022.113382.

Rajendran, K., Kankanala, H.R., Lundin, M., Taherzadeh, M.J., 2014. A novel process simulation model (PSM) for anaerobic digestion using Aspen Plus. Bioresource Technology 168, 7–13. Available at https://doi.org/10.1016/j.biortech.2014.01.051.

Raketh, M., Kongjan, P., O-Thong, S., Mamimin, C., Jariyaboon, R., Promnuan, K., 2024. Life cycle assessment (LCA) and economic analysis of two-stage anaerobic process of co-digesting palm oil mill effluent (POME) with concentrated latex wastewater (CLW) for biogas production. Process Safety and Environmental Protection 192, 450–459. Available at https://doi.org/10.1016/j.psep.2024.10.079.

Richards, D., Yabar, H., 2023. Promoting energy and resource recovery from livestock waste: case study Yuge Farm, Japan. Case Studies in Chemical and Environmental Engineering 7. Available at https://doi.org/10.1016/j.cscee.2023.100299, 100299.

Rojas Arboleda, M., Pfeiffer, A., Bezama, A., Thrän, D., 2021. Anticipatory study for identifying the key influential factors of the biogas system in Germany contributing to the energy system of 2050. Futures 128. Available at https://doi.org/10.1016/j.futures.2021.102704, 102704.

Ruan, D., Zhou, Z., Pang, H., Yao, J., Chen, G., Qiu, Z., 2019. Enhancing methane production of anaerobic sludge digestion by microaeration: enzyme activity stimulation, semi-continuous reactor validation and microbial community analysis. Bioresource Technology 289, 121643. Available at https://doi.org/10.1016/j.biortech.2019.121643.

Rupf, G.V., Bahri, P.A., de Boer, K., McHenry, M.P., 2016. Development of a model for identifying the optimal biogas system design in Sub-Saharan Africa. Computer Aided Chemical Engineering 38, 1533–1538. Available at https://doi.org/10.1016/B978-0-444-63428-3.50260-5.

Rutland, H., You, J., Liu, H., Bowman, K., 2025. Application of machine learning for FOS/TAC soft sensing in bio-electrochemical anaerobic digestion. Molecules (Basel, Switzerland) 30 (5), 1092. Available at https://doi.org/10.3390/molecules30051092.

Saggi, S.K., Gupta, G., Dey, P., 2017. Biological pretreatment of lignocellulosic biomaterials. In: Singh, L.K., Chaudhary, G. (Eds.), Advances in Biofeedstocks and Biofuels: Biofeedstocks and Their Processing, I. Wiley, pp. 97–119. Available at https://doi.org/10.1002/9781119117322.ch5.

Saikat, B., Sivamani, S., Prasad, B.S.N., 2023. Biogas as a value generation in industrial wastewater—a review. In: Shah, M.P. (Ed.), Biorefinery for Water and Wastewater Treatment. Springer, Cham, pp. 49–69. Available at https://doi.org/10.1007/978-3-031-20822-5_3.

Sesini, M., Cretì, A., Massol, O., 2024. Unlocking European biogas and biomethane: policy insights from comparative analysis. Renewable and Sustainable Energy Reviews 199. Available at https://doi.org/10.1016/j.rser.2024.114521, 114521.

Shaw, K.M., Poh, P.E., Ho, Y.K., Chen, Z.Y., Chew, I.M.L., 2024. Modeling the anaerobic digestion of palm oil mill effluent via physics-informed deep learning. Chemical Engineering Journal 485. Available at https://doi.org/10.1016/j.cej.2024.149826.

Sime, G., Jin, Z., 2020. Technical and socioeconomic constraints to the domestication and functionality of biogas technology in rural areas of southern Ethiopia. Cogent Engineering 7 (1). Available at https://doi.org/10.1080/23311916.2020.1765686.

Sinsuw, A.A.E., Chen, T.-H., Dokmaingam, P., Suriandjo, H.S., Chu, C.-Y., 2024. Life cycle assessment of environmental impacts for two-stage anaerobic biogas plant between commercial and pilot scales. International Journal of Hydrogen Energy 52, 58–70. Available at https://doi.org/10.1016/j.ijhydene.2023.06.331.

Skovsgaard, L., Jacobsen, H.K., 2017. Economies of scale in biogas production and the significance of flexible regulation. Energy Policy 101, 77–89. Available at https://doi.org/10.1016/j.enpol.2016.11.021.

Sunar, S.L., Kumara, M.K., Oruganti, R.K., Khadka, K.K., Panda, T.K., Bhattacharyya, D., 2025. Pretreatment and anaerobic co-digestion of lignocellulosic biomass: recent developments. Bioresource Technology Reports 30. Available at https://doi.org/10.1016/j.biteb.2025.102133.

Tabatabaei, M., Aghbashlo, M., Valijanian, E., Kazemi Shariat Panahi, H., Nizami, A.S., Ghanavati, H., et al., 2020. A comprehensive review on recent biological innovations to improve biogas production, part 1: upstream strategies. Renewable Energy 146, 1204–1220. Available at https://doi.org/10.1016/j.renene.2019.07.037.

Tembo, A., Rahman, M.M., Jerin, T., 2023. Barriers to development and adoption of biogas in Mokambo peri-urban of Mufulira, Zambia: how does local government fail to provide renewable energy? Zambia Biofuels 14 (6), 583–594. Available at https://doi.org/10.1080/17597269.2022.2156055.

Thompson, T.M., Young, B.R., Baroutian, S., 2019. Advances in the pretreatment of brown macroalgae for biogas production. Fuel Processing Technology 195. https://doi.org/10.1016/j.fuproc.2019.106151.

Tian, H., Wang, X., Lim, E.Y., Lee, J.T.E., Ee, A.W.L., Zhang, J., et al., 2021. Life cycle assessment of food waste to energy and resources: centralized and decentralized anaerobic digestion with different downstream biogas utilization. Renewable and Sustainable Energy Reviews 150. Available at https://doi.org/10.1016/j.rser.2021.111489.

Tian, T., Qiao, S., Li, X., Zhang, M., Zhou, J.z, 2017. Nano-graphene induced positive effects on methanogenesis in anaerobic digestion. Bioresource Technology 224, 41–47. Available at https://doi.org/10.1016/j.biortech.2016.10.058.

Vu, H.P., Nguyen, L.N., Vu, M.T., Johir, M.A.H., McLaughlan, R., Nghiem, L.D., 2020. A comprehensive review on the framework to valorise lignocellulosic biomass as biorefinery feedstocks. Science of the Total Environment 743. Available at https://doi.org/10.1016/j.scitotenv.2020.140630.

Wahid, R., Horn, S.J., 2021. Impact of operational conditions on methane yield and microbial community composition during biological methanation in in situ and hybrid reactor systems. Biotechnology for Biofuels 14 (1). Available at https://doi.org/10.1186/s13068-021-02019-4.

Wang, J., 2014. Decentralized biogas technology of anaerobic digestion and farm ecosystem: opportunities and challenges. Frontiers in Energy Research 2. Available at https://doi.org/10.3389/fenrg.2014.00010.

Wang, L., Long, F., Liao, W., Liu, H., 2020. Prediction of anaerobic digestion performance and identification of critical operational parameters using machine learning algorithms. Bioresource Technology 298. Available at https://doi.org/10.1016/j.biortech.2019.122495, 122495.

Wang, Y.N., Hou, Q.D., Liu, L., Ju, M.T., Li, W.Z., Chang, C.C., 2018. Anaerobic process. Water Environment Research 90 (10), 928–946. Available at https://doi.org/10.2175/106143018X15289915807083.

Westerholm, M., Isaksson, S., Sun, L., Schnürer, A., 2017. Microbial community ability to adapt to altered temperature conditions influences operating stability in anaerobic digestion. Energy Procedia 105, 895–900. Available at https://doi.org/10.1016/j.egypro.2017.03.408.

Wu, Y., 2025. Thermoeconomic modeling of new energy system based on biogas upgrading to multigenerational purpose. Desalination 593, 118215. Available at https://doi.org/10.1016/j.desal.2024.118215.

Wynn, J.P., Hanchar, R., Kleff, S., Senyk, D., Tiedje, T., 2016. Biobased technology commercialization: The importance of lab to pilot scale-up. In: Van Dien, S. (Ed.), Metabolic Engineering for Bioprocess Commercialization. Springer, Cham, pp. 101–119. Available at https://doi.org/10.1007/978-3-319-41966-4_6.

Xie, S., Li, X., Wang, C., Kulandaivelu, J., Jiang, G., 2020. Enhanced anaerobic digestion of primary sludge with additives: performance and mechanisms. Bioresource Technology 316. Available at https://doi.org/10.1016/j.biortech.2020.123970, 123970.

Xie, Z., Cao, Q., Chen, Y., Luo, Y., Liu, X., Li, D., 2022. The biological and abiotic effects of powdered activated carbon on the anaerobic digestion performance of cornstalk. Bioresource Technology 343. Available at https://doi.org/10.1016/j.biortech.2021.126072, 126072.

Yalcinkaya, S., Yucel, O., 2025. Enhancing biogas production from municipal wastewater sludge and grease trap waste: explainable machine learning models for prediction and parameter identification. Fuel 391, 134787. Available at https://doi.org/10.1016/j.fuel.2025.134787.

Yousuf, A., Khan, M.R., Pirozzi, D., Ab Wahid, Z., 2016. Financial sustainability of biogas technology: barriers, opportunities, and solutions. Energy Sources, Part B: Economics, Planning and Policy 11 (9), 841–848. Available at https://doi.org/10.1080/15567249.2016.1148084.

Zhao, Y., Xu, C., Ai, S., Wang, H., Gao, Y., Yan, L., et al., 2019. Biological pretreatment enhances the activity of functional microorganisms and the ability of methanogenesis during anaerobic digestion. Bioresource Technology 290, 121660. Available at https://doi.org/10.1016/j.biortech.2019.121660.

Zhong, L., Lin, F., Wo, D., Yang, X., Sun, Y., Feng, X., et al., 2024. The dominant-substrate driven the enhanced performance in co-digestion of Pennisetum hybrid and livestock waste. Journal of Environmental Management 366, 121700. Available at https://doi.org/10.1016/j.jenvman.2024.121700.

Ziganshin, A.M., Wintsche, B., Seifert, J., Carstensen, M., Born, J., Kleinsteuber, S., 2019. Spatial separation of metabolic stages in a tube anaerobic baffled reactor: reactor performance and microbial community dynamics. Federation Applied Microbiology and Biotechnology 103 (9), 3915–3929. Available at https://doi.org/10.1007/s00253-019-09767-2.

Assessment of biogas fertilisers: efficacy and application

6

6.1 Introduction

The growth of agricultural biogas plants and anaerobic treatment facilities for organic waste generates substantial amounts of digestate, a by-product of anaerobic digestion that is increasingly valued as an organic fertiliser (Logan and Visvanathan, 2019; Haque et al., 2023). Digestate is a mixture of liquid and solid residues remaining after digestion. It is rich in nutrients, mainly nitrogen (N), phosphorus (P), and potassium (K), as well as organic matter and micronutrients (Martín-Sanz-Garrido et al., 2025). Unlike traditional manures, digestate shows a high degree of mineralization, and a considerable share of nitrogen occurs in the ammonium form that is readily available to plants (Martín-Sanz-Garrido et al., 2025; Möller and Müller, 2012). For this reason, it is regarded as a valuable fertiliser that can partly replace mineral fertilisers and supports circular economy principles by recycling nutrients from waste streams (Skrzypczak, et al., 2023b). Demand for alternative sources of N and P in agriculture continues to rise due to the depletion of raw materials, natural gas for N production and phosphate rock for P, as well as increasing prices of synthetic fertilisers (García-López et al., 2023). Nutrient recycling from waste, including the use of digestate, has therefore become an important element of strategies aimed at ensuring food security and promoting sustainable agriculture (Rosemarin et al., 2020). According to the European Biogas Association, in 2022 Europe produced approximately 31 million tonnes of digestate on a dry-matter basis; this volume could cover about 15% of EU demand for nitrogen fertilisers and avoid around 10 million tonnes of CO_2-equivalent emissions by reducing synthetic fertiliser production and improving waste management (European Biogas Association, 2023; European Biogas Association, 2024). The fertiliser potential of digestate is therefore substantial. However, to use it fully, it is necessary to understand its composition and properties, its effects on soil and yields, and the conditions required for safe application.

This chapter presents the characteristics of digestate based fertilisers and assesses their effectiveness and potential use in agriculture. It discusses the content and bioavailability of nutrients in digestates derived from different feedstocks, the effects of digestate application, both untreated and processed, on soil properties and plant growth, and a comparison of the fertiliser performance of digestate with conventional mineral fertilisers. It also addresses digestate processing, for example phase separation, composting, and pelletizing, and explains how these approaches influence nutrient availability.

Sustainable Biogas Production. DOI: https://doi.org/10.1016/B978-0-443-40557-0.00003-2

6.2 Nutrient content and bioavailability in digestate based fertilisers

6.2.1 *Chemical composition of digestate*

Digestate from biogas plants is characterised by a high concentration of macronutrients essential for plant growth. The total nitrogen content (N_{tot}) in digestate is similar to the nitrogen content in the original feedstock. However, due to anaerobic digestion, most organic nitrogen is converted into mineral forms, mainly ammonium, that are readily taken up by plants. Typically, up to about 80% of total nitrogen in digestate is present as ammonium nitrogen (N-NH^{4+}), whereas in manure this share may be only 10%–15% (Martín-Sanz-Garrido et al., 2025; Möller and Müller, 2012). For example, in digestate from pig slurry about 80% of N occurs as NH^{4+} (Pampillón-González et al., 2017), while in digestate derived from food waste the proportion can reach about 99% (Slepetiene et al., 2023). This means that nitrogen in digestate can be highly available to plants, with availability comparable to mineral fertilisers containing ammonium or urea. Digestate also contains substantial amounts of phosphorus, often present as calcium or magnesium phosphates, which become available to plants at rates comparable to those of conventional phosphorus fertilisers, and potassium in ionic form that is fully plant available. The typical range of macronutrient concentrations in digestates is broad due to the diversity of feedstocks (Samoraj et al., 2022; Skrzypczak et al., 2023a). In a study by Jurgutis et al. (2021), concentrations reached up to 73 g N/kg fresh matter and 25 g K/kg in digestates from selected agricultural biogas plants (Jurgutis, Šlepetienė, Šlepetys, et al., 2021). On average, the nitrogen content of liquid digestate corresponds to about 2%– 10% of the nitrogen content in typical mineral nitrogen fertilisers. This implies that higher application rates by mass are needed, while at the same time supplying other nutrients and organic matter.

6.2.2 *Variability depending on feedstock and processing*

The fertiliser composition of digestate strongly depends on the type and origin of the substrates subjected to anaerobic digestion. Digestates derived from animal slurries, including pig and cattle manure, are characterized by relatively high concentrations of ammonium nitrogen and potassium and somewhat lower phosphorus content. This reflects the composition of manure and slurry, although differences in animal feeding regimes and housing systems may affect elemental proportions (Ragályi et al., 2025; Kovács et al., 2016). For example, digestate from poultry manure contains higher concentrations of total and ammonium nitrogen than digestate from cattle or pig slurry, due to the original composition of poultry litter (Holm-Nielsen, Al Seadi, and Oleskowicz-Popiel, 2009; Jurgutis, Šlepetienė, Šlepetys, et al., 2021).

Digestates produced from food waste, including household and catering residues, usually contain large amounts of easily degradable organic matter derived from food leftovers. This translates into high methane potential and also affects fertiliser composition. Such digestates are rich in ammonium nitrogen, phosphorus, and potassium, and may also contain elevated levels of sodium and chloride ions originating from table salt (Jurgutis,

Šlepetienė, Šlepetys, et al., 2021; Pawlett and Tibbett, 2015). Consequently, digestate from protein rich meat waste tends to be more nitrogen rich, whereas digestate from plant based residues, such as vegetable waste, is generally lower in nitrogen (Ragályi et al., 2025).

Digestates from sewage sludge are also used as fertilisers and typically contain 1%–6% nitrogen in dry matter along with substantial amounts of phosphorus. Approximately 50% of sludge after digestion consists of stable organic matter (Ragályi et al., 2025). Due to their municipal origin, such materials may contain heavy metals or persistent organic contaminants, including pharmaceuticals and other chemical compounds, which are not removed during digestion (Cao and Pawłowski, 2012).

Digestates obtained from plant biomass or green waste, such as maize silage, grasses, or garden residues, generally have higher fibre and lignin contents. This results in a greater proportion of the solid fraction and a higher C:N ratio. Their nitrogen content may be lower than that of slurry-based digestates, but they provide more recalcitrant organic matter that is valuable for improving soil properties (Alburquerque et al., 2012).

Post digestion processing methods also influence digestate composition. A common practice is the separation of digestate into liquid and solid fractions. The liquid fraction usually contains most of the ammonium nitrogen and potassium, while the solid fraction is richer in phosphorus, calcium, and organic carbon (Bolzonella et al., 2018; Tambone et al., 2017; Vázquez-Rowe et al., 2015). This separation allows better adaptation of the product to specific needs. The liquid fraction can serve as a fast acting nitrogen fertiliser, whereas the solid fraction can be used as a slower acting soil amendment or as feedstock for further processing, such as drying, pelletizing, or composting.

Drying and pelletising increase nutrient concentration relative to mass, facilitate transport and application, and extend shelf life by reducing water content (Martín-Sanz-Garrido et al., 2025). However, drying and pelletizing must be carefully controlled. Intensive heating can lead to significant nitrogen losses due to ammonia volatilisation. The literature indicates that uncontrolled drying may result in nitrogen losses exceeding 95% of the digestate nitrogen content (Szymańska et al., 2022). For this reason, technologies such as digestate acidification prior to drying or closed systems designed to minimize ammonia losses are applied (Pantelopoulos et al., 2016).

Composting of digestate leads to partial degradation of easily biodegradable residues and to humification of organic matter, thereby stabilising the material (Kovačić et al., 2022; Huang et al., 2019). Studies have shown that composting digestate from food waste significantly reduces CO_2 emissions after soil incorporation, which indicates lower carbon mineralisation. Emissions decreased from 2850 mg CO_2/kg soil for fresh digestate to 625 mg CO_2/kg soil for composted material (Grigatti et al., 2020). At the same time, composting did not significantly reduce the fertiliser availability of the main nutrients. Both phosphorus fertilization efficiency and nitrogen use from composted digestate remained similar to values observed for fresh digestate (Grigatti et al., 2020). This indicates that processed digestate largely retains its fertiliser value while becoming more stable and easier to handle.

6.2.3 Nutrient bioavailability and its determining factors

Due to the high proportion of mineral forms, nutrients in digestate are generally well available to plants. Ammonium nitrogen NH^{4+} from digestate is directly taken up by plant roots in the same way as nitrogen from mineral fertilisers (Li et al., 2023). A portion of nitrogen remains in organic compounds that become available more slowly, as they undergo gradual mineralisation in the soil and release nitrogen over an extended period. As a result, digestate acts partly as a fast acting fertiliser through its ammonium fraction and partly as a slow release fertiliser through its organic fraction. This dual effect can support prolonged crop nutrition and reduce nitrogen losses through leaching after the initial period following application (Ragályi et al., 2025; Tambone and Adani, 2017).

Phosphorus in digestate is mainly present in inorganic forms, often associated with calcium and magnesium as Ca and Mg phosphates, and depending on composition, as struvite, which is magnesium ammonium phosphate. The solubility and plant availability of these forms strongly depend on soil pH and are usually gradual rather than immediate compared with highly soluble mineral phosphorus fertilisers (Mazzini et al., 2020; Regelink et al., 2021; Talboys et al., 2016). Potassium, present in ionic form (Barłóg et al., 2020), is immediately soluble in soil water and is taken up by plants as readily as potassium from synthetic fertiliser salts.

It is important to note that the quantitative ratio of N:P:K in digestate often differs from that of conventional mineral fertilisers designed primarily to supply nitrogen. Digestate frequently shows an unfavourable N:P balance in relation to crop requirements. When application rates are based on nitrogen demand, relatively large amounts of phosphorus may be added, increasing the risk of excessive phosphorus accumulation in soil (García-López et al., 2023). Therefore, application rates should be adjusted with regard to both nitrogen and phosphorus, and nitrogen deficiencies may need to be supplemented with mineral fertiliser to avoid phosphorus over fertilization. Similarly, digestates with high potassium content may supply more potassium than is removed with the harvested yield (Möller and Müller, 2012; Jurgutis, Šlepetienė, Šlepetys, et al., 2021). This may be beneficial on potassium deficient soils, but on potassium rich soils it can represent a loss or lead to an unfavourable K:Mg ratio, causing antagonism in magnesium uptake by plants (Xie et al., 2021; Dölger et al., 2024).

Digestate composition should therefore be evaluated comprehensively. Key parameters include total and ammonium nitrogen content, the N_{org}/N_{tot} ratio, the N:P:K ratio, as well as pH and the content of salts such as sodium and chloride, and any potential contaminants. All these factors influence fertilisation efficiency and should be considered when planning digestate application rates and methods.

6.2.4 Contaminants and quality aspects of nutrients

Although digestate is valued as a source of nutrients, it may also contain undesirable substances. The most frequently cited include heavy metals such as Cd, Pb, Ni, Zn, and Cu. Their presence and concentrations depend on the type of substrates and on contaminants in the feedstock sent to anaerobic digestion (Lamolinara et al., 2022; Zheng et al., 2022). In many cases, heavy metal concentrations in digestates meet the standards for organic fertilisers. For example, studies of food waste digestates in Sweden reported cadmium at 0.34–0.37 mg/kg dry matter, lead at 2–7.4 mg/kg, nickel at 7.2–15 mg/kg dry matter, mercury below 0.1 mg/kg dry matter, zinc at 180–540 mg/kg dry matter, and copper at 41–100 mg/kg dry matter. These values are well below the EU limits for fertilisers, namely 1.5 mg/kg dry matter for Cd, 120 mg/kg dry matter for Pb, 50 mg/kg dry matter for Ni, 1 mg/kg dry matter for Hg, 300 mg/kg dry matter for Cu, and 800 mg/kg dry matter for Zn (Dincă et al., 2025; Golovko et al., 2022). In digestates derived from sewage sludge or certain industrial wastes, concentrations of specific metals may be higher, which is why chemical testing is required before agricultural use.

Another potential concern involves organic contaminants and residues of pharmaceuticals or plant protection products. Anaerobic digestion reduces some pathogens and decreases the phytotoxicity of waste, but not all chemical compounds are degraded. The literature reports trace amounts of antibiotics, hormones, feed additives, and microplastic particles in digestates, originating from contaminants in the feedstock (Sobhi et al., 2024). At present, specific legal limits for antibiotics and microplastics in digestate-based organic fertilisers are generally not harmonised; therefore, these contaminants should be discussed as emerging quality risks requiring feedstock control and monitoring rather than as routinely safe at typical concentrations (Czatzkowska et al., 2025; Sobhi et al., 2024). Nevertheless, these issues are under active investigation, and future regulations may impose stricter requirements for digestate purity. Quality assurance schemes, including digestate certification programs, already emphasize control of input materials to minimise contamination risks.

In summary, digestate is a nutrient rich source of readily available macro and micronutrients for plants, but its composition can vary depending on origin and processing. Understanding these differences is essential for optimizing its fertiliser use and avoiding potential environmental problems.

6.3 Effects of digestate based fertilisers on soil and plant growth

6.3.1 Impact on soil chemical properties and fertility

The application of digestate affects soil in multiple ways. It supplies nutrients and organic matter and influences soil reaction and biological activity. Due to the presence of slowly degradable organic matter, digestate can improve humus content and soil organic carbon under long-term use (Sobhi et al., 2024). Although most readily available carbon is consumed during anaerobic digestion, the remaining organic fraction, such as lignin, cellulose, and humic compounds, is relatively stable and undergoes gradual humification in soil (Kovačić et al., 2022; Huang et al., 2019). Studies indicate that long-term digestate application can increase soil organic carbon stocks and the proportion of more stable organic matter fractions, particularly when materials with higher structural carbon content are used, such as the solid fraction or aerobically processed products (Mayerová et al., 2023).

For example, field experiments on low fertility and eroded soils in Lithuania showed that digestate application increased soil organic carbon and mobile fractions of humified carbon, including humic acid fractions, compared with unfertilized controls. This effect is practically important because an increase in stable organic matter supports improvements in soil structure and physical properties, including water retention capacity, and thus enhances overall soil fertility (Slepetiene et al., 2020, 2022).

Digestate usually has a neutral to slightly alkaline reaction. Its incorporation into soil can therefore help neutralize excessive acidity. In studies on acidic soils, digestate application reduced hydrolytic acidity and increased soil buffering capacity (Rolka et al., 2024). On acidic soils, digestate fertilization typically raises soil pH due to its alkalizing effect associated with the presence of basic and ammonium compounds. At the same time, it may slightly increase soil electrical conductivity as a result of the input of readily soluble mineral salts (Möller and Müller, 2012; Rolka et al., 2024). In most cases, moderate digestate rates do not cause harmful salinity. However, on poorly drained soils or when very high doses are applied, a temporary increase in sodium and chloride concentrations may occur. Digestates derived from food waste, which often contain higher sodium levels, may increase soil salinity and electrical conductivity (Pawlett and Tibbett, 2015). For this reason, dilution or application on irrigated fields is recommended to prevent salt accumulation.

6.3.2 Effects on nutrient availability in soil

As a fertiliser, digestate usually increases soil levels of mineral nitrogen and plant available phosphorus and potassium. Shortly after application, an increase in ammonium nitrogen is observed in the topsoil, followed by a rise in nitrate nitrogen as nitrification proceeds (Doyeni et al., 2021; Möller and Müller, 2012; Petraityte et al., 2022). García López et al. (2023) showed that digestate application, either alone or combined with mineral nitrogen, significantly increased soil mineral nitrogen, mainly as NH_4^+, as well as available phosphorus. At higher rates, soil organic matter content also increased (García López et al., 2023).

Digestate also supplies a range of micronutrients required by plants, including Zn, Cu, Mn, Fe, and B. Their availability in soil may increase after digestate application, depending on feedstock composition, soil properties and digestate processing (Przygocka-Cyna and Grzebisz, 2018; Dincă et al., 2025). In many studies, no excessive increase in heavy metal mobility was observed in soils fertilised with digestate, even though trace amounts of these elements are introduced. Analyses of soluble and operationally defined metal fractions have shown that a large proportion of heavy metals remains in poorly available forms, with less mobile fractions dominating and the exchangeable and soluble fraction remaining low (Baldasso et al., 2023; Guo et al., 2020). As a result, any increase in the bioavailable pool is often limited, although it may depend on time, soil properties, and digestate composition, including the share of dissolved organic matter (Baldasso et al., 2023; Guo et al., 2020). In addition, initial metal concentrations in digestate are often several times lower than regulatory limits, so even long term use does not lead to exceedance of soil standards (Dincă et al., 2025; Golovko et al., 2022). Nevertheless, monitoring of soils fertilized with digestate over extended periods is recommended, especially when the digestate originates from municipal or industrial waste.

6.3.3 Effects on soil biological activity

Digestate application influences soil microbial and enzymatic activity, and these effects depend on dose and composition (Karimi et al., 2022; van Midden et al., 2023). Digestate introduces carbon fractions with different susceptibility to decomposition, and both the quantity and quality of organic matter, as well as the chemical composition of the material, shape the response of soil microorganisms (van Midden et al., 2023). Literature reviews indicate that under certain conditions, such as high concentrations of ammonium compounds or salts, digestate may have less favorable effects on soil organisms. The direction and magnitude of these effects depend on application rate and on the properties of both digestate and soil (van Midden et al., 2023). In a field experiment, García López et al. (2023) reported that a high digestate rate of up to 340 kg N/ha was associated with reduced activity of selected soil enzymes, including phosphatase and β glucosidase, as well as lower microbial biomass indicators for bacteria and fungi compared with the control or mineral fertilization. The authors linked the decrease in phosphatase activity to high phosphorus availability in soil following digestate application (García López et al., 2023). Digestate may also contain microorganisms originating from the anaerobic digestion process, including communities involved in hydrolysis, fermentation of organic compounds, and methanogenesis (Lim et al., 2020). For this reason, digestate is sometimes considered not only a nutrient source but also a material that introduces specific microbial groups and metabolic potential into the soil environment (Lim et al., 2020). The literature also describes the use of digestate as an amendment supporting the bioremediation of soils contaminated with petroleum hydrocarbons. In one study using sewage sludge digestate, application to hydrocarbon contaminated soil increased the initial density and diversity of bacteria and promoted the presence and abundance of hydrocarbon degradation markers, including alkB genes (Gielnik et al., 2021). Digestate has also been tested as an amendment in experiments focused on remediation of hydrocarbon pollution (Jumbo et al., 2022).

From a sanitary safety perspective, anaerobic digestion can effectively reduce certain pathogens, including indicator faecal bacteria such as *Escherichia coli*. The efficiency of reduction depends on process temperature and operational parameters, with thermophilic conditions generally being more effective (Zhao and Liu, 2019; Álvarez- Fraga et al., 2025). For example, in a study of thermophilic anaerobic digestion, *Salmonella* spp. and *E. coli* were reduced below detection limits, while spores such as *Clostridium perfringens* persisted (Lloret et al., 2013). It has also been shown that the process can eliminate the germination capacity of weed seeds, particularly under thermophilic conditions (Johansen et al., 2013). Moreover, levels of indicator microorganisms in digestate may further decline during post digestion storage, as reported in studies on hygienization and storage scenarios (Nag et al., 2021). Consequently, when the process is properly managed and quality control is maintained, the sanitary risk associated with digestate application can be substantially reduced (Zhao and Liu, 2019; Álvarez Fraga et al., 2025).

6.3.4 Effects on yield and plant growth

Digestate based fertilisers can effectively support plant growth, and in many field experiments yields are comparable to those obtained with mineral fertilization (Curadelli et al., 2023). Digestate supplies nitrogen mainly in mineral form, with a substantial share of N NH_4^+, as well as potassium, which is largely present in dissolved form as K^+. This promotes rapid fertiliser action (Möller and Müller, 2012). In a meta analysis including data from 35 studies with a total of 106 observations, Curadelli et al. (2023) showed that digestate fertilization resulted in average yields about 80% higher than the unfertilized control, while yields were statistically similar to those achieved with mineral fertilisers, on average about 2% lower (Curadelli et al., 2023). In a field experiment on silage maize, Verdi et al. (2019) reported no significant differences in dry matter yield between the liquid fraction of pig slurry digestate and urea at the same rate of 150 kg N/ha, with values of 13.63 and 13.24 t dry matter/ha, respectively (Verdi et al., 2019). Additional evidence comes from field trials conducted under various soil and climatic conditions and on different crops. In a study on switchgrass grown on very acidic soil, a lower digestate rate of 30 m^3/ha produced a biomass yield of 4.30 ± 0.20 t/ha, practically identical to mineral NPK fertilization at 4.33 ± 0.22 t/ha. A higher rate of 60 m^3/ha increased yield to 5.15 ± 0.26 t/ha (Głowacka et al., 2020). This indicates that, when properly dosed, digestate can not only match but even exceed mineral fertilization in terms of biomass production.

Similar patterns were observed in winter wheat, where all fertilization treatments, including digestate, NPK, cattle slurry, and digestate combined with straw, significantly increased grain and straw yields compared with the unfertilized control. Grain yield increases ranged from 39.6% to 47.4%, with the highest values recorded for the digestate treatment (Šimon et al., 2015). In a three year field study including maize, spinach, and potato, bio based fertilisers such as the liquid fraction of digestate were compared with synthetic fertilization using calcium ammonium nitrate. No significant differences were found in yield or plant nitrogen uptake between digestate based treatments and mineral fertiliser, with p values above 0.05 (Reuland et al., 2024). These results indicate that, under long-term field conditions, digestate can sustain crop production at levels equivalent to conventional nitrogen fertilisers. Regarding crop quality, some studies suggest that digestate fertilization may improve selected quality traits, particularly those related to pigments and antioxidant compounds. For example, Panuccio et al. (2019) found that the application of two types of digestate in cucumber cultivation increased the content of phenolic and flavonoid compounds with antioxidant activity in fruits. Certain compounds, such as neohesperidin, hesperetin, naringin, and narirutin, were detected only in digestate treatments (Panuccio et al., 2019). In lettuce grown under strong salt stress at 7.5 dS/m NaCl, Li et al. (2023) observed that digestate application increased fresh and dry mass, enhanced photosynthetic pigment content and photosynthesis, and reduced oxidative stress indicators compared with mineral fertilization. These findings suggest that, under comparable nutrient supply, digestate may in specific conditions help maintain or improve selected physiological quality parameters of crops.

However, quality effects depend on digestate composition and nutrient balance. For example, deficiencies of certain elements may occur when digestate is used as the sole fertiliser source in soilless systems. Therefore, interpretation of quality responses should consider specific cultivation conditions and fertilization strategies (Weimers et al., 2022).

Increasing attention is also given to modifying yield responses through combination of digestate with sorptive materials or by processing it into more concentrated and standardized forms, which influence nutrient availability dynamics in the root zone. In a 2-year field experiment on processing tomato, treatments included liquid digestate, biochar, pelletized digestate, and a combined treatment of liquid digestate with biochar. The highest marketable yield was obtained in the digestate plus biochar treatment at 72 t/ha, compared with 47 t/ha in the unfertilized control. Fertilized treatments also improved selected technological quality parameters, including Brix per hectare, and reduced the Bostwick index compared with the control (Ronga et al., 2020). On the other hand, results from a long-term field experiment conducted on highly fertile soil showed that biochar applied alone or in combination with digestate did not necessarily cause significant changes in yield or nitrogen use efficiency. When digestate was mixed with biochar without additional mineral nitrogen, only a non-significant tendency toward reduced nitrogen uptake and lower nitrogen use efficiency was observed (Sixt et al., 2025). Among fertiliser products derived from digestate, struvite also plays an important role, as it allows part of the phosphorus to be transferred into a more concentrated form. Due to the low N:P ratio in struvite, nitrogen supplementation is required. Studies have shown that increasing nitrogen supply to a level comparable with a reference fertiliser enhances the yield effect of struvite to a level similar to that achieved with commercial ammonium phosphate (Szymańska et al., 2019).

6.3.5 Potential risks for plants and soil

When applied responsibly, digestate is generally beneficial for both crops and soil. However, risks may arise in cases of excessive rates or improper use. Fresh digestate can exhibit phytotoxic effects, mainly associated with high ammonium nitrogen content, the presence of organic acids, and in some cases elevated salinity expressed as electrical conductivity (Möller and Müller, 2012; Abbas et al., 2023). For this reason, fresh and undiluted digestate should not be applied in direct contact with seeds or young seedlings. Safer practice involves pre sowing application or techniques that reduce contact with emerging plants, while ensuring that rates are aligned with crop demand. In relation to organic residues, materials with relatively higher carbon content may temporarily cause nitrogen immobilization in soil, meaning short term binding of nitrogen in microbial biomass. Although digestate often has a lower C:N ratio than raw manure or slurry, which reduces immobilization risk, this effect may be more likely when applying the solid fraction or carbon rich materials (van Midden et al., 2024; Abbas et al., 2023). According to good fertilization practice, digestate rates should be adjusted to the farm nitrogen balance and environmental requirements. Under the Nitrates Directive, the limit of

170 kg N/ha/year applies specifically to nitrogen from livestock manure, including processed manure; whether and how digestate is counted depends on its feedstock origin and national implementation, with higher limits possible only under specific derogations and conditions (European Union, 2008). While digestate generally has a positive impact on soil, proper management is essential. Rate, timing, and application method should be selected to minimize phytotoxicity risk and prevent temporary nitrogen shortages when using carbon richer materials.

6.4 Digestate and conventional fertilisers

6.4.1 Fertiliser efficiency

When comparing digestate with mineral fertilisers, crop yield responses are often similar. A large meta analysis showed that average yields under digestate fertilization were statistically comparable to mineral fertilization, and about 80% higher than the unfertilized control (Curadelli et al., 2023). Mineral fertilisers have the advantage of high nutrient concentration. For example, Ammonium nitrate contains about 33%–34.5% nitrogen. In contrast, digestate typically contains around 2–5 kg N/m3 (Li et al., 2023). This means that larger volumes are required to supply the same nitrogen dose, which increases logistical demands. On the other hand, digestate simultaneously supplies phosphorus, potassium, micronutrients, and organic matter. Möller and Müller (2012) describe digestate as a source of both macro and micronutrients, and indicate that separated fractions can function as N K fertilisers comparable to mineral N K products. From a practical perspective, nutrient use efficiency can be high under certain conditions with digestate fertilization. Some studies report comparable crop production combined with reduced nitrogen losses relative to mineral fertilisers (Rossi et al., 2023; Verdi et al., 2019). In a field experiment on silage maize, Verdi et al. (2019) compared the liquid fraction of pig slurry digestate with urea at 150 kg N/ha. Cumulative nitrogen emissions over 25 days were lower for digestate at 2.87 kg N/ha/25 days than for urea at 3.76 kg N/ha/25 days. Emission profiles differed, with higher N_2O emissions from digestate and higher NH_3 emissions from urea. Regarding nitrogen leaching, Rossi et al. (2023) conducted rainfall simulation experiments and found that the mineral treatment using ammonium sulfate resulted in the highest cumulative nitrogen leaching losses. Digestate derived from municipal organic waste reduced leaching compared with mineral fertilization. These findings suggest that, under certain conditions, digestate may limit part of the nitrogen losses to the environment, although outcomes depend on fertiliser composition, soil type, and weather conditions. Application technique is a key factor influencing nitrogen losses. Field studies show that significant amounts of total ammoniacal nitrogen may volatilize as ammonia when digestate is applied using trailing hose systems (Efosa et al., 2025). In contrast, open disc injection significantly reduces NH_3 emissions compared with trailing hose application (Quilez et al., 2025). These results highlight the importance of low emission techniques and appropriate management to minimize ammonia losses.

6.4.2 Effects on the environment and soil

From an environmental perspective, digestate offers several potential advantages over mineral fertilisers. Its use reduces demand for industrial fertilisers, whose production is associated with a substantial carbon footprint. For instance, a report by Fertilisers Europe estimated the carbon footprint of ammonium nitrate production at about 3.319 t CO_2e/t N as the EU average, and up to 10.875 t CO_2e/t N in coal based production systems (Hoxha and Christensen, 2018). Replacing part of mineral nitrogen with nitrogen from digestate can therefore reduce emissions linked to fertiliser manufacturing, although the scale depends on the reference fertiliser and energy mix. According to the European Biogas Association, Europe produced 31 Mt of digestate on a dry matter basis in 2022, an amount that could cover about 15% of EU nitrogen fertiliser demand. The same source estimates that this could avoid around 10 Mt CO_2e and save approximately 2 billion cubic meters of natural gas (European Biogas Association (EBA) 2023). Digestate may also contribute to building soil carbon stocks. The European Biogas Association reports that digestate contains significant amounts of stable carbon compounds with high humification potential, which can increase humus content and support carbon sequestration (European Biogas Association (EBA) 2024). Grigatti et al. (2024) showed that composting raw digestate improved stability and promoted greater conservation of soil organic carbon compared with raw digestate application. Quality and contamination issues must also be considered. The presence of contaminants depends on feedstock and process control, so long term use requires monitoring. Both digestate and mineral fertilisers may contribute to eutrophication if applied in excess or at inappropriate times.

6.4.3 Economic and practical aspects

Compared with mineral fertilisers, digestate can be economically advantageous, especially when available locally. In some regions with nutrient surpluses, digestate is provided to farmers free of charge due to disposal costs and logistical constraints (European Biogas Association (EBA) 2024). However, transport and application costs increase significantly with distance, so digestate is most often used near the biogas plant (Drosg et al., 2015). Its economic advantage is therefore greatest where transport distances are short, particularly for the liquid fraction with low dry matter content. To address volume and handling challenges, various technological solutions are implemented. Separation into solid and liquid fractions facilitates storage and management and can reduce handling costs (Nowak and Czekała, 2024). Technologies for concentrating and recovering nutrients from the liquid fraction, such as membrane processes or evaporation, are also being developed to produce more concentrated fertiliser products. Nitrogen recovery through stripping and scrubbing can produce ammonium sulfate solution, while the remaining digestate with reduced nitrogen content is used locally. Phosphorus recovery as struvite is also expanding. Field studies have shown that yields under struvite, especially when supplemented with nitrogen, were similar to or higher than those obtained with commercial ammonium phosphate (Szymańska et al., 2019). Such products increase distribution possibilities and may enable competition with mineral fertilisers beyond the immediate vicinity of biogas plants. Mineral fertilisers are generally more convenient due to high nutrient concentration and ease of spreading. In contrast, digestate requires appropriate equipment and application techniques. Development of soil injection systems is important not only for nutrient efficiency but also for reducing odour emissions. Field studies have shown that anaerobic digestion lowers odour emission potential compared with untreated slurry, and injection of digestate further reduces odor emissions after field application (Orzi et al., 2018). Overall, the economic advantage of digestate mainly results from local availability and its capacity to replace part of mineral fertilisers. Its main limitation lies in transport and application costs associated with large volumes. Therefore, continued development of separation, concentration, nutrient recovery, and low emission application technologies is crucial for wider adoption in agricultural practice.

6.5 Conclusion

Digestate is a key tool for closing nutrient cycles in agriculture, combining high availability of mineral forms with the supply of organic matter. Its substantial degree of mineralization, including the dominance of ammonium nitrogen, allows fertiliser effects comparable to mineral fertilisers while also providing phosphorus, potassium, and micronutrients. However, its chemical composition is variable and depends on feedstock type and processing technology. Each batch therefore requires quality assessment and rate adjustment according to crop needs and soil conditions. In the context of rising synthetic fertiliser costs and pressure to reduce emissions associated with their production, digestate can significantly substitute mineral fertilisers. Research indicates that, when properly dosed, yields are usually comparable to mineral fertilization and may exceed them under certain conditions. This is linked to high bioavailability of nitrogen and potassium and beneficial effects on selected soil properties, which can support stable crop production even under environmental stress. Long-term use may increase soil organic carbon stocks, enhance stable organic matter fractions, and exert a liming effect on acidic soils. Potential risks must nevertheless be considered. The input of readily soluble salts can increase soil electrical conductivity, and salinity risk rises when using materials with elevated sodium content or at high rates. Processing method is also important. Fraction separation enables more precise targeting of fertilization goals, composting stabilizes the material and reduces rapid carbon mineralization after application, and drying or pelletizing improves handling and transport properties but may cause nitrogen losses through ammonia volatilization if not properly controlled. From an environmental safety perspective, systematic quality control is essential, as digestate may contain heavy metals or organic contaminants depending on feedstock origin. Periodic analysis of the material and monitoring of soils are recommended, especially when using municipal or industrial inputs. Nitrogen losses and environmental impacts are also influenced by timing and application technique, which supports the use of low emission practices consistent with regulatory requirements. In conclusion, digestate is a valuable fertiliser with high potential in sustainable agriculture. Effective and safe use requires quality standardization, appropriate processing, balanced nutrient management, and application technologies that minimize losses and environmental risks.

AI disclosure

During the preparation of this work the author(s) used DEEPL, ChatGPT, Grammarly, Mendeley in order to Translation, language correction, bibliography management. After using this tool/service, the author(s) reviewed and edited the content as needed and take(s) full responsibility for the content of the publication.

References

Abbas, A.M., Abdelazeem, M., Novak, S.J., 2023. Anaerobic Digestion Reduces Seed Germination and Viability of Six Plant Species from the Upper Nile Valley, Egypt. Agronomy 13 (2), 396. Available at https://doi.org/10.3390/agronomy13020396.

Ahmed, M.B., Zhou, J.L., Ngo, H. Hao, Guo, W., Thomaidis, N.S., Xu, J., 2017. Progress in the biological and chemical treatment technologies for emerging contaminant removal from wastewater: A critical review. Journal of Hazardous Materials 323, 274–298. Available at https://doi.org/10.1016/j.jhazmat.2016.04.045.

Alburquerque, J.A., de la Fuente, C., Bernal, M.P., 2012. Chemical properties of anaerobic digestates affecting C and N dynamics in amended soils. Agriculture, Ecosystems & Environment 160, 15–22. Available at https://doi.org/10.1016/j.agee.2011.03.007.

Alburquerque, J.A., de la Fuente, C., Campoy, M., Carrasco, L., Nájera, I., Baixauli, C., et al., 2012. Agricultural use of digestate for horticultural crop production and improvement of soil properties. European Journal of Agronomy 43, 119–128. Available at https://doi.org/10.1016/j.eja.2012.06.001.

Álvarez-Fraga, L., Capson-Tojo, G., Sanglier, M., Hamelin, J., Escudié, R., Wéry, N., et al., 2025. A meta-analysis of pathogen reduction data in anaerobic digestion. Renewable and Sustainable Energy Reviews 207 (3), 114982. Available at https://doi.org/10.1016/j.rser.2024.114982.

Amon, B., Kryvoruchko, V., Amon, T., Zechmeister-Boltenstern, S., 2006. Methane, nitrous oxide and ammonia emissions during storage and after application of dairy cattle slurry and influence of slurry treatment. Agriculture, Ecosystems & Environment 112 (2-3), 153–162. Available at https://doi.org/10.1016/j.agee.2005.08.030.

Baldasso, V., Bonet-Garcia, N., Sayen, S., Guillon, E., Frunzo, L., Gomes, C.A.R., et al., 2023. Trace metal fate in soil after application of digestate originating from the anaerobic digestion of non-source-separated organic fraction of municipal solid waste. Frontiers in Environmental Science 10. Available at https://doi.org/10.3389/fenvs.2022.1007390.

Barłóg, P., Hlisnikovský, L., Kunzová, E., 2019. Yield, content and nutrient uptake by winter wheat and spring barley in response to applications of digestate, cattle slurry and NPK mineral fertilizers. Archives of Agronomy and Soil Science 66 (11), 1481–1496. Available at https://doi.org/10.1080/03650340.2019.1676890.

Bolzonella, D., Fatone, F., Gottardo, M., Frison, N., 2018. Nutrients recovery from anaerobic digestate of agro-waste: techno-economic assessment of full scale applications. Journal of Environmental Management 216, 111–119. Available at https://doi.org/10.1016/j.jenvman.2017.08.026.

Cao, Y., Pawłowski, A., 2012. Sewage sludge-to-energy approaches based on anaerobic digestion and pyrolysis: brief overview and energy efficiency assessment. Renewable and Sustainable Energy Reviews 16 (3). Available at https://doi.org/10.1016/j.rser.2011.12.014.

Curadelli, F., Alberto, M., Uliarte, E.M., Combina, M., Funes-Pinter, I., 2023. Meta-analysis of yields of crops fertilized with compost tea and anaerobic digestate. Sustainability 15 (2), 1357. Available at https://doi.org/10.3390/su15021357.

Czatzkowska, M., Rolbiecki, D., Korzeniewska, E., Harnisz, M., 2025. Heavy metal and antimicrobial residue levels in various types of digestate from biogas plants—a review. Sustainability 17 (2), 416. Available at https://doi.org/10.3390/su17020416.

Dincă, M.-N., Ferdeș, M., Zăbavă, B.-Ștefania, Ionescu, M., Moiceanu, G., Paraschiv, G., 2025. Effective valorization of anaerobic digestate—a sustainable approach to circular economy. Applied Sciences 15 (16), 8939. Available at https://doi.org/10.3390/app15168939.

Drosg, B., Fuchs, W., Al, T., Madsen, S.M., Linke, B. (2015). Nutrient Recovery by Biogas Digestate Processing.

Dölger, J.L., Pitann, B., Mühling, K.H., 2024. Potassium induced suppression of magnesium uptake and translocation is limited in oat (*Avena sativa* L.). Journal of Plant Nutrition and Soil Science 187 (4), 454–458. Available at https://doi.org/10.1002/jpln.202400151.

Doyeni, M.O., Stulpinaite, U., Baksinskaite, A., Suproniene, S., Tilvikiene, V., 2021. The effectiveness of digestate use for fertilization in an agricultural cropping system. Plants 10 (8), 1734. Available at https://doi.org/10.3390/plants10081734.

Głowacka, A., Szostak, B., Klebaniuk, R., 2020. Effect of Biogas Digestate and Mineral Fertilisation on the Soil Properties and Yield and Nutritional Value of Switchgrass Forage. Agronomy 10 (4), 490. Available at https://doi.org/10.3390/agronomy10040490.

Efosa, N., Häni, C., Kupper, T., Krause, H.M., Six, J., Bünemann, E.K., 2025. Ammonia emissions after trailing hose application of digestates and cattle slurry. Nutrient Cycling in Agroecosystems 130 (3), 419–426. Available at https://doi.org/10.1007/s10705-025-10407-7.

European Union, 2008. Consolidated text: Council Directive of 12 December 1991 concerning the protection of waters against pollution caused by nitrates from agricultural sources (91/676/EEC). https://eur-lex.europa.eu/legal-content/EN/TXT/?uri=CELEX%3A01991L0676-20081211.

European Biogas Association (EBA). (2023). 20% increase in biomethane production in Europe shows biogases industry report released today. European Biogas Association. Available at: https://www.europeanbiogas.eu/news/20-increase-in-biomethane-production-in-europe-shows-biogases-industry-report-released-today/.

European Biogas Association (EBA). (2024). Exploring digestate's contribution to healthy soils. European Biogas Association. Available at: https://www.europeanbiogas.eu/wp-content/uploads/2024/03/Exploring-digestate-contribution-to-health-soils_EBA-Report.pdf.

García-López, A.M., Delgado, A., Anjos, O., Horta, C., 2023. Digestate not only affects nutrient availability but also soil quality indicators. Agronomy 13 (5). Available at https://doi.org/10.3390/agronomy13051308.

Gielnik, A., Pechaud, Y., Huguenot, D., Cébron, A., Esposito, G., van Hullebusch, E.D., 2021. Functional potential of sewage sludge digestate microbes to degrade aliphatic hydrocarbons during bioremediation of a petroleum hydrocarbons contaminated soil. Journal of Environmental Management 280. Available at https://doi.org/10.1016/j.jenvman.2020.111648.

Golovko, O., Ahrens, L., Schelin, J., Sörengård, M., Bergstrand, K.-J., Asp, H., et al., 2022. Organic micropollutants, heavy metals and pathogens in anaerobic digestate based on food waste. Journal of Environmental Management 313, 114997. Available at https://doi.org/10.1016/j.jenvman.2022.114997.

Grigatti, M., Barbanti, L., Hassan, M.U., Ciavatta, C., 2020. Fertilizing potential and CO2 emissions following the utilization of fresh and composted food-waste anaerobic digestates. Science of the Total Environment 698. Available at https://doi.org/10.1016/j.scitotenv.2019.134198.

Grigatti, M., 2024. The Impact of Raw and Composted Food Waste Anaerobic Digestates on Soil Organic Carbon Management: A Pot Study. Waste and Biomass Valorization 15 (8), 4915–4925. Available at https://doi.org/10.1007/s12649-024-02501-8.

Guo, X., Xie, X., Liu, Y., Wang, C., Yang, M., Huang, Y., 2020. Effects of digestate DOM on chemical behavior of soil heavy metals in an abandoned copper mining areas. Journal of Hazardous Materials 393. Available at https://doi.org/10.1016/j.jhazmat.2020.122436 122436.

Haque, F., Fan, C., Lee, Y.-Y., 2023. From waste to value: Addressing the relevance of waste recovery to agricultural sector in line with circular economy. Journal of Cleaner Production 415 (14), 137873. Available at https://doi.org/10.1016/j.jclepro.2023.137873.

Holm-Nielsen, J.B., Al Seadi, T., Oleskowicz-Popiel, P., 2009. The future of anaerobic digestion and biogas utilization. Bioresource Technology 100 (22), 5478–5484. Available at https://doi.org/10.1016/j.biortech.2008.12.046.

Hoxha, A., Christensen, B. (2018). International Fertiliser Society The carbon footprint of fertilizer production: regional reference values. www.fertilizer-society.org.

Huang, J., Han, L., Huang, G., 2019. Characterization of digestate composting stability using fluorescence EEM spectroscopy combining with PARAFAC. Waste Management & Research: The Journal for a Sustainable Circular Economy 37 (5), 486–494. Available at https://doi.org/10.1177/0734242x19828181.

Johansen, A., Nielsen, H.B., Hansen, C.M., Andreasen, C., Carlsgart, J., Hauggard-Nielsen, H., et al., 2013. Survival of weed seeds and animal parasites as affected by anaerobic digestion at meso- and thermophilic conditions. Waste Management 33 (4), 807–812. Available at https://doi.org/10.1016/j.wasman.2012.11.001.

Jumbo, R.B., Coulon, F., Cowley, T., Azuazu, I., Atai, E., Bortone, I., et al., 2022. Evaluating different soil amendments as bioremediation strategy for wetland soil contaminated by crude oil. Sustainability 14 (24), 16568. Available at https://doi.org/10.3390/su142416568.

Jurgutis, L., Šlepetienė, A., Amalevičiūtė-Volungė, K., Volungevičius, J., Šlepetys, J., 2021. The effect of digestate fertilisation on grass biogas yield and soil properties in field-biomass-biogas-field renewable energy production approach in Lithuania. Biomass and Bioenergy 153. Available at https://doi.org/10.1016/j.biombioe.2021.106211 106211.

Jurgutis, L., Šlepetienė, A., Šlepetys, J., Cesevičienė, J., 2021. Towards a full circular economy in biogas plants: sustainable management of digestate for growing biomass feedstocks and use as biofertilizer. Energies 14 (14), 4272. Available at https://doi.org/10.3390/en14144272.

Karimi, B., Sadet-Bourgeteau, S., Cannavacciuolo, M., Chauvin, C., Flamin, C., Haumont, A., et al., 2022. Impact of biogas digestates on soil microbiota in agriculture: a review. Environmental Chemistry Letters 20 (5), 3265–3288. Available at https://doi.org/10.1007/s10311-022-01451-8.

Kovačić, Đ., Lončarić, Z., Jović, J., Samac, D., Popović, B., Tišma, M., 2022. Digestate management and processing practices: a review. Applied Sciences 12 (18), 9216. Available at https://doi.org/10.3390/app12189216.

Kovács, A.B., Kremper, R., Kincses, I., Leviczky, Á., 2016. Influences of different organic fertilizers on nutrients of humic sandy soil and on the growth of Spinach (Spinacia oleracea L.). Acta Agraria Debreceniensis 70, 23–28. Available at https://doi.org/10.34101/actaagrar/70/1812.

Lamolinara, B., Pérez-Martínez, A., Guardado-Yordi, E., Guillén Fiallos, C., Diéguez-Santana, K., Ruiz-Mercado, G.J., 2022. Anaerobic digestate management, environmental impacts, and techno-economic challenges. Waste Management 140, 14–30. Available at https://doi.org/10.1016/j.wasman.2021.12.035.

Lim, J.W., Park, T., Tong, Y.W., Yu, Z., 2020. The microbiome driving anaerobic digestion and microbial analysis. Advances in Bioenergy 5, 1–61. Available at https://doi.org/10.1016/bs.aibe.2020.04.001.

Li, F., Yuan, Y., Imazumi, Y., Gong, P., Na, R., Shimizu, N., 2023). Comparative effects of mineral fertilizer and digestate on growth, antioxidant system, and physiology of lettuce under salt stress. Horticulture Environment and Biotechnology 64 (3), 379–391. Available at https://doi.org/10.1007/s13580-022-00492-w.

Lloret, E., Pastor, L., Pradas, P., Pascual, J.A., 2013. Semi full-scale thermophilic anaerobic digestion (TAnD) for advanced treatment of sewage sludge: stabilization process and pathogen reduction. Chemical Engineering Journal 232, 42–50. Available at https://doi.org/10.1016/j.cej.2013.07.062.

Logan, M., Visvanathan, C., 2019. Management strategies for anaerobic digestate of organic fraction of municipal solid waste: current status and future prospects. Waste Management & Research 37 (1), 27–39. Available at https://doi.org/10.1177/0734242X18816793.

Martín-Sanz-Garrido, C., Revuelta-Aramburu, M., Santos-Montes, A.M., Morales-Polo, C., 2025. A review on anaerobic digestate as a biofertilizer: characteristics, production, and environmental impacts from a life cycle assessment perspective. Applied Sciences 15 (15), 8635. Available at https://doi.org/10.3390/app15158635.

Mayerová, M., Šimon, T., Stehlík, M., Madaras, M., Koubová, M., Smatanová, M., 2023. Long-term application of biogas digestate improves soil physical properties. Soil and Tillage Research 231. Available at https://doi.org/10.1016/j.still.2023.105715 105715.

Mazzini, S., Borgonovo, G., Scaglioni, L., Bedussi, F., D'Imporzano, G., Tambone, F., et al., 2020. Phosphorus speciation during anaerobic digestion and subsequent solid/liquid separation. Science of The Total Environment 734, 139284. Available at https://doi.org/10.1016/j.scitotenv.2020.139284.

van Midden, C., Harris, J., Shaw, L., Sizmur, T., Morgan, H., Pawlett, M., 2024. Immobilisation of anaerobic digestate supplied nitrogen into soil microbial biomass is dependent on lability of high organic carbon materials additives. Frontiers in Sustainable Food Systems 8. Available at https://doi.org/10.3389/fsufs.2024.1356469.

van Midden, C., Harris, J., Shaw, L., Sizmur, T., Pawlett, M., 2023. The impact of anaerobic digestate on soil life: a review. Applied Soil Ecology 191, 105066. Available at https://doi.org/10.1016/j.apsoil.2023.105066.

Möller, K., Müller, T., 2012. Effects of anaerobic digestion on digestate nutrient availability and crop growth: a review. Engineering in Life Sciences 12 (3), 242–257. Available at https://doi.org/10.1002/elsc.201100085.

Nag, R., Auer, A., Nolan, S., Russell, L., Markey, B.K., Whyte, P., et al., 2021. Evaluation of pathogen concentration in anaerobic digestate using a predictive modelling approach (ADRISK). Science of the Total Environment 800. Available at https://doi.org/10.1016/j.scitotenv.2021.149574.

Nowak, M., Czekała, W., 2024. Sustainable Use of Digestate from Biogas Plants: Separation of Raw Digestate and Liquid Fraction Processing. Sustainability 16 (13), 5461. Available at https://doi.org/10.3390/su16135461.

Orzi, V., Riva, C., Scaglia, B., D'Imporzano, G., Tambone, F., Adani, F., 2018. Anaerobic digestion coupled with digestate injection reduced odour emissions from soil during manure distribution. Science of The Total Environment 621, 168–176. Available at https://doi.org/10.1016/j.scitotenv.2017.11.249.

Panuccio, M.R., Papalia, T., Attinà, E., Giuffrè, A., Muscolo, A., 2019. Use of digestate as an alternative to mineral fertilizer: effects on growth and crop quality. Archives of Agronomy and Soil Science 65 (5), 700–711. Available at https://doi.org/10.1080/03650340.2018.1520980.

Pampillón-González, L., Luna-Guido, M., Ruíz-Valdiviezo, V.M., Franco-Hernández, O., Fernández-Luqueño, F., Paredes-López, O., et al., 2017. Greenhouse Gas Emissions and Growth of Wheat Cultivated in Soil Amended with Digestate from Biogas Production. Pedosphere 27 (2), 318–327. Available at https://doi.org/10.1016/s1002-0160(17)60319-9.

Pantelopoulos, A., Magid, J., Jensen, L.S., 2016. Thermal drying of the solid fraction from biogas digestate: Effects of acidification, temperature and ventilation on nitrogen content. Waste Management 48, 218–226. Available at https://doi.org/10.1016/j.wasman.2015.10.008.

Pawlett, M., Tibbett, M., 2015. Is sodium in anaerobically digested food waste a potential risk to soils? Sustainable Environment Research 25 (4), 235–239.

Petraityte, D., Arlauskiene, A., Ceseviciene, J., 2022. Use of Digestate as an Alternative to Mineral Fertilizer: Effects on Soil Mineral Nitrogen and Winter Wheat Nitrogen Accumulation in Clay Loam. Agronomy 12 (2), 402. Available at https://doi.org/10.3390/agronomy12020402.

Przygocka-Cyna, K., Grzebisz, W., 2018. Biogas digestate – benefits and risks for soil fertility and crop quality – an evaluation of grain maize response. Open Chemistry 16 (1), 258–271. Available at https://doi.org/10.1515/chem-2018-0027.

Quilez, D., Balcells, M., Herrero, E., 2025. Reducing Ammonia Emissions from Digested Animal Manure: Effectiveness of Acidification, Open Disc Injection, and Fertigation in Mediterranean Cereal Systems. AgriEngineering 7 (10), 352. Available at https://doi.org/10.3390/agriengineering7100352.

Ragályi, P., Szécsy, O., Uzinger, N., Magyar, M., Szabó, A., Rékási, M., 2025. Factors influencing the impact of anaerobic digestates on soil properties. Soil Systems 9 (3). Available at https://doi.org/10.3390/soilsystems9030078.

Ronga, D., Caradonia, F., Parisi, M., Bezzi, G., Parisi, B., Allesina, G., et al., 2020. Using Digestate and Biochar as Fertilizers to Improve Processing Tomato Production Sustainability. Agronomy 10 (1), 138. Available at 10.3390/agronomy10010138.

Regelink, I.C., Egene, C.E., Tack, F.M.G., Meers, E., 2021. Speciation of P in solid organic fertilizers from digestate and biowaste. Agronomy 11 (11). Available at https://doi.org/10.3390/agronomy11112233.

Reuland, G., Van de Sande, T., Dekker, H., Sigurnjak, I., Meers, E., 2024. Digestate in replacement of synthetic fertilizers: a comparative 3–year field study of the crop performance and soil residual nitrates in West-Flanders. European Journal of Agronomy 161, 127380. Available at https://doi.org/10.1016/j.eja.2024.127380.

Rolka, E., Wyszkowski, M., Żołnowski, A.C., Skorwider-Namiotko, A., Szostek, R., Wyżlic, K., et al., 2024. Digestate from an agricultural biogas plant as a factor shaping soil properties. Agronomy 14 (7). Available at https://doi.org/10.3390/agronomy14071528.

Rosemarin, A., Macura, B., Carolus, J., Barquet, K., Ek, F., Järnberg, L., et al., 2020. Circular nutrient solutions for agriculture and wastewater – a review of technologies and practices. Current Opinion in Environmental Sustainability 45, 78–91. Available at https://doi.org/10.1016/j.cosust.2020.09.007.

Rossi, G., Beni, C., Benedetti, A., Felici, B., Neri, U., 2023. Effect of mineral or OFMSW digestate fertilization on ryegrass and nitrogen leaching. Agronomy 13 (5). Available at https://doi.org/10.3390/agronomy13051316.

Samoraj, M., Mironiuk, M., Izydorczyk, G., Witek-Krowiak, A., Szopa, D., Moustakas, K., et al., 2022. The challenges and perspectives for anaerobic digestion of animal waste and fertilizer application of the digestate. Chemosphere 295, 133799. Available at https://doi.org/10.1016/j.chemosphere.2022.133799.

Skrzypczak, D., Trzaska, K., Gil, F., Chawla, Y., Mikula, K., Izydorczyk, G., et al., 2023a. Towards anaerobic digestate valorization to recover fertilizer nutrients: Elaboration of technology and profitability analysis. Biomass and Bioenergy 178, 106967. Available at https://doi.org/10.1016/j.biombioe.2023.106967.

Skrzypczak, D., Trzaska, K., Mikula, K., Gil, F., Izydorczyk, G., Mironiuk, M., et al., 2023b. Conversion of anaerobic digestates from biogas plants: Laboratory fertilizer formulation, scale-up and demonstration of applicative properties on plants. Renewable Energy 203, 506–517. Available at https://doi.org/10.1016/j.renene.2022.12.080.

Šimon, T., Kunzová, E., Friedlová, M., 2015. The effect of digestate, cattle slurry and mineral fertilization on the winter wheat yield and soil quality parameters. Plant, Soil and Environment 61 (11), 522–527. Available at https://doi.org/10.17221/530/2015-pse.

Slepetiene, A., Ceseviciene, J., Amaleviciute-Volunge, K., Mankeviciene, A., Parasotas, I., Skersiene, A., et al., 2023. Solid and liquid phases of anaerobic digestate for sustainable use of agricultural soil. Sustainability 15 (2), 1345. Available at https://doi.org/10.3390/su15021345.

Slepetiene, A., Volungevicius, J., Jurgutis, L., Liaudanskiene, I., Amaleviciute-Volunge, K., Slepetys, J., et al., 2020. The potential of digestate as a biofertilizer in eroded soils of Lithuania. Waste Management 102, 441–451. Available at https://doi.org/10.1016/j.wasman.2019.11.008.

Sixt, T., Amann, M., Schuster, J., Hülsbergen, K.-J., 2025. Does biochar mixed with digestate affect crop yield, nitrogen use efficiency, and nitrate leaching? European Journal of Agronomy 169 (1), 127698. Available at https://doi.org/10.1016/j.eja.2025.127698.

Sobhi, M., Elsamahy, T., Zakaria, E., Gaballah, M.S., Zhu, F., Hu, X., et al., 2024. Characteristics, limitations and global regulations in the use of biogas digestate as fertilizer: a comprehensive overview. Science of the Total Environment 957 , 177855. Available at https://doi.org/10.1016/j.scitotenv.2024.177855.

Szymańska, M., Ahrends, H.E., Srivastava, A.K., Sosulski, T., 2022. Anaerobic digestate from biogas plants—nuisance waste or valuable product. Applied Sciences 12 (8), 4052. Available at https://doi.org/10.3390/app12084052.

Szymańska, M., Szara, E., Wąs, A., Sosulski, T., van Pruissen, G.W.P., Cornelissen, R.L., 2019. Struvite—An Innovative Fertilizer from Anaerobic Digestate Produced in a Bio-Refinery. Energies 12 (2), 296. Available at https://doi.org/10.3390/en12020296.

Talboys, P.J., Heppell, J., Roose, T., Healey, J.R., Jones, D.L., Withers, P.J.A., 2016. Struvite: a slow-release fertiliser for sustainable phosphorus management? Plant and Soil 401 (1-2), 109–123. Available at https://doi.org/10.1007/s11104-015-2747-3.

Tambone, F., Adani, F., 2017. Nitrogen mineralization from digestate in comparison to sewage sludge, compost and urea in a laboratory incubated soil experiment. Journal of Plant Nutrition and Soil Science 180 (3), 355–365. Available at https://doi.org/10.1002/jpln.201600241.

Tambone, F., Orzi, V., D'Imporzano, G., Adani, F., 2017. Solid and liquid fractionation of digestate: mass balance, chemical characterization, and agronomic and environmental value. Bioresource Technology 243, 1251–1256. Available at https://doi.org/10.1016/j.biortech.2017.07.130.

Vabio, 2025. The production of biomethane in Europe increased by 20%—the new EBA reporthttps://uabio.org/en/materials/15629/

Vázquez-Rowe, I., Golkowska, K., Lebuf, V., Vaneeckhaute, C., Michels, E., Meers, E., et al., 2015. Environmental assessment of digestate treatment technologies using LCA methodology. Waste Management 43, 442–459. Available at https://doi.org/10.1016/j.wasman.2015.05.007.

Verdi, L., Kuikman, P.J., Orlandini, S., Mancini, M., Napoli, M., Dalla Marta, A., 2019. Does the use of digestate to replace mineral fertilizers have less emissions of N_2O and NH_3? Agricultural and Forest Meteorology 269-270, 112–118. Available at https://doi.org/10.1016/j.agrformet.2019.02.004.

Weimers, K., Bergstrand, K.J., Hultberg, M., Asp, H., 2022. Liquid anaerobic digestate as sole nutrient source in soilless horticulture—or spiked with mineral nutrients for improved plant growth. Frontiers in Plant Science 13. Available at https://doi.org/10.3389/fpls.2022.770179.

Xie, K., Cakmak, I., Wang, S., Zhang, F., Guo, S., 2021. Synergistic and antagonistic interactions between potassium and magnesium in higher plants. The Crop Journal 9 (2), 249–256. Available at https://doi.org/10.1016/j.cj.2020.10.005.

Zhao, Q., Liu, Y., 2019. Is anaerobic digestion a reliable barrier for deactivation of pathogens in biosludge? Science of The Total Environment 668, 893–902. Available at https://doi.org/10.1016/j.scitotenv.2019.03.063.

Zheng, X., Zou, D., Wu, Q., Wang, H., Li, S., Liu, F., et al., 2022. Review on fate and bioavailability of heavy metals during anaerobic digestion and composting of animal manure. Waste Management 150, 75–89. Available at https://doi.org/10.1016/j.wasman.2022.06.033.

Case studies: global perspectives on biogas projects

7

Biogas projects convert organic wastes into renewable energy and nutrient-rich digestate. Global assessments report significant greenhouse-gas (GHG) reductions, better waste management, and livelihood gains when biogas systems are competently designed and operated (Hijazi et al., 2016; Alengebawy et al., 2022). Performance hinges on assured feedstock, robust process control, and judicious upgrading choice; hydrogenotrophic and in-situ biological routes are advancing alongside established absorption and membrane systems, and methane slip requires strict monitoring and mitigation (Antukh et al., 2022; Paolini et al., 2019; Knoll, 2024; Hurtig et al., 2025; Zhao et al., 2021).

Deployment outcomes reflect policy stability and local context. Municipal leadership, local trust, and inclusive planning clearly aid biogas adoption, whereas seasonal feedstock gaps and logistics in small plants can undermine reliability; economies of scale improve unit costs when policy support and steady offtake are in place (Skovsgaard and Jacobsen, 2017; Smith and Culas, 2025). Moreover, carbon revenues and circular use of waste streams (e.g., gate fees and fertiliser value) significantly enhance project economics.

When properly treated and applied, digestate can replace synthetic fertilisers and enhance soil fertility, though its use must be quality-controlled; conversely, emissions from certain biogas combined heat-and-power (CHP) engines (e.g., formaldehyde) require regulatory monitoring (Sürmen and Kara, 2021; Neumann et al., 2011).

Key findings highlight the need for an integrated approach: Successful projects align technological innovation with supportive policies and financing, engage communities for social acceptance, and close the resource loop by utilising digestate in agriculture. Biogas initiatives can reduce GHG emissions, enhance waste management, and improve rural livelihoods, but outcomes depend on context-specific factors such as local policy frameworks, economic viability, and cultural acceptance.

7.1 Introduction and framing

Biogas technology is a cornerstone of sustainable development, contributing to renewable energy targets, better waste management, and rural development (Hijazi et al., 2016; Ncube et al., 2021). Global biogas capacity has expanded rapidly, roughly doubling over the past decade, underscoring the growing importance of anaerobic digestion (AD) in renewable-energy transitions and circular waste management (Ncube et al., 2021; Uranga-Valencia et al., 2025). Through AD, organic wastes are transformed into biogas and nutrient-rich digestate, thereby linking energy recovery with circular nutrient recycling and improved sanitation. Life-cycle studies and regional case analyses confirm that well-designed biogas

Sustainable Biogas Production. DOI: https://doi.org/10.1016/B978-0-443-40557-0.00005-6

systems can achieve substantial GHG reductions, avoid landfill emissions, and deliver co-benefits, provided methane leakage is minimised and upgrading methods are well-chosen (Alengebawy et al., 2022; Hurtig et al., 2025). This chapter synthesises case studies across Europe, Asia, Africa, and the Americas to identify common success factors, constraints, and impacts.

Scope and significance: Biogas projects contribute to multiple sustainable development goals. Key roles include the following:

- Renewable energy generation: Biogas can displace fossil fuels via electricity, heat, or biomethane for grid injection and transport, contributing to energy security and climate goals, provided methane leakage is kept low and there is reliable offtake (Hijazi et al., 2016; Alengebawy et al., 2022)
- Waste management: AD valorises agricultural residues, manures and food wastes, preventing uncontrolled decomposition (e.g., in heaps or landfills). Actual performance depends on collection logistics, system boundaries, and feedstock catchment design (Xu et al., 2018; Natividad Pérez-Camacho et al., 2019).
- Sustainable development: Well-designed biogas programmes simultaneously improve rural livelihoods, enable cleaner cooking and sanitation, and embed circular farming practices. Community participation, user training, and local financing are key enablers, and integrating biogas into farm systems can boost food production without harming the environment (Clemens et al., 2018; Fatima et al., 2023).

The analysis is structured around four interacting dimensions: technological options (from mature processes to emerging biological and hybrid upgrades), economic and policy factors (incentives, permitting, carbon credit schemes), social acceptance determinants, and agronomic use of digestate. Each dimension co-evolves in specific contexts, so a systems perspective is used to interpret outcomes and transfer lessons (Skovsgaard and Jacobsen, 2017; Mancini and Raggi, 2022). Assessing biogas sustainability requires an integrated approach that jointly evaluates economic, environmental, and social criteria (Hijazi et al., 2016; Börjesson and Berglund, 2007). In practice, combinations of life-cycle assessment, cost-benefit analysis and social research are used to quantify GHG reductions, economic viability and community impacts (Mancini and Raggi, 2022; O'Keeffe et al., 2019; Natividad Pérez-Camacho et al., 2019). This chapter synthesises peer-reviewed literature, programme evaluations and policy documents and uses cross-regional comparisons to interpret technical, economic, environmental, and social performance (Skovsgaard and Jacobsen, 2017; Mancini and Raggi, 2022).

7.2 Success stories of biogas implementation across different regions

Biogas projects have been implemented worldwide, with each region exhibiting unique drivers, challenges, and innovations. In this section, we present global perspectives through representative case studies and regional experiences, focusing on Europe, Asia, Africa, and the Americas. These examples illustrate how technology, policy, economics, and social factors interplay in different contexts.

7.2.1 Europe

Europe has led the biogas expansion, spurred by EU renewable-energy targets and national incentives (e.g., feed-in tariffs for biogas power, capital subsidies for biomethane upgrading) (O'Keeffe et al., 2019). The region today produces a large share of the world's biogas (O'Keeffe et al., 2019). Germany alone now has approximately 9000 biogas plants— mostly farm-based units launched under its Renewable Energy Sources Act—which benefited from guaranteed electricity buy-back rates. German projects have since broadened their feedstocks (from manures and energy crops to various organic wastes) and increasingly involve local stakeholders: Farmers often form cooperatives or partner with municipalities to ensure community support (Prager and Freese, 2009). Denmark, by contrast, invested in centralised biogas digesters co-owned by farmer collectives and municipalities. These hubs aggregate manure from many farms and coordinate utilisation of the resulting heat; municipalities play multiple roles (facilitator, regulator, and anchor energy consumer) in this model (Lybaek and Kjaer, 2015). The well-known Solrød Biogas project exemplifies such public-private collaboration, guided by careful lifecycle planning from inception through operation (Nielsen et al., 2022).

Outcomes still vary by region. Local context, public perception, feedstock supply and spatial planning all influence success, meaning even strong national policies must be adapted and trusted locally (Mancini and Raggi, 2022; Bourdin et al., 2020). For example, some Eastern European biogas projects faltered due to weaker community buy-in or support networks, whereas Western European projects tended to fit more easily into existing farming and energy systems. Overall, Europe's experience shows that coherent incentives, appropriate technologies and active stakeholder engagement must align to achieve robust projects with genuine environmental gains.

European projects pair technological innovation with stringent environmental standards. Germany, Denmark and Sweden deploy advanced upgrading to produce biomethane for grid injection and transport, using membranes and chemical absorption alongside emerging hydrogenotrophic and hybrid membrane-VPSA routes (Paolini et al., 2019; Antukh et al., 2022; Azadi Tabar et al., 2025; Zhao et al., 2021; Natividad Pérez-Camacho et al., 2019). Sweden illustrates how strategic municipal leadership and cross-sector collaboration can reorient biogas from electricity toward transport fuel. In Linköping, municipal, agricultural, transit, and utility actors partnered to integrate waste collection, AD, and biomethane upgrading; by the late 2010s, Swedish biogas consumption exceeded 4 TWh (~14 PJ), with transport representing an important end-use pathway (O'Keeffe et al., 2019; Niu et al., 2021). Operators complement these technologies with methane-loss monitoring and control to safeguard climate performance (Knoll, 2024; Hurtig et al., 2025).

7.2.2 Asia

Asia presents a diverse biogas landscape, spanning millions of household digesters and an expanding cohort of industrial and municipal facilities. China has implemented one of the world's largest biogas programmes, historically promoting household digesters in rural areas and, more recently, medium to large plants for livestock farms and organic industrial wastes. Strong government support through capital subsidies, technical training, and dedicated research institutes enabled rapid expansion during the 2000s; medium- and large-scale projects have demonstrated the viability of supplying cooking fuel and electricity by utilising manures and crop residues within circular farming systems (Xu et al., 2018; Niu et al., 2021). Policies such as the Renewable Energy Law and agricultural waste regulations created an enabling environment, although challenges with maintenance of household digesters and seasonal feedstock availability have been noted (Niu et al., 2021).

In South Asia, India, Bangladesh, and Nepal have long promoted biogas for rural energy. India's National Biogas Programme has supported the installation of over 4 million family-type digesters to provide clean cooking fuel and organic fertiliser for rural households, with outcomes closely linked to user training and participation (Eswarlal et al., 2014). As early as the 1990s, trials of a simple batch-fed digester in India showed that improved designs could be built in days using unskilled labour and accept diverse biomass, significantly lowering costs compared to conventional digesters (Lichtman et al., 1996). Adoption studies in Bangladesh indicate that wealthier and better-educated farmers are more likely to adopt, and that microcredit and NGO outreach can improve uptake by addressing affordability and initial scepticism (Kabir et al., 2013). Indonesia has developed community biogas projects in which self-help groups manage communal digesters, increasing local buy-in and equitable access (Prayitno et al., 2021). Across these settings, integration with farming can strengthen livelihoods through fertiliser substitution and improved productivity (Fatima et al., 2023; Smith and Culas, 2025). Not all national efforts have scaled at the same pace. In Sri Lanka, diffusion has been slower despite technical potential, reflecting insufficient incentives and limited policy coordination (Bekchanov et al., 2019).

A recurring lesson across Asian experiences is the need to pair high-level policy with grassroots engagement. Programmes linked to training and user services deliver co-benefits for health and sanitation through cleaner cooking and better waste management, yet consistent stove and fuel use depends on reliability and alignment with cooking practices and preferences (Talevi et al., 2022; Moncada et al., 2019). Where social uplift and finance mechanisms accompany technology deployment, projects contribute to energy access and rural development, particularly in South-East Asia (Afridi et al., 2023).

7.2.3 Africa

Biogas development in Africa has been driven primarily by the goals of improving rural energy access and managing organic wastes, with explicit attention to social and economic co-benefits. Several countries in sub-Saharan Africa have launched national programmes, often with international partners. A prominent example is the Africa Biogas Partnership Programme, which since 2009 has supported the construction of tens of thousands of small household digesters across Kenya, Uganda, and Tanzania. The programme adopted a market-based approach that trained local masons and entrepreneurs, established financing schemes such as partial subsidies and micro-loans, and built awareness through regional marketing hubs (Clemens et al., 2018). Evaluation in East Africa indicates that combining modest subsidies with enterprise development catalysed a nascent biogas industry, delivered cleaner cooking fuel, provided slurry for fertiliser, and improved sanitation (Clemens et al., 2018).

At the national level, Rwanda and Ethiopia illustrate contrasting implementation contexts. Rwanda's National Domestic Biogas Programme aimed to install 15,000 family digesters and embedded a strong training component, including training women in operation and maintenance, to enhance acceptance. Reported outcomes include reduced woodfuel use and improved kitchen conditions, with affordability remaining a barrier for the poorest households (Kabera et al., 2016). In Ethiopia, government and NGO initiatives have promoted biogas within broader sustainable energy and agricultural agendas. Adoption is higher among relatively wealthier farmers with larger cattle holdings and among households exposed to demonstrations of benefits (Marie et al., 2021). Seasonal variability in dung and water availability constrains year-round operation, pointing to the need for technical and management strategies that bridge feedstock gaps (Wardle et al., 2021; Getaneh et al., 2024).

Across African case studies, key barriers include high upfront costs, limited access to credit, and gaps in technical support. Even where subsidies defray part of the capital cost, the remaining investment can be prohibitive for low-income households, slowing adoption (Afridi et al., 2023; Patinvoh and Taherzadeh, 2019; Mhlanga and Dzingirai, 2024; Kabir et al., 2013). Programmes have responded with partial subsidies, microfinance, and enterprise development to reduce entry barriers (Clemens et al., 2018). Maintenance and user support are decisive for sustained functionality, as systems without reliable after-sales service can fall into disuse (Clemens et al., 2018; Kabera et al., 2016). Social acceptance is generally positive once benefits are demonstrated, yet initial scepticism is common where familiarity with the technology is low. Projects that foreground local involvement and capacity building, including engagement with cooperatives and women's groups, achieve stronger uptake and create durable support networks (Kabera et al., 2016; Sari et al., 2022). In sum, Africa's experience underscores the importance of coupling technology deployment with micro-financing, user training, and after-sales service to secure lasting impact. In Kenya, donor-supported programmes and community innovations have yielded tens of thousands of domestic digesters and a handful of institutional systems. Projects such as the Homa Bay Prison digester demonstrate co-treatment of human and kitchen waste and show how targeted training and monitoring can mitigate common issues like gas leakage and under-performance; women often report time savings and cleaner kitchens as key benefits (Clemens et al., 2018; Sari et al., 2022).

7.2.4 Americas

Biogas activity in the Americas has been uneven yet is expanding as waste-to-energy and sustainability agendas gain traction. In North America, the United States and Canada have traditionally utilised biogas from wastewater treatment and landfills, with growing interest in farm-based and community digesters. A case study from Ontario, Canada demonstrates integration of biogas into municipal waste management within a circular-economy strategy: source-separated organics are co-digested with farm manures to produce electricity and heat, and digestate is returned to local farms as fertiliser (Ncube et al., 2021). Reported outcomes include substantial diversion of organics from landfill and net GHG savings through substitution for grid electricity, supported by provincial renewable-energy incentives and coordinated engagement among municipal waste managers, farmers, and energy offtakers (Ncube et al., 2021; O'Keeffe et al., 2019; Natividad Pérez-Camacho et al., 2019). In the United States, the biogas landscape is heterogeneous: large agricultural states deploy dairy digesters for biomethane used as vehicle fuel, while many cities recover landfill or sewage biogas for local energy services. Renewable natural gas credits under federal and state programmes have recently provided material economic drivers for new projects.

Across Latin America, biogas development aligns with both agricultural and urban waste priorities. Brazil, Argentina, and Mexico are scaling projects in sugarcane processing, livestock, and municipal organics. In Brazil, emerging agro-industrial biogas and biomethane projects in sugarcane mills, swine farms, and wastewater utilities point to substantial headroom; the national association estimates technical potential of approximately 41 billion m^3 of biogas annually, which stable policy could help convert from niche to mainstream (Uranga-Valencia et al., 2025; Sobczak et al., 2022).

At the policy level, support has often been inconsistent and incentives are generally less prevalent than in Europe or Asia, tempering the pace of scale-up. Even so, several municipalities are now integrating biogas into climate-action plans to curb methane from unmanaged organics while providing decentralised energy services (Uranga-Valencia et al., 2025).

Community-scale facilities can deliver health and environmental co-benefits when paired with training and user services. A project in an informal settlement in El Salvador reported improved indoor air quality and sanitation, while highlighting the need for continuous community engagement to sustain operation (Moncada et al., 2019).

7.3 Lessons learned and best practices

7.3.1 Technological foundations

AD is a microbiological process in which diverse microbial consortia break down organic matter under oxygen-free conditions, producing methane-rich biogas and nutrient-rich digestate. Common feedstocks range from animal manures and crop residues to source-separated municipal organics, depending on local waste streams. Systems vary in scale from small household or farm digesters to large continuously operated plants handling tens of thousands of tonnes per year (Afridi et al., 2023; Nielsen et al., 2022). AD operating parameters, temperature regime (mesophilic/thermophilic), hydraulic retention time and mixing intensity, are optimised to stabilise digestion, maximise methane yields and support hygienisation of the digestate (Ta and Babel, 2019; Sürmen and Kara, 2021). Raw biogas typically contains ~50%–70% CH_4 with CO_2 as the main balance; it can fuel CHP engines/boilers or be upgraded to approximately 97% CH_4 biomethane for grid injection or transport using mature absorption/membrane systems and newer biological/hybrid routes (Paolini et al., 2018, 2019; Antukh et al., 2022; Zhao et al., 2021; Natividad Pérez-Camacho et al., 2019).

7.3.2 System components and innovations

A biogas system integrates feedstock prep, the digester itself, gas cleaning/utilisation, and digestate post-treatment. In practice, co-digestion of multiple substrates can improve nutrient balance, and controlling feedstock-to-inoculum ratios and alkalinity helps stabilise methanogenesis and boost methane output (Ta and Babel, 2019). Tailored microbial consortia and other process-intensification techniques allow even high-strength wastes to be treated at low cost while still recovering energy (e.g., methane) (Jenkins et al., 2021). Biogas upgrading technologies range from mature chemical scrubbers (e.g., hot potassium carbonate absorption) to emerging methods. New approaches like in-situ biological methanation (adding H_2 to convert CO_2 to CH_4) and hybrid systems (e.g., membrane separation combined with vacuum PSA) are being piloted to attain high methane purity at lower cost (Paolini et al., 2019; Azadi Tabar et al., 2025). Table 7.1 summarises the cross-cutting success factors identified from global case studies.

7.3.3 Best practices and design checklist

Drawing on global case studies, the following principles support efficient, durable biogas systems.

- Holistic planning and systems approach: integrate economic, environmental, and social objectives from the outset; use feasibility studies that include life cycle assessment and cost-benefit analysis to verify net gains (Hijazi et al., 2016; Mancini and Raggi, 2022; Alengebawy et al., 2022; Natividad Pérez-Camacho et al., 2019). A holistic implementation approach has been suggested to improve the sustainability of biogas projects (Kshirsagar et al., 2019).
- Robust governance and stakeholder coordination: define clear roles and responsibilities; engage farmers, municipalities, and residents early through structured consultations and participatory planning; align energy, agriculture, and environment agencies to reduce friction and delays (Prager and Freese, 2009; Lybaek and Kjaer, 2015; Gedminaitė-Raudonė et al., 2025; Bourdin et al., 2020; Nielsen et al., 2022).
- Policy support and finance: combine capital subsidies or concessional loans with predictable offtake (e.g., feed-in tariffs or premium rates) and tax incentives; couple instruments with microfinance where household digesters are targeted; recognise carbon and renewable credit revenues in project structuring; prioritise waste- and residue-based feedstocks to avoid land-use conflicts (Lybaek and Kjaer, 2015; Skovsgaard and Jacobsen, 2017; Bekchanov et al., 2019; Zhang and Wu, 2023; O'Keeffe et al., 2019; Börjesson and Berglund, 2007; Clemens et al., 2018; Mhlanga and Dzingirai, 2024).
- Appropriate technology and emissions control: match reactor type and scale to feedstocks and heat/electricity demand; optimise temperature, retention time, and mixing; use

Table 7.1 Key insights from global biogas case studies.

Theme	Key insights	Supporting references
Integrated success factors	Durable performance is associated with stable policy frameworks, assured feedstock logistics, fit-for-purpose upgrading, and robust user services; Danish centralised cases and municipal-led models illustrate the value of lifecycle planning, clear roles, and anchor offtake; poor or non-uniform feeding depresses yields and economic outcomes.	Lybaek and Kjaer (2015), O'Keeffe et al. (2019), Hijazi et al. (2016), Mancini and Raggi (2022), Nielsen et al. (2022), Bourdin et al. (2020), Glivin et al. (2018)
Policy and regulatory frameworks	Clear incentives and permitting, methane monitoring requirements, and coordination across governance levels support deployment; structured stakeholder engagement improves agricultural planning; municipalities acting as facilitators, regulators, and customers accelerate diffusion; pollutant controls are required for certain combustion units emitting formaldehyde.	Patinvoh and Taherzadeh (2019), Prager and Freese (2009), Lybaek and Kjaer (2015), Neumann et al. (2011), Mancini and Raggi (2022), Gedminaitė-Raudonė et al. (2025), Nielsen et al. (2022)
Technological innovation	Biological hydrogenotrophic upgrading and in-situ routes advance alongside absorption and membrane options; hybrid membrane-vacuum pressure swing adsorption enables simultaneous biomethane and CO_2 recovery; hot potassium carbonate absorption is commercially established; leak detection and mitigation reduce methane losses; feedstock pretreatment and feedstock-to-inoculum ratios influence biomethane yields; engineered microbial communities support low-cost treatment of high-strength effluents; multi-criteria decision models aid siting and configuration.	Antukh et al. (2022), Paolini et al. (2019), Knoll (2024), Hurtig et al. (2025), Ta and Babel (2019), Yamaji et al. (2024), Zhao et al. (2021), Azadi Tabar et al. (2025), Jenkins et al. (2021)
Economic efficiency	Unit costs tend to fall with scale where logistics and regulation are favourable, yet on-farm and community plants are viable when gate fees, reliable manure or market wastes, fertiliser substitution, and carbon revenues are present; circular designs strengthen techno-economics and	Skovsgaard and Jacobsen (2017), Zhang and Wu (2023), Ncube et al. (2021), Sobczak et al. (2022), Garkoti and Thengane (2025),

(*Continued*)

Table 7.1 (Continued)

Theme	Key insights	Supporting references
	life-cycle performance; regional evidence links economic outcomes to policy instruments and waste availability.	Uranga-Valencia et al. (2025), Fatima et al. (2023), Smith and Culas (2025), Mhlanga and Dzingirai (2024), Ma et al. (2014)
Social dimensions	Adoption relates to factors such as income, education level, fuel preferences, after-sales service, and training. Market-based programmes in East Africa and national programmes in Rwanda highlight the importance of service delivery, financing and user training for sustained biogas use. Community participation and local champions support long-term adoption, although seasonal feedstock variability and supply reliability can undermine performance.	Clemens et al. (2018), Kabera et al. (2016), Eswarlal et al. (2014), Sari et al. (2022), Talevi et al. (2022), Marie et al. (2021), Wardle et al. (2021), Afridi et al. (2023), Prayitno et al. (2021), Kabir et al. (2013), Getaneh et al. (2024)
Environmental impacts	Life-cycle studies report lower greenhouse-gas emissions relative to fossil energy and landfilling when methane losses are controlled; system boundaries and catchment design shape outcomes in food-waste systems; crediting for displaced energy and nutrient recycling influences results; emissions monitoring is needed for pollutants such as formaldehyde; methane slip from upgrading can erode climate benefits without mitigation.	Börjesson and Berglund (2007), Hijazi et al. (2016), O'Keeffe et al. (2019), Alengebawy et al. (2022), Ishii et al. (2024), Paolini et al. (2018), Neumann et al. (2011), Knoll (2024), Hurtig et al. (2025), Ncube et al. (2021), Natividad Pérez-Camacho et al. (2019)
Fertiliser use of digestate	Agronomic value depends on feedstock, sanitisation, solid–liquid separation, and application practice; integration into circular farming systems supports nutrient cycling and can substitute mineral fertilisers; documented applications include slurry and sludge use in China, circular manure management in integrated systems, and production of high-quality fertilisers from digestate.	Xu et al. (2018), Sürmen and Kara (2021), Fatima et al. (2023), Smith and Culas (2025)

co-digestion to balance nutrients and stabilise operation; implement gas-tight storage and structured leak detection and repair (LDAR) to minimise methane losses; select upgrading routes suited to context, membrane/VPSA hybrids, chemical absorption such as hot potassium carbonate, or biological hydrogenotrophic/in-situ methanation; comply with safety and emissions standards, including controls for formaldehyde from engines (Ta and Babel, 2019; Sürmen and Kara, 2021; Antukh et al., 2022; Paolini et al., 2019; Hurtig et al., 2025; Knoll, 2024; Neumann et al., 2011; Azadi Tabar et al., 2025; Zhao et al., 2021; Nielsen et al., 2022).

- Feedstock strategy and logistics: secure long-term supply through contracts or cooperative pooling; design catchments to minimise transport; plan for seasonal variability in manures and crop residues with storage or co-feed arrangements; apply multi-criteria tools for siting and collection radius (Wardle et al., 2021; Ishii et al., 2024; Yamaji et al., 2024; Ncube et al., 2021; Nielsen et al., 2022).
- Energy offtake and integration: align plant design with reliable offtakers; plan heat use in CHP networks, grid injection of biomethane, or vehicle fuel supply; size upgrading units and interconnections to local market conditions (Paolini et al., 2019; Antukh et al., 2022; O'Keeffe et al., 2019; Ncube et al., 2021; Azadi Tabar et al., 2025).
- Monitoring, reporting, and verification: quantify methane losses and set performance thresholds; monitor air-pollutant emissions; define life-cycle assessment (LCA) system boundaries and credits transparently; publish operational metrics to maintain trust and eligibility for performance-based incentives (Knoll, 2024; Hurtig et al., 2025; Alengebawy et al., 2022; Natividad Pérez-Camacho et al., 2019).
- Digestate management and nutrient recycling: treat digestate as a product; deploy solid-liquid separation, composting/aerobic curing, and targeted nutrient recovery (e.g., ammonia stripping, struvite) to improve handling and fertiliser value; co-develop application guidelines with farmers to match crop demand and prevent losses; build local markets and logistics for distribution (Sürmen and Kara, 2021; Smith and Culas, 2025; Xu et al., 2018; Sobczak et al., 2022; Fatima et al., 2023).
- Social performance and capacity building: invest in user training, after-sales service, and demonstration sites; involve women and local leaders in decision-making; use cooperatives or user committees for shared ownership and maintenance; communicate tangible co-benefits such as time savings, indoor-air improvements, and fertiliser substitution (Clemens et al., 2018; Kabera et al., 2016; Eswarlal et al., 2014; Sari et al., 2022; Mancini and Raggi, 2022; Marie et al., 2021; Prayitno et al., 2021; Kabir et al., 2013; Getaneh et al., 2024).

Community training and social inclusion: Provide structured training for plant operators and end-users, and educate communities on the benefits and correct use of biogas and digestate (Eswarlal et al., 2014; Sari et al., 2022). Foster local ownership through cooperatives or user associations to build trust and accountability. Ensure gender and social inclusion by making women and marginalised groups explicit targets for access and training (Kabera et al., 2016; Clemens et al., 2018). A socially inclusive approach is associated with higher adoption and longer-term functionality.

Monitoring, evaluation, and adaptive management: Establish monitoring systems that track process performance, financial metrics, environmental emissions, and user satisfaction, enabling corrective action. Use Internet of Things sensors and supervisory control and data acquisition analytics for early fault detection and transparent reporting (Hurtig et al., 2025; Prager and Freese, 2009). Conduct periodic reviews using multi-criteria decision analysis to assess progress against sustainability objectives and adjust design or operations as required (Yamaji et al., 2024; Garkoti and Thengane, 2025). Continual improvement should be explicit in project governance.

Lifecycle planning tools such as multi-criteria decision-making frameworks can aid designers and policymakers in balancing cost, environmental impacts, and social outcomes; recent studies apply such models to optimise feedstock logistics, technology choice, and policy incentives (Yamaji et al., 2024; Ishii et al., 2024).

Adherence to this checklist increases the likelihood of creating systems that are resilient, efficient, and broadly beneficial. Technical competence must be matched with sound economics, enabling policy, and proactive community engagement; projects implemented on these principles have demonstrated stronger performance and sustainability in the case studies reviewed.

7.4 Environmental impacts of selected projects

Biogas systems are promoted for GHG mitigation and improved waste management. LCA commonly benchmarks these systems against fossil-energy or untreated-waste baselines and shows substantial GHG reductions, with magnitude contingent on feedstock, system efficiency, and the displaced energy or fertiliser pathway (Hijazi et al., 2016; Natividad Pérez-Camacho et al., 2019). Using biogas electricity in place of coal or diesel can cut GHG emissions per unit energy by more than half in many contexts (Alengebawy et al., 2022). In a regional analysis, small-scale manure-based systems achieved net-negative GHG outcomes on a per-kWh_el basis once avoided methane from manure storage was credited: farm-scale digesters offset 0.1–0.2 kg CO_2-eq per kWh_el, outperforming larger centralised plants co-digesting energy crops (O'Keeffe et al., 2019). The principal driver is avoidance of potent emissions from unmanaged decomposition in lagoons, dumps, or landfills through capture and use of methane (Hijazi et al., 2016; Alengebawy et al., 2022). Where feedstocks are purpose-grown energy crops rather than true wastes or residues, net benefits shrink and can turn negative in some scenarios because of cultivation and land-use change burdens (Börjesson and Berglund, 2007).

Methane leakage from digesters, storage, piping, and upgrading units is a pivotal determinant of climate performance. Even small fractional losses erode benefits, so engineering controls and monitoring are prioritised (Knoll, 2024; Hurtig et al., 2025). Modern European plants report methane losses well below 1% of throughput, supported by structured leak detection and repair, gas-tight digestate storage, and off-gas treatment on upgrading systems (Hurtig et al., 2025). With such measures in place, projects deliver benefits both by displacing fossil fuels and by avoiding uncontrolled emissions from waste streams (Natividad Pérez-Camacho et al., 2019).

Environmental gains extend beyond climate. Diverting organics from landfills, open dumping, or uncontrolled burning reduces local pollution and health risks; treating livestock manure in digesters lowers biochemical oxygen demand and pathogen loads before land application, protecting water quality compared with raw-manure spreading (Ncube et al., 2021). Urban digestion of food waste alleviates landfill pressure and mitigates leachate and odour issues (Uranga-Valencia et al., 2025). Digestate returns nutrients and organic matter to soils and can substitute for energy-intensive mineral fertilisers, improving soil carbon and microbial activity over time (Sürmen and Kara, 2021). When applied at agronomically appropriate rates and timings, and preferably using low-emission application techniques, digestate can contribute to nutrient recycling and reduce reliance on mineral fertilisers; however, ammonia volatilisation and nitrate leaching must be controlled through storage, timing and application method (Sürmen and Kara, 2021; Smith and Culas, 2025; Alengebawy et al., 2022).

7.4.1 Environmental performance (life-cycle assessment perspective)

Biogas projects are not without environmental trade-offs. One concern is air quality: combustion of biogas in engines or burners produces emissions such as nitrogen oxides (NO_x) and, if not completely combusted, can emit formaldehyde and other organics. Older biogas power plants in some areas were found to have notable formaldehyde emissions, which led to stricter emission standards (Neumann et al., 2011). Contemporary plants use catalytic converters and engine optimisations to keep these emissions in check, and levels are generally much lower than those from equivalent diesel generators. Another consideration is digestate management: if digestate is over-applied or stored improperly, it can create odour or runoff issues. These risks are manageable with good practice and regulation. Life cycle assessments consistently indicate that well-managed biogas systems offer net environmental gains in terms of GHG reduction, renewable energy output, and waste reduction (Hijazi et al., 2016; O'Keeffe et al., 2019). Maximising these benefits requires attention to system design, including appropriate feedstocks and integration of energy use with nutrient reuse, alongside deployment of best available technologies for emissions control. This also explains why policy frameworks often tie incentives to environmental performance: higher subsidies or credits for biogas achieving significant GHG savings or for upgrading to biomethane that displaces transport fuels (Skovsgaard and Jacobsen, 2017; O'Keeffe et al., 2019).

7.4.2 Emissions control and methane management

Even small methane leaks (from digesters, storage, or upgrading) can undermine biogas's climate benefit, so advanced projects use rigorous LDAR, frequent monitoring, and high-integrity equipment (Hurtig et al., 2025; Knoll, 2024). Indeed, modern plants can limit total methane losses to well below 1% of output by such measures, while upgrading systems often achieve ≥99% methane capture (Paolini et al., 2019). In practice, robust AD matched to local feedstocks, coupled with appropriately selected upgrading and disciplined emissions management, underpins efficient renewable-gas utilisation with credible climate performance.

7.5 Economic and efficiency analysis

The viability and scaling of biogas projects are strongly influenced by economics and policy. Costs and returns vary widely across contexts. Household digesters may cost only a few hundred dollars, whereas large industrial plants can require tens of millions; simple payback times range from 3 to 5 years in favourable settings with subsidies or carbon revenues (Skovsgaard and Jacobsen, 2017; Ma et al., 2014). Combined-heat-and-power plants typically convert about one third of biomass energy to electricity and another half to useful heat, enabling overall efficiencies above 80% when heat is utilised (Nielsen et al., 2022). Economic appraisal typically considers capital and operating costs, revenue streams from energy sales, tipping fees for waste treatment, carbon credits, and the agronomic value of digestate, alongside local employment and supply-chain effects. One clear finding from global case studies is reliance on subsidies or other financial support in early deployment. Capital costs for digesters and gas-utilisation equipment can be high, and without incentives the payback period may exceed what private investors or farmers consider acceptable. Analyses of large European plants indicate that capital grants or feed-in tariffs often cover the equivalent of 50%–80% of the initial investment or revenue, reducing payback to about 5–7 years rather than decades (Skovsgaard and Jacobsen, 2017; Nielsen et al., 2022). In China, a pig-farm project with government subsidies covering approximately 75% of installation costs achieved a payback under 5 years, whereas payback would otherwise have been far longer (Ma et al., 2014). Public financing mechanisms are therefore pivotal in making projects bankable (Skovsgaard and Jacobsen, 2017; Ma et al., 2014).

Additional policy instruments can strengthen margins. Carbon crediting and trading have improved the profitability of farm-scale biogas under certain market conditions, while robust crediting of avoided emissions contributes to regional performance assessments (Zhang and Wu, 2023; O'Keeffe et al., 2019). Circular-economy designs that secure gate fees for municipal organics and valorise digestate reinforce both techno-economics and life-cycle outcomes, as shown in Ontario and in broader assessments of circular biogas systems (Ncube et al., 2021; Garkoti and Thengane, 2025; Uranga-Valencia et al., 2025; Sobczak et al., 2022). Municipalities acting as facilitators, regulators, and anchor customers can de-risk projects by coordinating feedstock logistics, heat use, and offtake, exemplified by Danish centralised plants including Solrød Biogas (Lybaek and Kjaer, 2015; Nielsen et al., 2022). Territorial context and stakeholder processes influence delivery, so coordination across governance levels and structured engagement are recurrent features of successful programmes (Mancini and Raggi, 2022; Prager and Freese, 2009; Bourdin et al., 2020; Gedminaitė-Raudonė et al., 2025).

Economies of scale play a nuanced role. Larger centralised facilities can achieve lower specific costs, yet incur higher collection, transport, and distribution expenses. Decentralised on-farm and community systems can be cost-effective by using on-site wastes and meeting local energy needs, thereby avoiding logistics costs and capturing fertiliser substitution benefits (Skovsgaard and Jacobsen, 2017; Fatima et al., 2023; Smith and Culas, 2025). Flexible regulation that permits cooperative feedstock pooling and energy sharing helps both small and large projects to thrive, creating win–win arrangements in which farmers supply manures to a hub digester and the community benefits from heat and digestate redistribution (Lybaek and Kjaer, 2015; Nielsen et al., 2022). In lower-income settings, household digesters have relatively low absolute costs but remain a heavy burden for poor families; adoption is higher where microfinance, credit schemes, and coherent incentives are available (Bekchanov et al., 2019; Afridi et al., 2023; Patinvoh and Taherzadeh, 2019; Mhlanga and Dzingirai, 2024; Kabir et al., 2013).

Biogas projects deliver local economic benefits beyond direct financial returns. Construction, operation, and maintenance create jobs in rural areas, while the need for skilled technicians stimulates local capacity building (Lybaek and Kjaer, 2015; Sobczak et al., 2022). Reliable local energy can improve energy security and stabilise costs for communities. Farmers may earn additional income by processing neighbours' wastes or selling surplus electricity, and municipalities can lower waste-management expenditures when source-separated organics are diverted to energy production, creating new value chains for formerly low-value streams (Sobczak et al., 2022; Ncube et al., 2021). Public–private partnership models, including municipal equity stakes and municipalities acting as facilitators, regulators, and anchor customers, distribute benefits locally and de-risk projects (Lybaek and Kjaer, 2015; Nielsen et al., 2022).

Monetisation of environmental attributes has become an important revenue pillar. With expanding carbon markets and renewable-energy crediting, projects can improve profitability by selling offsets or renewable fuel credits. A farm case that incorporated carbon trading reported markedly higher net present value once credits for methane mitigation and renewable power were included (Zhang and Wu, 2023). European frameworks that recognise biomethane used as vehicle fuel create premium markets, while in the United States renewable identification credits have, at times, raised the value of biomethane from livestock wastes. These policy-driven markets reward the documented climate benefits of biogas (Zhang and Wu, 2023; O'Keeffe et al., 2019).

Policy settings are closely linked to these outcomes. Supportive instruments include capital grants, feed-in tariffs, tax credits, concessional finance, and mandates or targets for renewable gas. In Germany, guaranteed purchase prices for biogas electricity were a central driver of sector growth, reducing investment risk and accelerating deployment (Sobczak et al., 2022; Skovsgaard and Jacobsen, 2017). Denmark's success with centralised co-digestion was aided by policies that encouraged manure with organic industrial wastes, enabled tipping-fee revenues, and credited heat from combined heat and power toward municipal goals, with municipalities coordinating feedstock logistics and energy offtake (Lybaek and Kjaer, 2015; Nielsen et al., 2022). Where incentives are weak or inconsistent, diffusion slows despite technical potential, as observed in Sri Lanka with limited uptake under fragmented policy support (Bekchanov et al., 2019).

Effective biogas policy has several dimensions: incentive alignment, technical standards, and institutional coordination. Incentive alignment entails designing subsidies and tariffs that reflect societal benefits such as emissions reduction and waste treatment without creating perverse outcomes. Support that favours energy crops may satisfy renewable targets yet conflict with land-use and food-security objectives; many jurisdictions now tilt incentives toward waste- and residue-based feedstocks for this reason (O'Keeffe et al., 2019; Börjesson and Berglund, 2007). Technical standards are equally central, covering digester construction, gas quality, and emissions to secure safety and environmental performance. After reports of formaldehyde from some biogas engines in Germany, tighter emission norms accelerated adoption of control equipment (Neumann et al., 2011). Requirements linked to nutrient management plans for digestate application help prevent over-application and runoff, aligning fertiliser use with agronomic need (Nielsen et al., 2022).

Institutional coordination operates across ministries and tiers of government and hinges on structured stakeholder processes. Practical mechanisms include municipal facilitation of feedstock pooling and heat-use planning, standardised offtake contracts, and clear role allocation among agriculture, energy, and environment authorities. Danish experience shows municipalities acting simultaneously as facilitators, regulators, and anchor customers can de-risk centralised plants and secure manure mobilisation and heat utilisation (Lybaek and Kjaer, 2015; Nielsen et al., 2022). In Germany, local- and state-level forums involving farmer organisations and administrations improved the quality and acceptance of agri-environmental measures relevant to biogas (Prager and Freese, 2009). At national scale, stakeholder engagement in CAP (EU Common Agricultural Policy) strategic planning, illustrated by Lithuania, has been used to align renewable energy and agricultural incentives, including those affecting biogas (Gedminaitė-Raudonė et al., 2025). Territorial context still conditions outcomes, so participatory design at the regional level remains important (Bourdin et al., 2020).

Project economics depend on aligning support instruments with project configuration. Decentralised rural models use low-cost on-farm residues and local heat or power demand, delivering strong economic and environmental performance; scaling typically requires targeted support and regulatory flexibility (O'Keeffe et al., 2019; Skovsgaard and Jacobsen, 2017). Centralised schemes capture economies of scale, yet depend on coordinated feedstock logistics and offtake infrastructure for energy and digestate. Performance-based incentives that reward verified GHG savings and circular-resource outcomes, paired with regulations that secure quality and sustainability, underpin durable delivery; public-private partnerships frequently combine government support with private operational capability (Clemens et al., 2018; Mhlanga and Dzingirai, 2024).

7.5.1 Integrated sustainability assessment frameworks

Future biogas planning benefits from frameworks that evaluate environmental, economic, and social criteria in concert rather than in isolation. Multi-criteria decision-making models and integrated assessment tools are being developed to support policymakers and planners in optimising deployment pathways (Yamaji et al., 2024; Ishii et al., 2024). These approaches can couple LCA outputs with cost-benefit calculations and stakeholder-preference data to identify strategies that maximise net benefits; an integrated model may indicate an optimal combination of feed-in tariffs and fertiliser subsidies that achieves both emissions reduction and farmer-income goals (Garkoti and Thengane, 2025; Natividad Pérez-Camacho et al., 2019). Holistic planning becomes increasingly important as biogas scales and intersects with food systems, waste management, and climate policy.

Further development may explore new feedstock streams, including algae or aquatic plants linked to wastewater treatment, and hybrid energy systems that pair biogas with solar photovoltaics or battery storage to enhance reliability in rural contexts. Policy research is also needed on integrating biogas within carbon-neutrality strategies and circular-economy plans. The next generation of projects is likely to feature greater integration in biorefineries that, alongside energy and fertiliser, produce co-products such as fibres, pellets, or biochemicals, requiring interdisciplinary collaboration across engineering, agronomy, economics, and the social sciences to realise efficient biogas production in support of sustainable development.

7.6 Social aspects of the biogas production and fertilisers based on biogas wastes

7.6.1 Social acceptance and adoption

The social dimension of biogas projects spans community acceptance, user behaviour, equity of access, and the socio-cultural factors that shape adoption and sustained use. Case studies show that social acceptance can determine outcomes independently of technical merit or financial support (Mancini and Raggi, 2022). Acceptance depends on trust in the technology and implementing organisations, perceived benefits and risks, and the extent to which communities influence project design.

7.6.1.1 Community engagement

Early, transparent engagement reduces opposition and strengthens legitimacy. Public consultations, clear communication on environmental safeguards, and locally visible benefits such as discounted energy or access to fertiliser are associated with higher acceptance (Prager and Freese, 2009; Mancini and Raggi, 2022; Bourdin et al., 2020). Projects led by trusted local actors, farmer cooperatives, or municipal partners tend to secure durable support, while top-down schemes face resistance over odour, traffic, or operator credibility (Mancini and Raggi, 2022). In lower-income settings, respected community leaders and peer demonstration increase interest and trust; user committees and cooperatives foster ownership and routine upkeep (Eswarlal et al., 2014; Sari et al., 2022; Prayitno et al., 2021). Market-development approaches that pair enterprise training with modest subsidies have built service ecosystems and after-sales support in East Africa, improving functionality and user confidence (Clemens et al., 2018). National programmes that include targeted training, such as initiatives in Rwanda that trained women in operation and maintenance, report stronger acceptance and persistence in use (Kabera et al., 2016).

7.6.1.2 Training and user support

Sustained adoption depends on practical capability. Households and operators need simple routines for feedstock preparation, regular feeding, leak checks, and safe digestate use. Where programmes provided structured training with follow-up visits, utilisation and satisfaction improved; where support was absent, systems drifted into disuse (Kabera et al., 2016; Sari et al., 2022; Clemens et al., 2018). Education on health and safety reinforces benefits by encouraging consistent stove use and correct handling of residues; in an informal-settlement project, training correlated with better respiratory outcomes as users substituted biogas for wood fires (Moncada et al., 2019; Prayitno et al., 2021).

7.6.1.3 Determinants of adoption and sustained use

Adoption patterns reflect income, education, livestock holdings, and access to credit, alongside exposure to demonstrations of biogas performance (Marie et al., 2021; Kabir et al., 2013; Getaneh et al., 2024). Reliability and fit with cooking practices are decisive for continued use, so projects that match fuel supply to household routines and appliance preferences see higher persistence (Talevi et al., 2022). Seasonal variability in dung or water supply constrains household digesters, indicating needs for storage, co-digestion, or operational adjustments during dry periods (Wardle et al., 2021). Where finance, user services, and livelihood benefits are aligned, such as reduced fuel spending or fertiliser substitution, biogas contributes to social uplift and resilience, reinforcing acceptance over time (Afridi et al., 2023; Fatima et al., 2023).

7.6.1.4 Adoption drivers and barriers

Socio-economic studies in Bangladesh, Ethiopia, and elsewhere identify consistent drivers of adoption: higher income or asset bases, larger households with greater energy needs and labour availability, and greater livestock holdings or feedstock access increase the likelihood of installing a digester (Marie et al., 2021; Kabir et al., 2013; Getaneh et al., 2024). Farming households perceive compound benefits, clean cooking fuel alongside fertiliser, so willingness to invest is higher where manure and crop residues are readily available (Marie et al., 2021; Kabir et al., 2013). Barriers recur across settings: high upfront costs, constrained access to credit, limited awareness or information about system operation, and context-specific fuel preferences (Patinvoh and Taherzadeh, 2019; Kabir et al., 2013). Where subsidised LPG or freely collected firewood meet cooking needs, biogas adoption is harder to justify unless positioned as a complementary, reliability-enhancing option (Afridi et al., 2023). Gender dynamics matter because women typically manage cooking; programmes that involve women in decision-making and training report stronger persistence in use (Kabera et al., 2016). Seasonal variability in dung and water supply can interrupt operation in small systems, reinforcing the need for storage or co-digestion strategies (Wardle et al., 2021).

Targeted interventions address these constraints. Microfinance, instalment credit, and revolving funds reduce entry barriers; partial subsidies further improve affordability, especially for poorer households (Bekchanov et al., 2019; Clemens et al., 2018). Demonstration units and "biogas ambassadors" build trust where familiarity is low, while user committees or cooperatives organise labour-sharing and routine maintenance (Eswarlal et al., 2014; Sari et al., 2022; Prayitno et al., 2021). Training and follow-up services improve day-to-day operation and safe digestate use, aligning perceived benefits with lived experience and supporting sustained adoption (Kabera et al., 2016; Sari et al., 2022).

7.6.1.5 Social co-benefits

Biogas projects yield social and health benefits alongside energy and waste services. A major gain is reduced indoor air pollution when biogas replaces wood or charcoal for cooking, improving health among women and children who are most exposed to smoke; studies in East Africa and South Asia document declines in respiratory ailments and eye irritation following fuel switching (Talevi et al., 2022; Moncada et al., 2019). Sanitation improves where toilets are linked to digesters, safely treating human waste while producing gas. Employment and income generation follow from training technicians and maintenance providers, particularly for youth (Lybaek and Kjaer, 2015). Community-scale systems that power milling equipment or light schools can strengthen social cohesion by creating a shared, locally managed resource. Equity requires deliberate design so benefits reach poorer households as well as wealthier early adopters, through subsidy criteria tailored to smallholders or women-headed households. In Rwanda, specific efforts to include women in user training reflected their central role in cooking and household energy management (Kabera et al., 2016). In Indonesia, collective ownership models extend access to families unable to finance individual units (Prayitno et al., 2021).

Cultural acceptance of digestate use can be a constraint, especially where human waste is a feedstock. Hesitation has been reduced through extension and demonstration, by showing yield gains on plots fertilised with properly treated digestate (Eswarlal et al., 2014).

Social sustainability rests on trust built through transparency and engagement, inclusivity so benefits are broadly shared, and practical knowledge that enables users to operate and maintain systems effectively (Mancini and Raggi, 2022; Prayitno et al., 2021). Projects aligned with local customs, that empower communities through user associations, and that address concrete needs such as lighting or cooking convenience, prove most durable; technocratic deployments without grassroots support tend to falter.

7.6.2 Digestate as fertiliser

Digestate, whether used as a slurry or separated into solids, retains most of the feedstock's nitrogen, phosphorus, potassium and micronutrients, much of it in crop-available forms. As such, digestate is a valuable biofertiliser that can substitute for synthetic fertilisers on farms (Smith and Culas, 2025). For example, China has decades of experience applying biogas slurry and sludge to fields to improve soil fertility, and Indonesian smallholders with digesters report spending less on chemical fertiliser and seeing improved crop yields.

Realising digestate's full agronomic value while avoiding pitfalls requires good management. Because nutrient content and quality depend on feedstock and process conditions, digestate is often post-treated for use as fertiliser (Sürmen and Kara, 2021). Typical treatments include solid–liquid separation (yielding a P-rich solid and

N-rich liquid), composting or curing (to stabilise organics and cut odour), and targeted nutrient recovery steps (e.g., ammonia stripping or struvite precipitation) to produce usable fertiliser products. Such measures improve storability and can significantly reduce ammonia loss during field spreading (Sürmen and Kara, 2021). Furthermore, co-digestion can help optimise nutrient balance: for instance, adding carbon-rich crop residues raises the C:N ratio, yielding a more stable soil amendment (Ta and Babel, 2019).

When used appropriately, digestate closes the nutrient loop in circular farming: nutrients taken up by crops (or fed to animals) are returned to the soil, helping sustain soil organic matter and fertility (Sürmen and Kara, 2021; Sobczak et al., 2022). Studies show digestate can maintain or even raise crop yields and soil health over multiple seasons while cutting the need for chemical N and P fertilisers, shrinking agriculture's environmental footprint (Sürmen and Kara, 2021). In practice, best management of digestate includes testing for contaminants (especially if feedstocks include sewage or MSW) and matching application rates to crop needs and soil conditions to prevent nutrient leaching or runoff (Sürmen and Kara, 2021).

7.7 Comparative analysis: decentralised rural versus centralised urban systems

To contextualise regional insights, we juxtapose two archetypes: decentralised rural systems (India) and centralised/large-scale systems (Germany). The comparison highlights trade-offs in feedstock logistics, technology configuration, economics, and environmental performance. Table 7.2 compares decentralised rural systems in India with centralised urban systems in Germany across feedstocks, technology configuration, economics, environmental performance, methane management, integration, digestate use, social outcomes, policy instruments, and key challenges.

Decentralised systems excel where waste streams are nearby and there is local demand for heat and fertiliser. Centralised models leverage economies of scale, higher utilisation factors, and access to stable energy markets. In both approaches, rigorous methane control, predictable support instruments, and genuine stakeholder engagement are essential.

7.8 Future directions

Addressing identified knowledge gaps, including comparative urban and rural performance and validation of novel technologies in operational settings, is critical for scale-up (Garkoti and Thengane, 2025; Glivin et al., 2018). Independent verification of methane emissions from upgrading units and gas networks remains a priority to safeguard climate performance (Knoll, 2024; Hurtig et al., 2025). A growing policy focus on climate action and the circular economy provides a favourable context for expansion, with biogas increasingly positioned alongside mainstream waste and energy strategies (O'Keeffe et al., 2019; Hijazi et al., 2016;

Table 7.2 Comparison of decentralised rural systems in India and centralised/large-scale systems in Germany.

Aspect	Decentralised rural (India)	Centralised urban (Germany)
Primary feedstocks and logistics	On-farm manures and crop residues used near source; co-digestion and basic pre-treatment stabilise operation; integration within farming systems supports closed nutrient loops (Eswarlal et al., 2014; Ta and Babel, 2019; Fatima et al., 2023).	Manure, energy crops, agricultural residues and selected organic wastes are supplied to larger farm-based or regional plants; contracted supply and quality control are important where substrates are aggregated across several farms or municipalities (O'Keeffe et al., 2019; Sobczak et al., 2022).
Technology configuration	Household- to farm-scale digesters supply cooking gas and local heat/power; their performance is sensitive to routine feeding and operator training (Eswarlal et al., 2014; Talevi et al., 2022).	Large continuous reactors with gas cleaning and upgrading for grid injection or CHP; widespread use of membranes and chemical scrubbing, with emerging hybrid membrane-VPSA and biological methanation options (Paolini et al., 2019; Antukh et al., 2022; Azadi Tabar et al., 2025; Zhao et al., 2021).
Economic structure and returns	Lower capital intensity and avoided transport; returns realised via displaced purchased fuels and fertiliser substitution; viability improves with targeted support and credit (Eswarlal et al., 2014; Skovsgaard and Jacobsen, 2017).	Higher specific capex for reception, pre-treatment, compliance, and upgrading; revenues rely on grid sales, gate fees, and stable policy instruments (O'Keeffe et al., 2019; Sobczak et al., 2022).
Environmental performance (LCA)	High GHG abatement when manures/residues are used and polluting fuels are displaced; local digestate use supports soil health (Alengebawy et al., 2022; Fatima et al., 2023; Natividad Pérez-Camacho et al., 2019).	Strong GHG performance when treating municipal organics with tight methane control; benefits shrink where energy crops dominate due to cultivation burdens (O'Keeffe et al., 2019; Börjesson and Berglund, 2007; Natividad Pérez-Camacho et al., 2019).

(*Continued*)

Table 7.2 (Continued)

Aspect	Decentralised rural (India)	Centralised urban (Germany)
Methane management and air emissions	Outcomes depend on operating practice; programmes including follow-up support improve functionality (Eswarlal et al., 2014).	Structured LDAR and high-integrity equipment reduce methane slip to well below 1% in modern plants; emission standards and controls address formaldehyde from engines (Hurtig et al., 2025; Knoll, 2024; Neumann et al., 2011).
Energy offtake and integration	Predominantly household cooking and local heat/power; matching production with household routines is critical for persistence (Eswarlal et al., 2014; Talevi et al., 2022).	Grid injection and CHP for district heat and power are common; configuration tailored to local markets and interconnection (Paolini et al., 2019; O'Keeffe et al., 2019).
Digestate management	On-farm application is typical; separation/curing improve handling and acceptance; agronomic gains documented in integrated Indian systems (Sürmen and Kara, 2021; Fatima et al., 2023).	Post-treatment and compliance frameworks support safe reuse; regional practice includes market development for fertiliser products (Sürmen and Kara, 2021; Sobczak et al., 2022).
Social outcomes	Clean-cooking benefits, time savings, and access to fertiliser; acceptance increases with participatory design and user training (Eswarlal et al., 2014; Talevi et al., 2022).	Contribution to urban waste management and energy services; acceptance linked to transparent siting, engagement, and trust in operators (Prager and Freese, 2009; Mancini and Raggi, 2022; Bourdin et al., 2020).
Policy and enabling instruments	Household/farm-level support (subsidies, credit) plus extension and after-sales service underpin uptake (Eswarlal et al., 2014).	Long-term tariffs/premia, clear permitting, and municipal roles as facilitator/offtaker underpin bankability and scale (O'Keeffe et al., 2019; Sobczak et al., 2022).
Key challenges	Access to finance and reliable after-sales service; ensuring regular feeding and maintenance; aligning fuel use with cooking practices (Eswarlal et al., 2014; Talevi et al., 2022).	Complex feedstock logistics and cost exposure; dependence on stable policy; compliance with air-emission and odour rules (O'Keeffe et al., 2019; Neumann et al., 2011; Sobczak et al., 2022).

Ncube et al., 2021; Uranga-Valencia et al., 2025; Natividad Pérez-Camacho et al., 2019). Realising this potential depends on applying global lessons: adopting best practice, securing supportive and flexible policies, and giving social dimensions equal weight with technical delivery (Skovsgaard and Jacobsen, 2017; Mancini and Raggi, 2022; Prager and Freese, 2009; Clemens et al., 2018). Well-designed projects convert residual organics into clean energy and fertiliser while empowering communities (Sürmen and Kara, 2021; Smith and Culas, 2025). Continued advances in upgrading and digestate valorisation, coupled with policy learning through structured stakeholder processes, will determine the pace and quality of scale-up (Sürmen and Kara, 2021; Azadi Tabar et al., 2025; Gedminaitė-Raudonė et al., 2025). The case studies and analyses in this chapter provide cautionary evidence and workable precedents to inform the next generation of biogas projects worldwide. Looking ahead, research and development in the biogas sector are focusing on enhancing performance, broadening applications, and integrating AD within sustainable energy and agricultural frameworks.

Urban–rural comparative analyses: There is a need for systematic comparisons of biogas projects in urban versus rural settings across countries. Studies should evaluate levelised cost of energy, GHG abatement per unit of energy, soil-health outcomes from digestate use, and socio-economic impacts. Evidence from India and Germany points to meaningful contrasts, with rural farm-based systems often yielding greater GHG benefits per unit while urban systems address municipal organics (Eswarlal et al., 2014; O'Keeffe et al., 2019; Sobczak et al., 2022; Smith and Culas, 2025). Rigorous multi-country analyses with consistent system boundaries and crediting rules would guide context-appropriate models.

Long-term technology validation: Innovations such as membrane-VPSA hybrid upgrading, in-situ biological methanation, and advanced digestate treatment have shown promise in pilot trials. Longitudinal studies are required to quantify real-world performance, reliability, and cost trajectories (Paolini et al., 2019; Azadi Tabar et al., 2025). Independent monitoring of methane emissions from upgrading units and gas distribution networks is critical to verify low leakage at scale (Knoll, 2024; Hurtig et al., 2025). Shared operational datasets, ideally anonymised and quality-assured, could accelerate learning and technology refinement.

Digital tools for operations and engagement: The sector can leverage digital technologies for both operational efficiency and stakeholder engagement. On the operations side, artificial intelligence and machine-learning methods can analyse sensor and SCADA data to optimise digester performance and anticipate maintenance needs. On the social side, mobile and online platforms can support continuous user feedback, issue reporting, and coordination of feedstock supply across networks of small digesters, improving transparency and trust in community projects (Prager and Freese, 2009; Mancini and Raggi, 2022; Kabera et al., 2016; Talevi et al., 2022; Prayitno et al., 2021).

Global case studies indicate that durable performance emerges when policy instruments, technology choices, and finance are configured as a single system. Predictable support and carbon-related revenues enable bankability; high-integrity digestion, upgrading and methane-loss control protect climate performance;

disciplined digestate management turns residues into a fertiliser asset (Skovsgaard and Jacobsen, 2017; Uranga-Valencia et al., 2025; Hurtig et al., 2025; Sürmen and Kara, 2021).

Decentralised rural models can be highly cost-effective where on-farm residues and local energy uses are assured, with additional livelihood gains and fertiliser substitution benefits (Eswarlal et al., 2014; Smith and Culas, 2025). Many such projects report shorter payback periods and higher GHG savings per unit energy when displacing polluting household or farm fuels. Targeted support and capacity building remain prerequisites for scale.

Social acceptance underwrites longevity. Projects built around participation and knowledge sharing, with visible local benefits and credible service provision, sustain use and expand over time (Mancini and Raggi, 2022; Prayitno et al., 2021).

References

Afridi, Z.U.R., Ullah, K., Mustafa, M.F., Saleem, H., Shaker, B., Ashraf, N., et al., 2023. Biogas as sustainable approach for social uplift in South East Asian Region. Energy Reports 10, 4808–4818. Available at https://doi.org/10.1016/j.egyr.2023.11.037.

Alengebawy, A., Mohamed, B.A., Ghimire, N., Jin, K., Liu, T., Samer, M., et al., 2022. Understanding the environmental impacts of biogas utilization for energy production through life cycle assessment: An action towards reducing emissions. Environmental Research 213, 113632. Available at https://doi.org/10.1016/j.envres.2022.113632.

Antukh, T., Lee, I., Joo, S., Kim, H., 2022. Hydrogenotrophs-based biological biogas upgrading technologies. Frontiers in Bioengineering and Biotechnology 10. Available at https://doi.org/10.3389/fbioe.2022.833482.

Azadi Tabar, M., Maghsoudi, H., Denayer, J.F.M., 2025. Hybrid membrane-vacuum pressure swing adsorption: Low-cost technology for simultaneous biomethane and carbon dioxide production from biogas. Bioresource Technology 422, 132241. Available at https://doi.org/10.1016/j.biortech.2025.132241.

Bekchanov, M., Mondal, M.A.H., de Alwis, A., Mirzabaev, A., 2019. Why adoption is slow despite promising potential of biogas technology for improving energy security and mitigating climate change in Sri Lanka? Renewable and Sustainable Energy Reviews 105, 378–390. Available at https://doi.org/10.1016/j.rser.2019.02.010.

Bourdin, S., Raulin, F., Josset, C., 2020. On the (un)successful deployment of renewable energies: Territorial context matters. A conceptual framework and an empirical analysis of biogas projects. Energy Studies Review 24 (1), 4088. Available at https://doi.org/10.15173/ESR.V24I1.4088.

Börjesson, P., Berglund, M., 2007. Environmental systems analysis of biogas systems-Part II: The environmental impact of replacing various reference systems. Biomass and Bioenergy 31 (5), 326–344. Available at https://doi.org/10.1016/j.biombioe.2007.01.004.

Clemens, H., Bailis, R., Nyambane, A., Ndung'u, V., 2018. Africa biogas partnership program: a review of clean cooking implementation through market development in East Africa. Energy for Sustainable Development 46, 23–31. Available at https://doi.org/10.1016/j.esd.2018.05.012.

Eswarlal, V.K., Vasudevan, G., Dey, P.K., Vasudevan, P., 2014. Role of community acceptance in sustainable bioenergy projects in India. Energy Policy 73, 333–343. Available at https://doi.org/10.1016/j.enpol.2014.04.019.

Fatima, A., Singh, V.K., Babu, S., Singh, R.K., Upadhyay, P.K., Rathore, S.S., et al., 2023. Food production potential and environmental sustainability of different integrated farming system models in northwest India. Frontiers in Sustainable Food Systems 7. Available at https://doi.org/10.3389/fsufs.2023.959464.

Garkoti, P., Thengane, S.K., 2025. Techno-economic and life cycle assessment of circular economy-based biogas plants for managing organic waste. Journal of Cleaner Production 504. Available at https://doi.org/10.1016/j.jclepro.2025.145412.

Gedminaitė-Raudonė, Ž., Simonaitytė, V., Lankauskienė, R., 2025. Unfolding stakeholders' engagement in the design process of National Common Agricultural Policy (CAP) strategic plans: case of Lithuania. Agriculture 15 (5), 538. Available at https://doi.org/10.3390/agriculture15050538.

Getaneh, A., Eba, K., Tucho, G.T., 2024. Assessment of biomass energy potential for biogas technology adoption and its determinant factors in rural district of Limmu Kossa, Jimma, Ethiopia. Energies 17 (9). Available at https://doi.org/10.3390/en17092176, Article 2176.

Glivin, G., Edwin, M., Sekhar, S.J., 2018. Techno-economic studies on the influences of nonuniform feeding in the biogas plants of educational institutions. Environmental Progress and Sustainable Energy 37 (6), 2156–2164. Available at https://doi.org/10.1002/ep.12892.

Hijazi, O., Munro, S., Zerhusen, B., Effenberger, M., 2016. Review of life cycle assessment for biogas production in Europe. Renewable and Sustainable Energy Reviews 54, 1291–1300. Available at https://doi.org/10.1016/j.rser.2015.10.013.

Hurtig, O., Buffi, M., Besseau, R., Scarlat, N., Carbone, C., Agostini, A., 2025. Mitigating biomethane losses in European biogas plants: a techno-economic assessment. Renewable and Sustainable Energy Reviews 210, 115187. Available at https://doi.org/10.1016/j.rser.2024.115187.

Ishii, K., Yamanaka, M., Ochiai, S., Sato, M., 2024. Multicriteria evaluation of biogas plants receiving food waste from a broad area considering different spatial system boundaries and stakeholders. Journal of Solid Waste Technology and Management 50 (2), 512–526. Available at https://doi.org/10.5276/jswtm/iswmaw/502/2024.512.

Jenkins, S.N., Weerasekara, M.L.M.A.W., Zavahir, J.S., 2021. Role of microbial communities in the low-cost, sustainable treatment of pig effluent waste. Role of Microbial Communities for Sustainability. Microorganisms for Sustainability, vol 29. Springer, Singapore, pp. 289–316. Available at https://doi.org/10.1007/978-981-15-9912-5_11.

Kabera, T., Nishimwe, H., Imanantirenganya, I., Mbonyi, M.K., 2016. Impact and effectiveness of Rwanda's National Domestic Biogas programme. International Journal of Environmental Studies 73 (3), 402–421. Available at https://doi.org/10.1080/00207233.2016.1165480.

Kabir, H., Yegbemey, R.N., Bauer, S., 2013. Factors determinant of biogas adoption in Bangladesh. Renewable and Sustainable Energy Reviews 28, 881–889. Available at https://doi.org/10.1016/j.rser.2013.08.046.

Knoll, L., 2024. ETA-Florence Renewable Energies Germany METHANE EMISSIONS FROM BIOGAS UPGRADING PLANTS, European Biomass Conference and Exhibition Proceedings, 528-530, http://www.etaflorence.it/proceedings/

Kshirsagar, V.S., Pawar, P.M., Mehetre, S.T., 2019. Holistic approach for biogas technology implementation to improve sustainability. Current Science 116 (2), 249–255. Available at https://doi.org/10.18520/cs/v116/i2/249-255.

Lichtman, R., Ellegård, A., Lal, S., Sharma, N., 1996. The improved biogas systems project: results and future work. Energy for Sustainable Development 3 (4), 28–42. Available at https://doi.org/10.1016/S0973-0826(08)60202-1.

Lybaek, R., Kjaer, T., 2015. Municipalities as facilitators, regulators and energy consumers: enhancing the dissemination of biogas technology in Denmark. International Journal of Sustainable Energy Planning and Management 8, 17–30. Available at https://doi.org/10.5278/ijsepm.2015.8.3.

Ma, W.L., Li, X.Q., Deng, B.L., 2014. Case study on economic evaluation of biogas project in large-scale pig farms. Applied Mechanics and Materials 448-453, 4412–4418. Available at https://doi.org/10.4028/www.scientific.net/AMM.448-453.4412.

Mancini, E., Raggi, A., 2022. Out of sight, out of mind? The importance of local context and trust in understanding the social acceptance of biogas projects: a global scale review. Energy Research & Social Science 91. Available at https://doi.org/10.1016/j.erss.2022.102697, 102697.

Marie, M., Yirga, F., Alemu, G., Azadi, H., 2021. Status of energy utilization and factors affecting rural households' adoption of biogas technology in north-western Ethiopia. Heliyon 7 (3). Available at https://doi.org/10.1016/j.heliyon.2021.e06487, e06487.

Mhlanga, D., Dzingirai, M., 2024. Sustainable finance and business in sub-saharan africa: a conclusion. In: Mhlanga, D., Dzingirai, M. (Eds.), Sustainable Finance and Business in Sub-Saharan Africa. Springer Nature, Cham, pp. 493–505. Available at https://doi.org/10.1007/978-3-031-74050-3_24.

Moncada, S., Bambrick, H., Briguglio, M., 2019. The health impacts of a community biogas facility in an informal urban settlement: does training matter? Journal of Development Effectiveness 11 (2), 189–202. Available at https://doi.org/10.1080/19439342.2019.1638434.

Natividad Pérez-Camacho, M., Curry, R., Cromie, T., 2019. Life cycle environmental impacts of biogas production and utilisation substituting for grid electricity, natural gas grid and transport fuels. Waste Management 95, 90–101. Available at https://doi.org/10.1016/j.wasman.2019.05.045.

Ncube, A., Cocker, J., Ellis, D., Fiorentino, G., 2021. Biogas from source separated organic waste within a circular and life cycle perspective. A case study in Ontario, Canada. Environmental and Sustainability Indicators 11. Available at https://doi.org/10.1016/j.indic.2021.100134, 100134.

Neumann, T., Zikoridse, G., Gamer, P., Moczigemba, T., 2011. Formaldehyd aus biogas-BHKW: Ist-zustand für einige in Sachsen und Thüringen installierte anlagen. Gefahrstoffe Reinhaltung der Luft 71 (5), 227–230.

Nielsen, M.B., Lybæk, R., Kjær, T., 2022. Successfully navigating the project lifecycle for deployment of centralized biogas projects, the case of Solrød Biogas. Energies 15 (16), 5879. Available at https://doi.org/10.3390/en15165879.

Niu, S., Dai, R., Zhong, S., Wang, Y., Qiang, W., Dang, L., 2021. Multiple benefit assessment and suitable operation mechanism of medium- and large-scale biogas projects for cooking fuel in rural Gansu, China. Sustainable Energy Technologies and Assessments 46. Available at https://doi.org/10.1016/j.seta.2021.101285, 101285.

O'Keeffe, S., Franko, U., Oehmichen, K., Daniel-Gromke, J., Thrän, D., 2019. Give them credit-the greenhouse gas performance of regional biogas systems. GCB Bioenergy 11 (6), 791–808. Available at https://doi.org/10.1111/gcbb.12603.

Paolini, V., Petracchini, F., Segreto, M., Tomassetti, L., Naja, N., Cecinato, A., 2018. Environmental impact of biogas: A short review of current knowledge. Journal of Environmental Science and Health - Part A Toxic/Hazardous Substances and Environmental Engineering 53 (10), 899–906. Available at https://doi.org/10.1080/10934529.2018.1459076.

Paolini, V., Torre, M., Giacopini, W., Pastori, M., Segreto, M., Tomassetti, L., et al., 2019. CO2/CH4 separation by hot potassium carbonate absorption for biogas upgrading.

International Journal of Greenhouse Gas Control 83, 186–194. Available at https://doi.org/10.1016/j.ijggc.2019.02.011.
Patinvoh, R.J., Taherzadeh, M.J., 2019. Challenges of biogas implementation in developing countries. Current Opinion in Environmental Science and Health 12, 30–37. Available at https://doi.org/10.1016/j.coesh.2019.09.006.
Prager, K., Freese, J., 2009. Stakeholder involvement in agri-environmental policy making - Learning from a local- and a state-level approach in Germany. Journal of Environmental Management 90 (2), 1154–1167. Available at https://doi.org/10.1016/j.jenvman.2008.05.005.
Prayitno, G., Hakim, A.N., Meidiana, C., 2021. Community participation on the self-help group of methane gas (biogas) management as renewable energy in Indonesia. International Journal of Energy Economics and Policy 11 (1), 200–211. Available at https://doi.org/10.32479/ijeep.10595.
Sari, A.I., Suwarto, S., Suminah, S., Purnomo, S.H., 2022. Empowering the community in the use of livestock waste biogas as a sustainable energy source. Sustainability (Switzerland) 14 (21). Available at https://doi.org/10.3390/su142114121.
Skovsgaard, L., Jacobsen, H.K., 2017. Economies of scale in biogas production and the significance of flexible regulation. Energy Policy 101, 77–89. Available at https://doi.org/10.1016/j.enpol.2016.11.021.
Smith, A., Culas, R.J., 2025. The economic and agricultural impacts of biogas technology on small-scale farms in Indonesia: a case study in Bali and Lombok. In: Gorawala, W.P., Mandhatri, S. (Eds.), Agricultural Research Updates. Volume 49. Nova Science Publishers, Hauppauge, NY, pp. 1–48. Available at https://doi.org/10.52305/JXIX3014.
Sobczak, A., Chomać-Pierzecka, E., Kokiel, A., Różycka, M., Stasiak, J., Soboń, D., 2022. Economic conditions of using biodegradable waste for biogas production, using the example of Poland and Germany. Energies 15 (14), 5239. Available at https://doi.org/10.3390/en15145239.
Sürmen, M., Kara, E., 2021. High-quality fertilizers from biogas digestate. In: Galanakis, C.M. (Ed.), Environment and Climate-smart Food Production. Springer, Cham, pp. 319–347. Available at https://doi.org/10.1007/978-3-030-71571-7_10.
Ta, A.T., Babel, S., 2019. Utilization of green waste from vegetable market for biomethane production: influences of feedstock to inoculum ratios and alkalinity. Journal of Material Cycles and Waste Management 21 (6), 1391–1401. Available at https://doi.org/10.1007/s10163-019-00898-2.
Talevi, M., Pattanayak, S.K., Das, I., Lewis, J.J., Singha, A.K., 2022. Speaking from experience: preferences for cooking with biogas in rural India. Energy Economics 107. Available at https://doi.org/10.1016/j.eneco.2021.105796.
Uranga-Valencia, L.P., Pérez-Álvarez, S., Gabriel-Parra, R., Chávez-Medina, J.A., Magallanes-Tapia, M.A., Sánchez-Chávez, E., et al., 2025. Biogas production from organic waste in the forestry and agricultural context: challenges and solutions for a sustainable future. Energies 18 (12). Available at https://doi.org/10.3390/en18123174.
Wardle, J.M., Fischer, A., Tesfaye, Y., Smith, J., 2021. Seasonal variability of resources: The unexplored adversary of biogas use in rural Ethiopia. Current Research in Environmental Sustainability 3. Available at https://doi.org/10.1016/j.crsust.2021.100072.
Yamaji, D.M., Amâncio-Vieira, S.F., Fidelis, R., Contani, E.Ad.R., 2024. Proposal of multicriteria decision-making models for biogas production. Energies 17 (4). Available at https://doi.org/10.3390/en17040806.

Xu, X., Ma, Z., Chen, Y., Gu, X., Liu, Q., Wang, Y., et al., 2018. Circular economy pattern of livestock manure management in Longyou, China. Journal of Material Cycles and Waste Management 20 (2), 1050–1062. Available at https://doi.org/10.1007/s10163-017-0667-4.

Zhang, C., Wu, L., 2023. Farm biogas project considering carbon trading indicates promising economic results—a case study. Environmental Science and Pollution Research 30 (24), 65865–65878. Available at https://doi.org/10.1007/s11356-023-27016-y.

Zhao, J., Li, Y., Dong, R., 2021. Recent progress towards in-situ biogas upgrading technologies. Science of the Total Environment 800, 149667. Available at https://doi.org/10.1016/j.scitotenv.2021.149667.

Sustainability assessment: carbon footprint and life cycle assessment of biogas production

8

8.1 Sustainability assessment: carbon footprint and life cycle assessment of biogas production

Biogas systems are increasingly recognised as a climate mitigation option aligned with circular economy principles. They convert organic wastes into renewable energy and fertiliser (Ingrao et al., 2019). Anaerobic digestion (AD) of wastes such as animal manure, food scraps, sewage sludge, and crop residues can reduce greenhouse gas (GHG) emissions via two main mechanisms: (1) avoiding methane (CH_4) release from uncontrolled decomposition of organic waste, and (2) displacing fossil fuels using biogas (primarily CH_4) as a renewable energy source (Ingrao et al., 2019). Evidence from farm and municipal contexts indicates that avoided manure emissions and substitution of grid electricity or natural gas often dominate the benefits, conditional on controlling fugitive CH_4 along the chain (Battini et al., 2014; Bacenetti and Fiala, 2015; Fantin et al., 2015; van den Oever et al., 2021).

A rigorous life cycle assessment (LCA) or carbon footprint analysis is required to quantify these benefits and identify trade-offs or indirect effects (Earles and Halog, 2011). Method choices, system boundaries, functional unit, allocation versus system expansion, and attributional versus consequential framing, strongly influence results, which underscores the need for transparent reporting and sensitivity analysis (Ahlgren et al., 2015; Nematian and Scown, 2025; Vance et al., 2022). Concerns about misuse of LCA for marketing claims further motivate clear documentation of assumptions and data quality (Brandão et al., 2023).

LCA examines the entire biogas supply chain: feedstock provision, AD operations, biogas utilisation and digestate management, accounting for relevant emissions (CO_2, CH_4, N_2O) and credits for displaced products in cradle-to-grave or system-wide perspectives (Bjørn et al., 2017; Poeschl et al., 2012). The utilisation pathway, on-site heat and power, grid electricity, compressed biomethane for transport or grid injection, drive differences in both impacts and credits (Carnevale and Lombardi, 2015; Lombardi and Francini, 2020; Ahmed et al., 2021). Upgrading technology performance and energy demand, including recent advances in membrane-based and biological options, further shape outcomes (Lombardi and Francini, 2020; Gkotsis et al., 2023; Nashmin Elyasi et al., 2021). Accurate treatment of fugitive CH_4 is critical; plant-scale measurements using inverse dispersion and Lagrangian transport models show that leak rates vary across facilities and can erode expected savings if unmanaged (Hrad et al., 2021, 2022; Bühler et al., 2022).

Sustainable Biogas Production. DOI: https://doi.org/10.1016/B978-0-443-40557-0.00010-X

Digestate handling is another decisive factor. Nutrient substitution credits from digestate can constitute a material share of the carbon benefit of anaerobic digestion. Applying digestate as an organic fertiliser displaces fossil-fuel-intensive mineral fertiliser production and can improve soil organic matter and nutrient cycling. The magnitude of the credit depends on digestate nutrient content and local crop demand, and it must be balanced against direct field emissions of ammonia, methane, and nitrous oxide, which can erode gains if digestate is over-applied or poorly incorporated (Havukainen et al., 2018; Chen et al., 2021; Zeng et al., 2022; Dietrich et al., 2020). Robust LCAs therefore model nutrient substitution and field emissions together to avoid overstating climate benefits. Options for solid-liquid separation, composting, drying or nutrient recovery affect both energy use and emissions, while substitution credits hinge on realistic fertiliser displacement and field-emission factors (Havukainen et al., 2018; Chen et al., 2021; Zeng et al., 2022; Durdević and Hulenić, 2020; Đurđević et al., 2020; Dietrich et al., 2020). Feedstock mixes shape performance through both biogas yield and counterfactual emissions, which supports optimisation of co-digestion blends and regional supply logistics (González-Martínez et al., 2019; García-Gen et al., 2014; Dressler et al., 2012). Spatially explicit modelling and dynamic or ex-ante LCA capture heterogeneity in technology learning, grid mixes and land-based effects, improving decision relevance (Dresen and Jandewerth, 2012; El Akkari et al., 2023; Röder et al., 2022; Heidak et al., 2025; Slavkovic and Stephan, 2025; Beloin-Saint-Pierre et al., 2014). Alongside dynamic LCA, digital optimisation tools are emerging to support feedstock logistics and process design. Geographic information system models have been used to identify optimal co-digestion mixtures and supply chains, minimising transport distances and maximising biomethane potential (González-Martínez et al., 2019). Machine-learning methods can predict biogas yield from complex substrate combinations and inform process control and reactor design (Kim and Kim, 2025). These tools complement conventional LCA by offering spatial and predictive insights that can refine system boundaries and scenarios. Process-level simulators such as MCBioCH4 can also be coupled with LCA to forecast biogas and biomethane output and composition for inventory building and uncertainty reduction (Ravina et al., 2019).

Policy uses of LCA require methodological alignment with regulatory frameworks and attention to market-mediated effects. Applications of the Renewable Energy Directive (RED) formulas to biogas show both utility and limits when compared with comprehensive LCAs (Manninen et al., 2013). Consequential assessments illustrate how fuel-market feedbacks and leakage can shift net outcomes, including risks when crediting offsets under low-carbon fuel policies (Earles and Halog, 2011; Rajagopal and Zilberman, 2013; Fingerman et al., 2025). Broader sustainability appraisal, such as life cycle sustainability assessment, can integrate environmental, economic and social indicators where appropriate (Guinée, 2015; Mayer et al., 2025; Makepa and Chihambakwe, 2024; Kalkanis et al., 2024). In practice, life cycle sustainability assessment extends LCA by integrating economic and social metrics, for example evaluating operating costs, employment effects, and distributional impacts alongside environmental indicators (Kruse et al., 2009; Kalkanis et al., 2024). Early applications include studies on aquaculture and waste-to-energy technologies and show that incorporating socioeconomic indicators can reveal trade-offs not apparent in environmental LCAs alone (Kruse et al., 2009; Mayer et al., 2025). Such holistic approaches may help align biogas deployment with just transition objectives.

This chapter reviews the methodologies for calculating the carbon footprint of biogas projects, synthesises LCA results for biogas production and digestate utilisation, and discusses policy implications of these sustainability assessments in the context of climate-change mitigation. The focus is global, highlighting insights from Europe and other regions to understand how biogas LCA outcomes inform renewable-energy policies and GHG reduction strategies worldwide. Transitioning from waste to energy via biogas can yield significant GHG savings, but the magnitude of the benefit depends on project design, feedstock selection and management practices, as demonstrated by numerous studies over the past decade (Ingrao et al., 2019; Fantin et al., 2015).

8.1.1 Methodologies for carbon footprint calculation in biogas projects

Standardised methodologies enable consistent and transparent assessment of biogas carbon footprints. Most studies follow the ISO 14040/14044 LCA framework, which requires a defined goal and scope, explicit system boundaries, a clear functional unit, complete life cycle inventory and impact assessment. Carbon footprinting is a climate-focused LCA that reports results in CO_2-equivalents using 100-year global warming potential (GWP100) factors for greenhouse gases, with CH_4 typically assigned 28–34 and N_2O 265–298 relative to CO_2. In biogas applications, functional units commonly reference energy output, for example per kWh of electricity or per MJ of biomethane, or the quantity of feedstock treated, for example per tonne of manure or waste, to facilitate performance comparisons (Patterson et al., 2011; Bacenetti and Fiala, 2015). The system boundary must encompass all life cycle stages where significant emissions or credits arise. Fig. 8.1 illustrates the typical biogas LCA scope: feedstock collection and transport; pre-treatment; AD; biogas utilisation (e.g., combined heat and power [CHP]/boiler) or upgrading to biomethane; digestate handling and field application; and the avoided emissions or displaced products (e.g., grid electricity, mineral fertilisers, liquefied petroleum gas for cooking) attributable to the system. A cradle-to-grave perspective captures upstream emissions, for example from growing and transporting energy crops or the indirect effects of using crop residues, and downstream impacts, for example emissions from using digestate as fertiliser or CO_2 from biogas combustion, to determine the net GHG balance (Lijó et al., 2014; Poeschl et al., 2012). Methodological choices, including allocation versus system expansion and the inclusion of ILUC, can significantly influence results for crop-based biogas (Ahlgren et al., 2015). Beyond carbon and energy metrics, establishing clear LCA guidelines for biofuel systems demands careful treatment of boundary setting, allocation, and crediting choices, which can materially shape outcomes (Czyrnek-Delêtre et al., 2017). Selection of background data and counterfactuals requires regional specificity, for example grid mix and waste

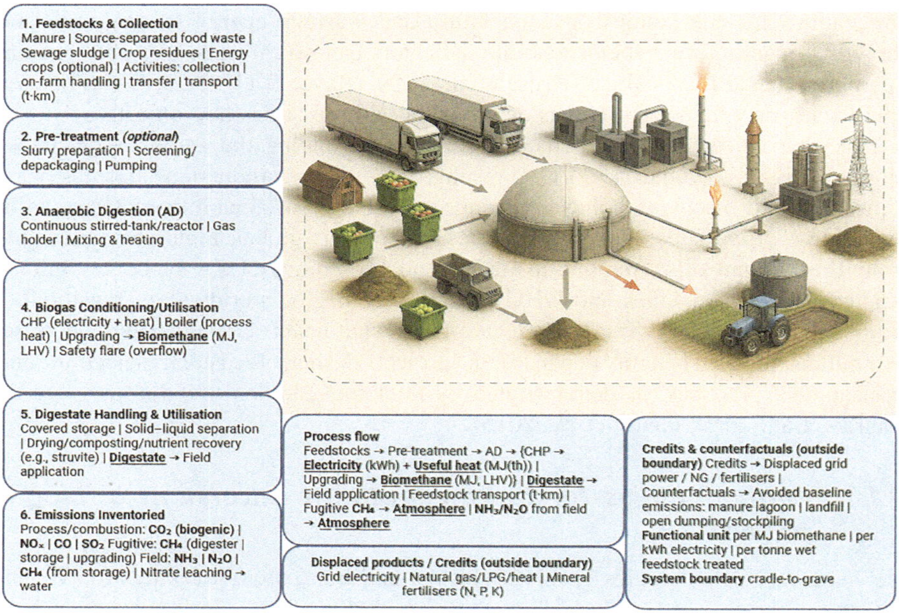

Figure 8.1 Schematic system boundary for a typical biogas life cycle assessment (LCA), covering feedstock collection and transport, pre-treatment, anaerobic digestion, biogas utilisation or upgrading to biomethane, digestate handling and field application, along with the associated emissions, avoided emissions, and displaced products.

management baselines, and can benefit from spatially explicit modelling and dynamic or prospective LCA when technologies and energy systems evolve (Dresen and Jandewerth, 2012; Heidak et al., 2025; Beloin-Saint-Pierre et al., 2014). Accurate treatment of fugitive CH_4 is central to credibility, with plant-scale measurements using inverse-dispersion and Lagrangian transport approaches demonstrating variable leak rates that should inform inventory factors (Hrad et al., 2021, 2022; Bühler et al., 2022).

Allocation and crediting methodologies are critical when waste feedstocks or co-products are involved. For wastes and residues, such as manure or food waste, many studies assign zero upstream burden and apply system expansion or displacement to credit the biogas system for avoiding conventional waste handling (Pérez-Camacho et al., 2018). A biogas plant treating manure is, for example, credited with the methane that would have been emitted if the manure were stored in a lagoon or spread raw on fields (Battini et al., 2014), with counterfactual emissions often parameterised using regulatory or IPCC methods (Makepa and Chihambakwe, 2024). Use of digestate as an organic fertiliser can claim credits for displacing energy-intensive synthetic fertilisers, while field emissions and nutrient substitution rates must be represented to avoid over- or under-crediting (Tidåker et al., 2014; Havukainen et al., 2018; Dietrich et al., 2020). In contrast, if purpose-grown energy crops (e.g., maize silage) are used as AD feedstock, their production carries an upstream GHG cost (farm fuel, fertiliser manufacture, N_2O from fields, etc). that must be included (Dressler et al., 2012). Methodological choices, allocation versus system expansion, selection of the GWP time horizon, and, in consequential LCA, the inclusion of indirect land-use change when land is diverted to energy crops, can materially shift results for crop-based biogas; therefore, sensitivity and scenario analyses should be performed to test robustness (Ahlgren et al., 2015; Nematian and Scown, 2025). A recent synthesis of biogas LCAs identifies the major influence factors to stress-test: feedstock choice and pre-processing, system boundaries and co-product crediting, digestate management and field emissions, energy for upgrading/utilisation, transport distances, and fugitive CH_4 treatment, as the dominant drivers of inter-study variability (Abu et al., 2020).

Another key methodological aspect is accounting for methane leakage and emissions during the biogas process. Not all methane produced is successfully captured and utilised. Even a small fraction escaping as fugitive emissions can have an outsized impact on the carbon footprint due to methane's high GWP (Bühler et al., 2022). Therefore, GHG accounting guidelines require including any measured or estimated CH_4 leakage along the biogas supply chain (Nematian and Scown, 2025). LCA models often include an assumed methane leak rate (e.g., approximately 1%–3% of biogas volume) if direct measurements are unavailable (Bühler et al., 2022). If actual leak measurements from a plant are available, they are used to improve accuracy (Bühler et al., 2022). It has been shown that even a small increase in methane leakage (e.g., from 1% to 3% of total biogas output escaping) can sharply reduce or even negate the GHG savings of a biogas system (Bühler et al., 2022). Thus, careful treatment of fugitive emissions is crucial in biogas LCA, often prompting analysts to run worst-case scenarios to ensure the system still offers net GHG mitigation (Ahlgren et al., 2015).

Carbon-footprint calculations for biogas follow general LCA principles while placing particular emphasis on boundary choices that capture counterfactual waste management and energy substitution, precise functional-unit definition, co-product handling for energy and digestate, and inclusion of all biogenic GHG flows. Methodological steps like these ensure biogas projects are evaluated on a fair, apples-to-apples basis against fossil energy or alternative waste scenarios (Brandão et al., 2023). It is generally recommended to perform sensitivity analyses under different assumptions to test the robustness of the carbon footprint results (Ahlgren et al., 2015). These practices establish a robust basis for judging how climate-friendly a given project is once all positive and negative contributions are integrated over the life cycle.

8.2 Life cycle assessment of biogas production and digestate utilisation

Life cycle studies consistently show that well-designed biogas systems can substantially reduce GHG emissions compared to conventional fossil energy systems or scenarios in which organic wastes are left untreated (Carnevale and Lombardi, 2015).

Table 8.1 Life cycle GHG intensities for selected biomethane pathways compared with fossil natural gas. Negative values indicate net GHG avoidance; positive values indicate residual emissions per unit energy (GWP100; g CO_2-eq/MJ).

Biogas pathway	Representative life cycle GHG intensity (g CO_2-eq/MJ)	GHG reduction versus fossil natural gas	Boundary notes and key drivers from LCA literature	References
Landfill-gas capture → biomethane	~ −50 to −30 (typical EU cases)	> 100% net-negative (i.e., greater than 100% reduction relative to the fossil baseline)	Avoided landfill CH_4 dominates; results hinge on gas-collection efficiency, utilisation (flare/engine/upgrade) and power for upgrading/compression. Regional waste policies and baseline landfill management are decisive.	Patterson et al. (2011), Huttunen et al. (2014), Del Borghi et al. (2007)
Dairy manure AD → biomethane	~ −40 to −20 (central ~ −30)	> 100% (net negative)	Counterfactual manure storage drives credits; outcomes depend on CH_4 leakage control, conversion efficiency and digestate field emissions with nutrient substitution.	Battini et al. (2014), van den Oever et al. (2021), Dressler et al. (2012)
Sewage-sludge AD → biomethane	~ −20 to −10 (system-dependent)	~ 80%–110%	Sludge thickening/dewatering, sidestream treatment and energy-recovery scheme drive results; grid mix and upgrading energy are influential.	Yoshida et al. (2013), Patterson et al. (2011)
Source-separated food-waste AD → biomethane	~ −15 to +5	~ 70%–105%	Avoided landfill or incineration credits are pivotal; pre-treatment energy, upgrading demand and digestate management can flip the sign near zero.	Pérez-Camacho et al. (2018), Pérez-Camacho et al. (2019), Jin et al. (2015), Chen et al. (2021), Zeng et al. (2022)

Co-digestion (manure + crop residues or food waste)	~ −20 to +10	~ 50%–110%	Higher specific gas yield combined with avoided-emission credits from the waste fraction; net result depends on mixing ratios, logistics and fertiliser substitution from digestate.	García-Gen et al. (2014), Martín et al. (2012), Kunatsa et al. (2022), van den Oever et al. (2021)
Energy-crop AD (e.g., maize silage) → biomethane	~ +30 to +60	~ 20%–60%	Upstream burdens from cultivation, fertiliser manufacture and direct N_2O dominate; rotation choices and land-use effects shape outcomes.	Bacenetti and Fiala (2015), Lijó et al. (2014), Tidåker et al. (2014), Dressler et al. (2012)
Fossil natural gas (baseline)	~ +70 to +90	0%	Well-to-end-use intensity including upstream supply plus combustion; regional leakage and supply chain losses affect the upper bound.	Carnevale and Lombardi (2015), Pérez-Camacho et al. (2019), Lombardi and Francini (2020), Ahmed et al. (2021)

The magnitude of GHG savings varies widely depending on feedstock type, scale of operation, and how the digestate is managed (Pérez-Camacho et al., 2018).

GHG emissions of biogas systems by feedstock: Feedstock selection is a primary driver of carbon performance. Waste-derived biogas generally offers the greatest GHG benefits, often even achieving net-negative emissions (preventing more emissions than it produces) (Pérez-Camacho et al., 2018). In contrast, crop-based biogas (e.g., using dedicated energy crops like maize) tends to have a more modest GHG advantage, as growing these crops entails emissions and can divert land from other uses (Dressler et al., 2012). Comparative European assessments report consistently lower GHG intensity for biogas than for the fossil references they replace, with wide dispersion in the degree of reduction across cases (Huttunen et al., 2014). Biogas from manure or food waste can even achieve net-negative GHG emissions in some cases, whereas biogas from energy crops might only deliver around a 10%–50% reduction relative to fossil energy (Pérez-Camacho et al., 2018; Dressler et al., 2012).

Table 8.1 presents representative life cycle carbon intensities for biomethane pathways. Manure-based and landfill-gas routes can fall below zero because the counterfactual releases substantial methane, while crop-based biomethane typically shows positive intensities in the order of tens of g CO_2-eq/MJ, lower than fossil fuel comparators, yet non-negligible (van den Oever et al., 2021; Fantin et al., 2015; Bacenetti and Fiala, 2015). Co-digestion strategies can improve outcomes by combining high-yield substrates with waste mitigation; for example, adding an energy-rich co-substrate (like crop residues or food waste) to manure can boost biogas yield while the manure portion provides a large emission credit, yielding a better overall GHG balance than either feedstock alone (Saghouri et al., 2024). Methane leakage remains a critical parameter across all feedstocks—even a small leak rate (on the order of a few percent of total biogas) can erode or eliminate the net GHG savings (Bühler et al., 2022). Therefore, technologies and maintenance practices that minimise fugitive emissions (e.g., gas-tight digesters, proper biogas storage, and emergency flares) are essential to realise the theoretical GHG mitigation potential of biogas (Bühler et al., 2022).

Notes. Ranges reflect cradle-to-gate or cradle-to-grave scopes typical of European studies and are sensitive to methodological choices, counterfactuals and measurement of fugitive CH_4. Measured leak rates at AD facilities vary across sites, and even low single-digit losses can erode savings, underscoring the need to base inventories on plant data where available (Hrad et al., 2021, 2022; Bühler et al., 2022). Upgrading technology and energy demand influence pathway intensities, with membrane and biological options offering improving performance in recent applications (Lombardi and Francini, 2020; Ahmed et al., 2021; Gkotsis et al., 2023; Nashmin Elyasi et al., 2021). Digestate processing and field application drive downstream emissions and credits; separation, drying/composting and nutrient-recovery choices, together with realistic fertiliser-substitution rates, are critical to avoid over- or under-crediting (Havukainen et al., 2018; Chen et al., 2021; Zeng et al., 2022; Durdević and Hulenić, 2020; Dietrich et al., 2020; Đurđević et al., 2020).

The data in Table 8.1 reinforce that using residual and waste feedstocks (landfill gas, manure, food waste) for biogas yields the greatest climate benefits, often resulting in carbon-negative energy (Vignali and Vitale, 2017). In one detailed analysis, converting dairy manure to biogas had an estimated life cycle carbon intensity (CI) around −30 g CO_2-eq/MJ, meaning the system avoids more GHG emissions than it emits because it captures methane that would otherwise escape from manure storage (Battini et al., 2014). By contrast, biogas from maize silage in the same context showed a positive CI of about +40 g CO_2-eq/MJ under typical assumptions, reflecting emissions from crop cultivation and transport (Fantin et al., 2015). Waste-based biogas pathways can show substantially lower carbon intensities than fossil natural gas, but the final result depends on the assumed counterfactual scenario, methane leakage, upgrading energy demand and digestate management. Fantin et al. (2015) reported that 1 kWh of electricity from an Italian co-digestion plant treating manure, maize and agro-industrial residues had 7%–20% lower GHG emissions than the average Italian grid mix, with larger savings when the manure share increased. Plants co-digesting a higher proportion of animal slurry achieved the best performance, whereas maize-heavy mixes exhibited higher net emissions (Fantin et al., 2015). Co-digestion that raises the share of waste substrates while maintaining specific gas yields typically improves CI relative to maize monoculture systems (van den Oever et al., 2021; García-Gen et al., 2014; Martín et al., 2012; Kunatsa et al., 2022). Recent evidence from a two-stage co-digestion of palm oil mill effluent with concentrated latex wastewater shows higher methane yields but a shift in LCA hotspots toward auxiliary chemicals and electricity, underscoring the need for process-specific inventories and sensitivity analysis (Raketh et al., 2024).

Methane control is a co-equal determinant of outcomes. Facility-scale measurements using inverse-dispersion and Lagrangian transport modelling show substantial variation in fugitive CH_4 across plants, with uncontrolled losses capable of eroding expected savings (Hrad et al., 2021, 2022; Bühler et al., 2022). Downstream management of digestate also shapes results: solid-liquid separation, composting or drying and nutrient-recovery options alter both emissions and credits, and realistic fertiliser-substitution rates are required to avoid over- or under-crediting (Havukainen et al., 2018; Chen et al., 2021; Zeng et al., 2022; Durdević and Hulenić, 2020; Dietrich et al., 2020; Đurđević et al., 2020). Technology choices in gas upgrading and utilisation influence energy demand and leak potential, with performance differences reported for water scrubbing, membrane and biological routes (Lombardi and Francini, 2020; Ahmed et al., 2021; Gkotsis et al., 2023; Nashmin Elyasi et al., 2021). Across diverse contexts, the clear LCA message is that feedstock choice and methane control determine whether a biogas project yields modest or very large GHG savings (Bühler et al., 2022; Pérez-Camacho et al., 2018). In practice, this means prioritising feedstocks like manure and organic wastes, and ensuring near-zero methane leakage, to maximise the climate mitigation value of biogas (Nematian and Scown, 2025).

Digestate management and environmental trade-offs: Digestate, the residual slurry after biogas extraction, is a pivotal determinant of life cycle impacts. Appropriate utilisation can enhance sustainability by recycling nutrients in agriculture and displacing synthetic fertilisers, whereas mismanagement can generate substantial emissions and air-quality burdens (Havukainen et al., 2018; Dietrich et al., 2020). LCA studies highlight that all emissions from digestate storage and field application must be included to accurately assess a biogas system's carbon footprint (Zeng et al., 2022). If raw digestate is stored in open lagoons or applied carelessly to fields, it can release methane (from residual anaerobic activity), ammonia (NH_3 volatilisation), and nitrous oxide (N_2O from soil nitrification/denitrification), which are potent GHGs or air pollutants (Zeng et al., 2022). Crediting for nutrient substitution must be balanced against these losses and based on realistic crop-available N, P and K to ensure carbon footprint integrity (Havukainen et al., 2018).

Mitigation options span storage, processing and agronomy. Covering digestate storage (tanks or lagoons) to capture any biogas produced during storage, and then flaring or using it, significantly improves the GHG balance of biogas plants, especially for manure-based systems (Zeng et al., 2022). Processing routes such as solid–liquid separation, composting of the solid fraction, low-temperature or solar drying using surplus heat and targeted nutrient recovery alter both energy demand and emission profiles; well-chosen configurations can lower life cycle impacts relative to storing and surface spreading raw digestate (Chen et al., 2021; Zeng et al., 2022; Durdević and Hulenić, 2020). Struvite precipitation provides a route to recover phosphorus while improving handling properties; coupling recovery with downstream use can reduce transport and enable precise nutrient placement (Durdević and Hulenić, 2020; Đurđević et al., 2020). Energy-intensive drying or concentration steps increase the processing burden and need surplus heat or efficient electricity supply to remain advantageous, a finding consistent across plant-scale assessments (Chen et al., 2021; Zeng et al., 2022). A consensus recommendation is to apply digestate to land in ways that minimise emissions. Subsurface injection or immediate soil incorporation can reduce NH_3 volatilisation, and applying digestate at agronomic rates avoids nitrate leaching and N_2O spikes (Nematian and Scown, 2025). When digestate is used efficiently as an organic fertiliser, it can substantially offset GHG emissions by displacing the production of synthetic fertilisers (Tidåker et al., 2014).

Beyond climate impacts, broader environmental trade-offs emerge in LCAs of biogas and digestate management. Several studies report higher acidification and eutrophication potentials when ammonia from digestate storage or land application and nitrogen oxides from combustion are insufficiently controlled (Fantin et al., 2015; Dressler et al., 2012). In Italy, electricity from biogas exhibited lower GHG intensity than the grid average yet higher acidification and photochemical smog formation, largely attributable to NH_3 losses and engine-out NO_x (Fantin et al., 2015). Mitigation relies on both process design and end-of-pipe measures. Selective catalytic or oxidation after-treatment on CHP units reduces NO_x and carbon monoxide; covered storage with abatement, for example biofilters, curbs ammonia and odour; and integrating heat use, for district heating or for low-temperature digestate drying, raises overall efficiency and lowers impacts per unit of delivered energy (Fantin et al., 2015; Carnevale and Lombardi, 2015; Lombardi and Francini, 2020; Poeschl et al., 2012; Pérez-Camacho et al., 2019). Beyond waste-heat use, integrating anaerobic digestion with other thermal technologies can further improve performance. Machine-learning-based assessments of AD coupled with a heat pump indicate that repurposing latent heat from the digestate stream reduces external energy inputs and overall GHG emissions relative to standalone digesters (Hajabdollahi Ouderji et al., 2023). This integration complements advances in upgrading: techno-economic analyses of water scrubbing, membrane and biological routes show declining energy demand and emissions (Lombardi and Francini, 2020; Ahmed et al., 2021; Gkotsis et al., 2023). Trade-offs are also linked to digestate handling. Nutrient recovery and processing options can lessen downstream burdens when aligned with local energy and land-use contexts, although energy-intensive treatments add to the inventory unless powered by surplus heat or low-carbon electricity (Havukainen et al., 2018; Chen et al., 2021; Zeng et al., 2022).

Context strongly conditions results. In industrialised regions, plants are typically larger, connected to electricity or gas markets and operate against baselines such as regulated landfills or tank storage that already suppress counterfactual emissions. In many low- and middle-income settings, biogas systems are smaller and used at the household or farm scale. Baseline practices often include open dumping of organic waste or reliance on fuelwood and charcoal for cooking, which amplifies the relative benefits of biogas adoption (Aziz et al., 2020; Hijazi et al., 2020). Field studies in Asia have found that small farm digesters can cut manure-management GHG emissions by around 20%–30%, yet this benefit is readily negated if digesters are not gas-tight or if gas is vented (Yoshida et al., 2013; Huttunen et al., 2014). Inadequate digestate handling can also elevate nutrient runoff and water-quality impacts, which supports coupling biogas deployment with training and infrastructure for composting, solid-liquid separation and targeted nutrient recovery so that fertiliser substitution is realised without shifting burdens to air or water (Durdević and Hulenić, 2020; Yoshida et al., 2013; Huttunen et al., 2014; Đurđević et al., 2020). LCAs across both contexts converge on a common conclusion: Biogas can deliver net sustainability gains when complemented by disciplined nutrient management, emission controls and efficient heat integration, with system design attentive to regional baselines and co-product use (Mayer et al., 2025; Fantin et al., 2015; Pérez-Camacho et al., 2019) (Fig. 8.2).

Life cycle GHG intensity of biogas pathways relative to fossil natural gas. Waste-based biogas routes often achieve net-negative emissions because the avoided CH_4 from baseline waste handling plus displaced fossil energy outweigh the process emissions. The net benefit is highly sensitive to methane leakage: a loss of roughly 5% of the biogas produced can flip a manure-based system from net-negative to net-positive emissions. In contrast, energy-crop AD typically yields a positive carbon intensity (~ +20–60 g CO_2-eq/MJ) because it incurs N_2O emissions from fertiliser use and field operations (e.g., silage production) and has no avoided-CH_4 credit. Only very low-input energy-crop systems with strong digestate substitution credits and minimal leakage come close to a neutral GHG balance. These comparisons highlight the importance of prioritising high-CH_4 waste feedstocks and ensuring leak-tight operation to maximise GHG reductions per unit energy, especially for waste-derived biomethane used in heat and transport (see Battini et al., 2014; van den Oever et al., 2021; Hrad et al., 2021, 2022; Bühler et al., 2022; Hijazi et al., 2020).

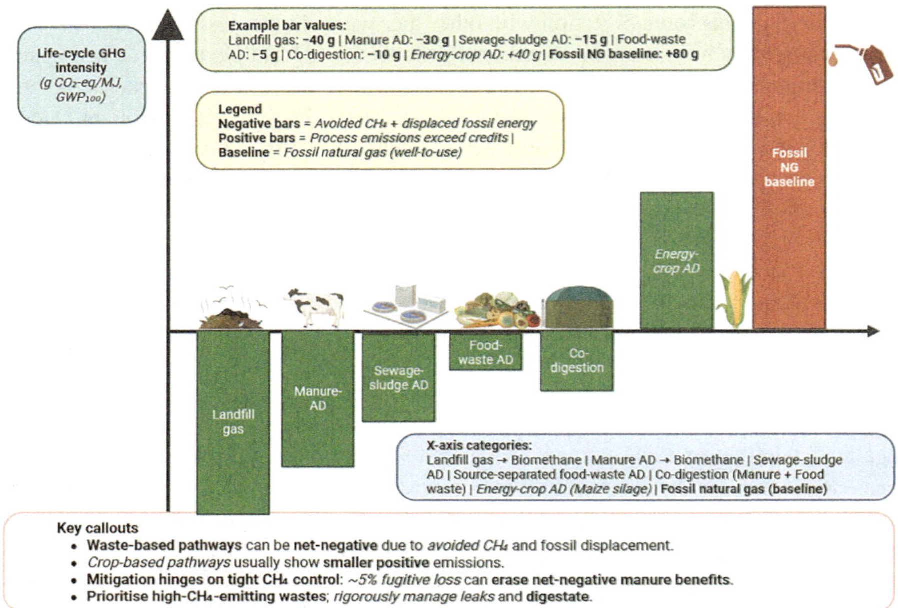

Figure 8.2 Contrasts biogas pathways with fossil gas: waste-based pathways can be net-negative, single-digit methane losses can substantially erode manure-system benefits, and energy-crop AD is typically positive (~+20–60 g CO_2-eq/MJ) due to fertiliser use, N_2O emissions, field operations and the absence of avoided-CH_4 credits.

8.3 Policy implications and contributions to climate change mitigation

Life cycle GHG assessments of biogas carry direct implications for energy and environmental policy. Governments are increasingly using LCA-based metrics to ensure that bioenergy (including biogas) contributes meaningfully to climate goals (Nematian and Scown, 2025). In the European Union, Directive (EU) 2018/2001 on the promotion of energy from renewable sources, as amended by Directive (EU) 2023/2413, sets sustainability and greenhouse-gas saving criteria for biofuels, bioliquids and biomass fuels, including biogas used in transport. For biogas consumed in the transport sector and produced in installations starting operation from 1 January 2021, the required GHG saving is at least 65% versus the fossil comparator (Nematian and Scown, 2025). Evidence reviewed in Section 8.2 indicates that waste-based pathways routinely meet and surpass this threshold, often with a wide margin (Pérez-Camacho et al., 2019). Typical manure or sewage-sludge biomethane exhibits CI values around −20 to −50 g CO_2-eq/MJ, well below the limit (van den Oever et al., 2021).

Policy design in the EU increasingly favours advanced biofuels from wastes and residues, reflecting LCA findings on superior climate performance. Accordingly, feedstocks such as manure, sewage sludge, and food waste receive policy incentives (e.g., double-counting towards targets), whereas crop-based biofuels face stricter limits or phased-out support due to concerns over their real-world GHG savings and land-use impacts (Nematian and Scown, 2025). Several Member States link support levels to the share of waste or residue in the feedstock mix, aligning plant economics with pathways that deliver larger avoided emissions (Chin et al., 2022). To operationalise GHG criteria, producers must complete a standardised calculation, using the methods specified in RED II annexes or approved tools such as BioGrace, as part of sustainability certification, which has made carbon footprint analysis a routine element of project development and auditing (Manninen et al., 2013; Nematian and Scown, 2025; Brandão et al., 2023).

Credible policy implementation depends on methodological clarity and data quality. Requirements to include measured or well-justified estimates of fugitive CH_4, to account for digestate storage and field emissions, and to document counterfactual baselines improve comparability and guard against overstated benefits (Hrad et al., 2021, 2022; Bühler et al., 2022; Havukainen et al., 2018; Dietrich et al., 2020). Attention to these details responds to practitioner concerns about LCA misuse and greenwashing, and supports robust certification decisions (Brandão et al., 2023). Policy instruments that reward efficient heat integration, effective leakage control, and best-practice digestate management reduce life cycle impacts per unit of delivered energy (Lombardi and Francini, 2020; Pérez-Camacho et al., 2019).

Experience from other regulatory contexts underlines additional design considerations. Credit-based schemes for transport decarbonisation need safeguards to avoid perverse incentives and over-crediting, particularly where carbon-offset claims depend on uncertain counterfactuals for manure management (Fingerman et al., 2025). Forward-looking policymaking benefits from prospective and dynamic LCA to reflect changing grids, technology learning and scale-up effects, thereby aligning targets and default values with likely future performance (Röder et al., 2022; Heidak et al., 2025; Wen and Ierapetritou, 2025; Vance et al., 2022). In emerging markets, supply-chain and institutional factors condition outcomes, so programmes that pair incentives with capacity-building for waste-based feedstock mobilisation and digestate valorisation can amplify mitigation while avoiding pollution displacement (Aziz et al., 2020; Chin et al., 2022).

In practice, LCA-based rules that set explicit GHG-reduction thresholds and prioritise wastes and residues in target accounting have redirected investment toward waste- and residue-derived biomethane that more reliably meets policy criteria. Certification schemes requiring standardised calculations, disclosure of counterfactual baselines and inclusion of fugitive CH_4, together with auditing, make climate-performance claims verifiable and comparable (Manninen et al., 2013; Nematian and Scown, 2025; Brandão et al., 2023).

In North America, California's Low Carbon Fuel Standard (LCFS) is a leading policy that leverages LCA to reduce GHG emissions. The LCFS assigns tradable credits to fuels based on their life cycle carbon intensity relative to a declining benchmark. Notably, dairy manure biogas receives a very low (even negative) carbon intensity because the methodology credits the capture of methane that would otherwise be released from manure (Nematian and Scown, 2025). As a result, using dairy biogas as a vehicle fuel in California yields substantial credits, and thus revenue, for project developers, spurring investment in renewable natural gas projects at large dairy operations (Fingerman et al., 2025).

The same crediting approach has raised questions of additionality and unintended consequences. If credit values are set too high, they could encourage expansion of industrial livestock operations, or reward facilities for methane emissions that may already be subject to control requirements (Fingerman et al., 2025). Some projects can "stack" credits by selling carbon offsets in multiple markets for the same volume of biogas; if not tightly regulated, this practice risks double-counting emissions reductions (Fingerman et al., 2025). Policymakers are therefore challenged to balance the clear climate benefits of manure biogas with safeguards that prevent perverse incentives or exaggerated net gains (Fingerman et al., 2025; Brandão et al., 2023). One proposal is to extend the system boundaries in policy accounting by evaluating whole-farm emissions when assessing manure biogas. This approach includes upstream emissions from producing animal feed and recognises that a digester does not eliminate all manure-related emissions (Fingerman et al., 2025; Rajagopal and Zilberman, 2013). Although more complex, this approach aligns with the LCA principle of avoiding problem shifting, ensuring that a credited reduction in one part of the system does not mask an increase elsewhere (Fingerman et al., 2025).

At the global level, biogas features in climate strategies because it mitigates emissions across waste, agriculture and energy. The Global Methane Pledge (launched 2021) commits countries to significant methane cuts by 2030, and anaerobic digestion is identified as a key measure to reduce emissions from waste and agriculture (Nematian and Scown, 2025). Nationally determined contributions in parts of Asia, Africa, and Latin America include biogas initiatives aimed at energy generation and methane abatement from organic wastes. Analyses suggest that large-scale deployment could displace fossil energy and mitigate a material share of global GHG emissions, provided best practice is adopted to minimise fugitive CH_4 and to realise fertiliser-substitution credits from digestate use (Mayer et al., 2025; Aziz et al., 2020; Chin et al., 2022). Delivering this potential requires disciplined control of leaks and evidence-based handling of digestate; field studies show that measured leakage varies across facilities, so monitoring and abatement should be embedded in programmes from the outset (Hrad et al., 2021, 2022; Bühler et al., 2022; Havukainen et al., 2018; Durdević and Hulenić, 2020). Policy levers include financial incentives, carbon finance and feed-in tariffs, together with capacity building and technology transfer to scale modern digesters in developing contexts and to integrate projects with waste-management improvement and energy-access agendas (Aziz et al., 2020; Chin et al., 2022).

Use of LCA in policy helps target improvement efforts and guide research. Extending national inventories and crediting schemes to explicitly include digestate storage and field emissions would sharpen incentives for nutrient-recovery technologies and low-emission spreading techniques, priorities that are already attracting R&D support (Havukainen et al., 2018; Durdević and Hulenić, 2020). Methodological advances, such as fuzzy-logic integrations for complex decision spaces, spatially explicit modelling to reflect regional counterfactuals and logistics, and prospective or dynamic LCA to capture technology learning and evolving grid mixes, can reduce uncertainty and improve planning (Choudhary et al., 2025; Dresen and Jandewerth, 2012; González-Martínez et al., 2019; El Akkari et al., 2023; Röder et al., 2022; Heidak et al., 2025; Wen and Ierapetritou, 2025; Vance et al., 2022). Consistent LCA methodologies also enable international comparison and certificate-based trading of biomethane, for example Guarantees of Origin for renewable gas in Europe, which depend on accepted accounting rules to compare pathways (Nematian and Scown, 2025).

As climate policies tighten, support will gravitate toward the most GHG-efficient pathways, reinforcing feedstock choices and technologies that maximise emission reductions while placing safeguards on crediting to avoid over-claiming and perverse incentives (Fingerman et al., 2025; Manninen et al., 2013; Brandão et al., 2023). Life cycle carbon footprint analysis has therefore become a practical instrument in policy design, shaping eligibility criteria, subsidy levels and environmental safeguards so that biogas deployment translates into verifiable climate-change mitigation (Nematian and Scown, 2025; Brandão et al., 2023).

Robust climate benefits from waste feedstocks: AD of manure, sewage sludge and source-separated food waste can achieve net-negative GHG emissions through avoided methane and fossil displacement, while crop-fed biogas delivers smaller savings that depend on cultivation practices and land-use effects (Battini et al., 2014; van den Oever et al., 2021; Bacenetti and Fiala, 2015).

Importance of system boundaries and credits: Results depend strongly on cradle-to-grave scope and consistent crediting for avoided emissions from waste management and fertiliser substitution, with transparent choices on allocation versus system expansion (Ahlgren et al., 2015; Pérez-Camacho et al., 2018; Poeschl et al., 2012).

Methane leakage can erode gains: Small, single-digit losses can negate expected benefits, which underscores the need for leak-tight designs, capture systems and regular monitoring informed by site-level measurements (Hrad et al., 2021, 2022; Bühler et al., 2022).

Digestate management is pivotal: open storage and surface spreading can generate CH_4, NH_3, and N_2O emissions that offset savings, whereas covered storage, digestate separation, and nutrient-recovery techniques—combined with low-volatilisation field application—markedly improve net outcomes (Rehl and Müller, 2011; Havukainen et al., 2018; Chen et al., 2021).

Comprehensive LCA to guide policy: Regulators employ LCA metrics, for example the EU's RED sustainability criteria, to ensure only projects with demonstrable GHG reductions receive support, which structurally favours waste-based pathways (Manninen et al., 2013; Nematian and Scown, 2025).

Mitigating unintended consequences: High incentives (such as LCFS credits for manure-derived biomethane) can drive rapid GHG reductions yet also raise risks of over-crediting and other market-mediated effects. Broadening the system boundaries and implementing safeguards can reduce the potential for such perverse outcomes (Fingerman et al., 2025; Rajagopal and Zilberman, 2013; Brandão et al., 2023).

Ongoing innovation and integration: Progress in membranes and biological upgrading, power-to-gas methanation that uses biogenic CO_2, and valorisation routes such as struvite recovery support further gains, particularly when integrated with regional waste and energy systems (Lombardi and Francini, 2020; Nashmin Elyasi et al., 2021; Gkotsis et al., 2023; Mayer et al., 2025; Đurđević et al., 2020).

Several avenues could further enhance the sustainability of biogas systems. First, innovation in leak detection and prevention, for example advanced continuous gas monitoring and improved seal materials, can drive fugitive emissions closer to zero and secure the climate advantage of biogas (Bühler et al., 2022; Hrad et al., 2021, 2022). Second, new digestate refinement technologies are being developed to manage nutrients more effectively and create higher-value outputs. These include nutrient stripping, producing concentrated bio-based fertilisers, and coupling anaerobic digestion with algae cultivation. Such innovations can reduce eutrophication potential while improving plant economics through added revenue or lower disposal costs (Rehl and Müller, 2011; Havukainen et al., 2018; Chen et al., 2021; Zeng et al., 2022; Durdević and Hulenić, 2020; Đurđević et al., 2020). Third, integrating biogas with other waste and energy systems can generate synergies that stand-alone applications miss. Power-to-gas schemes combine biogenic CO_2 with green hydrogen to produce additional methane, and agroecological farming uses digestate to build soil carbon and close nutrient loops (Mayer et al., 2025; Gkotsis et al., 2023; Ahmed et al., 2021; Lombardi and Francini, 2020). As the energy transition accelerates, biogas can also complement intermittent renewables by providing dispatchable renewable gas or by being upgraded to renewable CNG/LNG for uses that are difficult to electrify. All such applications should be guided by life cycle carbon criteria to ensure low net emissions (Nashmin Elyasi et al., 2021; Vance et al., 2022).

Sustainability assessment via carbon footprint analysis and LCA has proved indispensable in revealing the environmental performance of biogas production. It identifies where improvements are needed to realise the promise of waste-to-energy without unintended consequences. Continued research and real-world data collection (e.g., on long-term soil impacts of digestate, or region-specific LCAs in the Global South), alongside advances in LCA methodology (including dynamic and prospective analyses), will further refine understanding (Heidak et al., 2025). When policies heed LCA insights, biogas technology can be scaled up responsibly to transform organic wastes from liabilities into clean-energy assets, close loops in food-energy-waste systems and support a low-carbon, sustainable future (Brandão et al., 2023).

Robust climate benefits from waste feedstocks: AD of manure, sewage sludge and source-separated food waste can achieve net-negative GHG emissions through avoided methane and fossil displacement, while crop-fed biogas delivers smaller savings that depend on cultivation practices and land-use effects (Battini et al., 2014; van den Oever et al., 2021; Bacenetti and Fiala, 2015).

Importance of system boundaries and credits: Results depend strongly on cradle-to-grave scope and consistent crediting for avoided emissions from waste management and fertiliser substitution, with transparent choices on allocation versus system expansion (Ahlgren et al., 2015; Pérez-Camacho et al., 2018; Poeschl et al., 2012).

Methane leakage can erode gains: Small, single-digit losses can negate expected benefits, which underscores the need for leak-tight designs, capture systems and regular monitoring informed by site-level measurements (Hrad et al., 2021, 2022; Bühler et al., 2022).

Digestate management is pivotal: open storage and surface spreading can generate CH_4, NH_3, and N_2O emissions that offset savings, whereas covered storage, digestate separation, and nutrient-recovery techniques—combined with low-volatilisation field application—markedly improve net outcomes (Rehl and Müller, 2011; Havukainen et al., 2018; Chen et al., 2021).

Comprehensive LCA to guide policy: Regulators employ LCA metrics, for example the EU's RED sustainability criteria, to ensure only projects with demonstrable GHG reductions receive support, which structurally favours waste-based pathways (Manninen et al., 2013; Nematian and Scown, 2025).

Mitigating unintended consequences: High incentives (such as LCFS credits for manure-derived biomethane) can drive rapid GHG reductions yet also raise risks of over-crediting and other market-mediated effects. Broadening the system boundaries and implementing safeguards can reduce the potential for such perverse outcomes (Fingerman et al., 2025; Rajagopal and Zilberman, 2013; Brandão et al., 2023).

Ongoing innovation and integration: Progress in membranes and biological upgrading, power-to-gas methanation that uses biogenic CO_2, and valorisation routes such as struvite recovery support further gains, particularly when integrated with regional waste and energy systems (Lombardi and Francini, 2020; Nashmin Elyasi et al., 2021; Gkotsis et al., 2023; Mayer et al., 2025; Đurđević et al., 2020).

References

Abu, R., Aziz, M.A.A., Hassan, C.H.C., Noor, Z.Z., 2020. Renewable biogas from anaerobic digestion of biomass: Influence factors in life cycle assessment. IOP Conference Series: Materials Science and Engineering 808 (1), 012011. Available at https://doi.org/10.1088/1757-899X/808/1/012011.

Ahlgren, S., Björklund, A., Ekman, A., Karlsson, H., Berlin, J., Börjesson, P., et al., 2015. Review of methodological choices in LCA of biorefinery systems - key issues and recommendations. Biofuels, Bioproducts and Biorefining 9 (5), 606–619. Available at https://doi.org/10.1002/bbb.1563.

Ahmed, S.F., Mofijur, M., Tarannum, K., Chowdhury, A.T., Rafa, N., Nuzhat, S., et al., 2021. Biogas upgrading, economy and utilization: a review. Environmental Chemistry Letters 19 (6), 4137–4164. Available at https://doi.org/10.1007/s10311-021-01292-x.

El Akkari, M., Ben Fradj, N., Gabrielle, B., Njakou Djomo, S., 2023. Spatially-explicit environmental assessment of bioethanol from miscanthus and switchgrass in France. Cleaner and Circular Bioeconomy 6, 100059. Available at https://doi.org/10.1016/j.clcb.2023.100059.

Aziz, N.I.H.A., Hanafiah, M.M., Gheewala, S.H., Ismail, H., 2020. Bioenergy for a cleaner future: a case study of sustainable biogas supply chain in the Malaysian energy sector. Sustainability (Switzerland) 12 (8). Available at https://doi.org/10.3390/SU12083213.

Bacenetti, J., Fiala, M., 2015. Carbon footprint of electricity from anaerobic digestion plants in Italy. Environmental Engineering and Management Journal 14 (7), 1495–1502. Available at https://doi.org/10.30638/eemj.2015.161.

Battini, F., Agostini, A., Boulamanti, A.K., Giuntoli, J., Amaducci, S., 2014. Mitigating the environmental impacts of milk production via anaerobic digestion of manure: case study of a dairy farm in the Po Valley. Science of The Total Environment 481 (1), 196–208. Available at https://doi.org/10.1016/j.scitotenv.2014.02.038.

Beloin-Saint-Pierre, D., Heijungs, R., Blanc, I., 2014. The ESPA (enhanced structural path analysis) method: a solution to an implementation challenge for dynamic life cycle assessment studies. International Journal of Life Cycle Assessment 19 (4), 861–871. Available at https://doi.org/10.1007/s11367-014-0710-9.

Bjørn, A., Owsianiak, M., Molin, C., Laurent, A., 2017. Main characteristics of LCA. In: Hauschild, M., Rosenbaum, R., Olsen, S. (Eds.), Life Cycle Assessment: Theory and Practice. Springer International Publishing, Cham, pp. 9–16. Available at https://doi.org/10.1007/978-3-319-56475-3_2.

Del Borghi, A., Binaghi, L., Del Borghi, M., Gallo, M., 2007. The application of the environmental product declaration to waste disposal in a sanitary landfill: four case studies. International Journal of Life Cycle Assessment 12 (1), 40–49. Available at https://doi.org/10.1065/lca2005.08.224.

Brandão, M., Busch, P., Kendall, A., 2023. Life cycle assessment, quo vadis? Supporting or deterring greenwashing? A survey of practitioners. Environmental Science: Advances 3 (2), 266–273. Available at https://doi.org/10.1039/d3va00317e.

Bühler, M., Häni, C., Ammann, C., Brönnimann, S., Kupper, T., 2022. Using the inverse dispersion method to determine methane emissions from biogas plants and wastewater treatment plants with complex source configurations. Atmospheric Environment: 13. Available at https://doi.org/10.1016/j.aeaoa.2022.100161.

Carnevale, E., Lombardi, L., 2015. Comparison of different possibilities for biogas use by life cycle assessment. Energy Procedia 81, 215–226. Available at https://doi.org/10.1016/j.egypro.2015.12.088.

Chen, T., Qiu, X., Feng, H., Yin, J., Shen, D., 2021. Solid digestate disposal strategies to reduce the environmental impact and energy consumption of food waste-based biogas systems. Bioresource Technology 325. Available at https://doi.org/10.1016/j.biortech.2021.124706. 124706.

Chin, M.Y., Lee, C.T., Woon, K.S., 2022. Policy-driven municipal solid waste management assessment using relative quadrant eco-efficiency: a case study in Malaysia. Journal of Environmental Management 323. Available at https://doi.org/10.1016/j.jenvman.2022.116238.

Choudhary, R., Kumar, A., Narayan, S., Mishchenko, I., Choudhury, M., Bhardwaj, L.K., et al., 2025. Integrating fuzzy logic into life cycle assessment for sustainable municipal solid waste management in metro cities. Ecological Engineering and Environmental Technology 26 (7), 316–331. Available at https://doi.org/10.12912/27197050/205500.

Czyrnek-Delêtre, M.M., Smyth, B.M., Murphy, J.D., 2017. Beyond carbon and energy: The challenge in setting guidelines for life cycle assessment of biofuel systems. Renewable Energy 105, 436–448. Available at https://doi.org/10.1016/j.renene.2016.11.043.

Dietrich, M., Fongen, M., Foereid, B., 2020. Greenhouse gas emissions from digestate in soil. International Journal of Recycling of Organic Waste in Agriculture 9 (1), 1–19. Available at https://doi.org/10.30486/IJROWA.2020.1885341.1005.

Dresen, B., Jandewerth, M., 2012. Integration of spatial analyses into LCA-calculating GHG emissions with geoinformation systems. International Journal of Life Cycle Assessment 17 (9), 1094–1103. Available at https://doi.org/10.1007/s11367-011-0378-3.

Dressler, D., Loewen, A., Nelles, M., 2012. Life cycle assessment of the supply and use of bioenergy: Impact of regional factors on biogas production. International Journal of Life

Cycle Assessment 17 (9), 1104–1115. Available at https://doi.org/10.1007/s11367-012-0424-9.

Durdević, D., Hulenić, I., 2020. Anaerobic digestate treatment selection model for biogas plant costs and emissions reduction. Processes 8 (2). Available at https://doi.org/10.3390/pr8020142.

Đurđević, D., Hulenić, I., Kulišić, B., 2020. Degradation of lignocellulosic complex through production of struvite from digestate. Waste and Biomass Valorisation 11 (6), 2559–2566. Available at https://doi.org/10.1007/s12649-019-00637-6.

Earles, J.M., Halog, A., 2011. Consequential life cycle assessment: a review. International Journal of Life Cycle Assessment 16 (5), 445–453. Available at https://doi.org/10.1007/s11367-011-0275-9.

Fantin, V., Giuliano, A., Manfredi, M., Ottaviano, G., Stefanova, M., Masoni, P., 2015. Environmental assessment of electricity generation from an Italian anaerobic digestion plant. Biomass and Bioenergy 83, 422–435. Available at https://doi.org/10.1016/j.biombioe.2015.10.015.

Fingerman, K., Geronimo, C., Read, E., Wakeman, D., Martin, J.I., 2025. Risks of crediting carbon offsets in low carbon fuel standards: lessons learned from dairy biomethane. Energy Policy 206, 114738. Available at https://doi.org/10.1016/j.enpol.2025.114738.

Gkotsis, P., Kougias, P., Mitrakas, M., Zouboulis, A., 2023. Biogas upgrading technologies – recent advances in membrane-based processes. International Journal of Hydrogen Energy 48 (10), 3965–3993. Available at https://doi.org/10.1016/j.ijhydene.2022.10.228.

García-Gen, S., Rodríguez, J., Lema, J.M., 2014. Optimisation of substrate blends in anaerobic co-digestion using adaptive linear programming. Bioresource Technology 173, 159–167. Available at https://doi.org/10.1016/j.biortech.2014.09.089.

González-Martínez, A., de Simón-Martín, M., Diez Suárez, A.M., de la Puente-Gil, Á., Ramos Malmierca, M.Á., Álvarez-de Prado, L., 2019. GIS-based model for determining the optimal potential of co-digestion mixtures in the Spanish Iberian Peninsula. Renewable Energy and Power Quality Journal 17, 327–332. Available at https://doi.org/10.24084/repqj17.300.

Guinée, J., 2015. Life cycle sustainability assessment: what is it and what are its challenges? In: Clift, R., Druckman, A. (Eds.), Taking Stock of Industrial Ecology. Springer International Publishing, Cham, pp. 45–68. Available at https://doi.org/10.1007/978-3-319-20571-7_3.

Hajabdollahi Ouderji, Z., Gupta, R., Mckeown, A., Yu, Z., Smith, C., Sloan, W., et al., 2023. Integration of anaerobic digestion with heat pump: machine learning-based technical and environmental assessment. Bioresource Technology 369. Available at https://doi.org/10.1016/j.biortech.2022.128485. 128485.

Havukainen, J., Uusitalo, V., Koistinen, K., Liikanen, M., Horttanainen, M., 2018. Carbon footprint evaluation of biofertilizers. International Journal of Sustainable Development and Planning 13 (08), 1050–1060. Available at https://doi.org/10.2495/sdp-v13-n8-1050-1060.

Heidak, P., Isbert, A.M., Haas, S., Schmidt, M., 2025. Integration of recent prospective LCA developments into dynamic LCA of circular economy strategies for wind turbines. Energies 18 (10). Available at https://doi.org/10.3390/en18102509.

Hijazi, O., Mettenleiter, S., Bauerdick, J., Treiber, M., Gräff, A., Bernhardt, H., 2020. Sustainability of biogas production with small-sized plant in South America. Bio-economy and Agri-production: Concepts and Evidence. Elsevier, Germany, pp. 147–158. Available at https://doi.org/10.1016/B978-0-12-819774-5.00008-4.

Hrad, M., Huber-Humer, M., Reinelt, T., Spangl, B., Flandorfer, C., Innocenti, F., et al., 2022. Determination of methane emissions from biogas plants, using different quantification methods. Agricultural and Forest Meteorology 326, 109179. Available at https://doi.org/10.1016/j.agrformet.2022.109179.

Hrad, M., Vesenmaier, A., Flandorfer, C., Piringer, M., Stenzel, S., Huber-Humer, M., 2021. Comparison of forward and backward Lagrangian transport modelling to determine methane emissions from anaerobic digestion facilities. Atmospheric Environment: X 12, 100131. Available at https://doi.org/10.1016/j.aeaoa.2021.100131.

Huttunen, S., Manninen, K., Leskinen, P., 2014. Combining biogas LCA reviews with stakeholder interviews to analyse life cycle impacts at a practical level. Journal of Cleaner Production 80, 5–16. Available at https://doi.org/10.1016/j.jclepro.2014.05.081.

Ingrao, C., Bacenetti, J., Ioppolo, G., Messineo, A., 2019. Energy and environmental assessments of agro-biogas supply chains for energy generation: a comprehensive review. In: Basosi, R., Cellura, M., Longo, S., Parisi, M. (Eds.), Life Cycle Assessment of Energy Systems and Sustainable Energy Technologies. Green Energy and Technology. Springer, Cham, pp. 99–117. Available at https://doi.org/10.1007/978-3-319-93740-3_7.

Jin, Y., Chen, T., Chen, X., Yu, Z., 2015. Life-cycle assessment of energy consumption and environmental impact of an integrated food waste-based biogas plant. Applied Energy 151, 227–236. Available at https://doi.org/10.1016/j.apenergy.2015.04.058.

Kalkanis, K., Bourtsalas, A., Psomopoulos, C.S., 2024. Life cycle assessment applied to waste-to-energy technologies. In: Ong, H.W., Md Rizwanul Fattah, I., Indra Mahlia, T.M. (Eds.), Waste Valorisation for Bioenergy and Bioproducts: Biofuels, Biogas, and Value-Added Products. Woodhead Publishing, pp. 527–543. Available at https://doi.org/10.1016/B978-0-443-19171-8.00014-6.

Kim, D., Kim, J., 2025. Predicting biogas production from organic waste through anaerobic co-digestion. Journal of Cleaner Production 496. Available at https://doi.org/10.1016/j.jclepro.2025.145122. 145122.

Kruse, S.A., Flysjö, A., Kasperczyk, N., Scholz, A.J., 2009. Socioeconomic indicators as a complement to life cycle assessment - an application to salmon production systems. International Journal of Life Cycle Assessment 14 (1), 8–18. Available at https://doi.org/10.1007/s11367-008-0040-x.

Kunatsa, T., Zhang, L., Xia, X., 2022. Biogas potential determination and production optimisation through optimal substrate ratio feeding in co-digestion of water hyacinth, municipal solid waste and cow dung. Biofuels 13 (5), 631–641. Available at https://doi.org/10.1080/17597269.2020.1835452.

Lijó, L., González-García, S., Bacenetti, J., Fiala, M., Feijoo, G., Lema, J.M., et al., 2014. Life cycle assessment of electricity production in Italy from anaerobic co-digestion of pig slurry and energy crops. Renewable Energy 68, 625–635. Available at https://doi.org/10.1016/j.renene.2014.03.005.

Lombardi, L., Francini, G., 2020. Techno-economic and environmental assessment of the main biogas upgrading technologies. Renewable Energy 156, 440–458. Available at https://doi.org/10.1016/j.renene.2020.04.083.

Makepa, D.C., Chihambakwe, Z.J., 2024. 14 - Carbon emissions reduction assessment via biogas production and resource recovery: the IPCC methodology. In: Matambo, T.S., Christian, R. (Eds.), Innovations in the Global Biogas Industry: Applications of Green Principles. Woodhead Publishing, pp. 399–421. Available at https://doi.org/10.1016/B978-0-443-22372-3.00015-7.

Manninen, K., Koskela, S., Nuppunen, A., Sorvari, J., Nevalainen, O., Siitonen, S., 2013. The applicability of the renewable energy directive calculation to assess the sustainability of biogas production. Energy Policy 56, 549–557. Available at https://doi.org/10.1016/j.enpol.2013.01.040.

Martín, M.A., Siles Lopez, J.A., Martín, A., 2012. Anaerobic co-digestion of agro-industrial wastes. Biogas: Production, Consumption and Applications 2012, 117–140.

Mayer, F.D., Brondani, M., Carrillo, M.C.V., de Souza, L.L.P., Lora, E.E.S., 2025. Chapter Two - Life cycle analysis and sustainability of bioenergy systems. From Crops and Wastes to Bioenergy: Current Status and Challenges. Woodhead Publishing, pp. 33–66. Available at https://doi.org/10.1016/B978-0-443-16084-4.00013-4.

Nashmin Elyasi, S., He, L., Tsapekos, P., Rafiee, S., Khoshnevisan, B., Carbajales-Dale, M., et al., 2021. Could biological biogas upgrading be a sustainable substitution for water scrubbing technology? A case study in Denmark. Energy Conversion and Management 245, 114550. Available at https://doi.org/10.1016/j.enconman.2021.114550.

Nematian, M., Scown, C.D., 2025. Advances and emerging issues in life-cycle assessment for biofuel policy. GCB Bioenergy 17 (8). Available at https://doi.org/10.1111/gcbb.70061.

van den Oever, A.E.M., Cardellini, G., Sels, B.F., Messagie, M., 2021. Life cycle environmental impacts of compressed biogas production through anaerobic digestion of manure and municipal organic waste. Journal of Cleaner Production 306. Available at https://doi.org/10.1016/j.jclepro.2021.127156.

Patterson, T., Esteves, S., Dinsdale, R., Guwy, A., 2011. Life cycle assessment of biogas infrastructure options on a regional scale. Bioresource Technology 102 (15), 7313–7323. Available at https://doi.org/10.1016/j.biortech.2011.04.063.

Poeschl, M., Ward, S., Owende, P., 2012. Environmental impacts of biogas deployment: part I: Life cycle inventory for evaluation of production process emissions to air. Journal of Cleaner Production 24, 168–183. Available at https://doi.org/10.1016/j.jclepro.2011.10.039.

Pérez-Camacho, M.N., Curry, R., Cromie, T., 2018. Life cycle environmental impacts of substituting food wastes for traditional anaerobic digestion feedstocks. Waste Management 73, 140–155. Available at https://doi.org/10.1016/j.wasman.2017.12.023.

Pérez-Camacho, M.N., Curry, R., Cromie, T., 2019. Life cycle environmental impacts of biogas production and utilisation substituting for grid electricity, natural gas grid and transport fuels. Waste Management 95, 90–101. Available at https://doi.org/10.1016/j.wasman.2019.05.045.

Rajagopal, D., Zilberman, D., 2013. On market-mediated emissions and regulations on life cycle emissions. Ecological Economics 90, 77–84. Available at https://doi.org/10.1016/j.ecolecon.2013.03.006.

Raketh, M., Kongjan, P., O-Thong, S., Mamimin, C., Jariyaboon, R., Promnuan, K., 2024. Life cycle assessment (LCA) and economic analysis of two-stage anaerobic process of co-digesting palm oil mill effluent (POME) with concentrated latex wastewater (CLW) for biogas production. Process Safety and Environmental Protection 192, 450–459. Available at https://doi.org/10.1016/j.psep.2024.10.079.

Ravina, M., Castellana, C., Panepinto, D., Zanetti, M.C., 2019. MCBioCH$_4$: a computational model for biogas and biomethane evaluation. Journal of Cleaner Production 227, 739–747. Available at https://doi.org/10.1016/j.jclepro.2019.04.224.

Rehl, T., Müller, J., 2011. Life cycle assessment of biogas digestate processing technologies. Resources, Conservation and Recycling 56 (1), 92–104. Available at https://doi.org/10.1016/j.resconrec.2011.08.007.

Röder, H., Kumar, K., Füchsl, S., Sieber, V., 2022. Ex-ante life cycle assessment and scale up: a protein production case study. Journal of Cleaner Production 376. Available at https://doi.org/10.1016/j.jclepro.2022.134329. 134329.

Saghouri, M., Abdi, R., Ebrahimi-Nik, M., Rohani, A., Maysami, M., 2024. Modeling and optimization of biomethane production from solid-state anaerobic co-digestion of organic fraction municipal solid waste and other co-substrates. Energy Sources, Part A: Recovery,

Utilization and Environmental Effects 46 (1), 8564–8580. Available at https://doi.org/10.1080/15567036.2020.1767728.

Slavkovic, K., Stephan, A., 2025. Dynamic life cycle assessment of buildings and building stocks – a review. Renewable and Sustainable Energy Reviews 212. Available at https://doi.org/10.1016/j.rser.2024.115262. 115262.

Tidåker, P., Sundberg, C., Öborn, I., Kätterer, T., Bergkvist, G., 2014. Rotational grass/clover for biogas integrated with grain production - a life cycle perspective. Agricultural Systems 129, 133–141. Available at https://doi.org/10.1016/j.agsy.2014.05.015.

Vance, C., Sweeney, J., Murphy, F., 2022. Space, time, and sustainability: the status and future of life cycle assessment frameworks for novel biorefinery systems. Renewable and Sustainable Energy Reviews 159. Available at https://doi.org/10.1016/j.rser.2022.112259. 112259.

Vignali, G., Vitale, G., 2017. Life cycle assessment of bio-methane obtained by valorization of food waste. Proceedings of the Summer School Francesco Turco. AIDI - Italian Association of Industrial Operations Professors Italy, pp. 136–142. Available at http://www.summerschool-aidi.it/.

Wen, C.-M., Ierapetritou, M., 2025. Improving life cycle assessment consistency for biomass-derived processes: a case study on triacetic acid lactone production with CO_2 recycling. Computers & Chemical Engineering 201, 109244. Available at https://doi.org/10.1016/j.compchemeng.2025.109244.

Yoshida, H., Christensen, T.H., Scheutz, C., 2013. Life cycle assessment of sewage sludge management: a review. Waste Management and Research 31 (11), 1083–1101. Available at https://doi.org/10.1177/0734242X13504446.

Zeng, Q., Zhen, S., Liu, J., Ni, Z., Chen, J., Liu, Z., et al., 2022. Impact of solid digestate processing on carbon emission of an industrial-scale food waste co-digestion plant. Bioresource Technology 360, 127639. Available at https://doi.org/10.1016/j.biortech.2022.127639.

Economic viability and market analysis of biogas production

9

9.1 Introduction

Biogas constitutes a renewable energy source generated through the anaerobic digestion of organic matter, including agricultural residues, municipal waste, and sewage sludge. This process yields a gaseous mixture primarily composed of methane and carbon dioxide. The growing importance of biogas derives from its combined environmental and economic benefits: its production enables the simultaneous management of organic waste streams and the reduction of greenhouse gas emissions (Klimek et al., 2021).

In addition, biogas offers a relatively stable and dispatchable energy supply. It can be upgraded to biomethane and injected into the natural gas grid, while the digestion process produces digestate as a by-product with recognised fertilising value (Lamolinara et al., 2022; Purkus et al., 2018; Keogh et al., 2022). Consequently, biogas systems align with the principles of the circular economy by transforming waste and residual biomass into valuable energy carriers and nutrient-rich fertilisers, thereby fostering the creation of new markets and employment opportunities (Klimek et al., 2021).

The economic performance of electricity generation from biogas is strongly influenced by prevailing economic conditions and the design of support mechanisms. The profitability of projects may differ substantially depending on whether biogas is utilised in combined heat and power (CHP) systems or upgraded to biomethane within alternative plant configurations (Budzianowski and Budzianowska, 2015). In practice, a major limitation of the CHP approach often lies in the insufficient utilisation of the heat generated, which reduces overall system efficiency and constrains the economic value derived from the thermal output (Weinand et al., 2019). For this reason, increasing attention is being directed toward pathways other than electricity generation, including upgrading biogas to biomethane for use as a transport fuel and, where technically and economically feasible, for injection into the gas grid (Patrizio et al., 2015; Sulewski et al., 2023).

Global biogas production currently amounts to approximately 40 billion cubic metres per year (in natural gas equivalent). Biomethane production reaches roughly 10 billion cubic metres annually, corresponding to only about 0.2% of global natural gas demand; however, the biomethane segment has been expanding at an estimated annual growth rate of approximately 20% (IEA, 2025). Within the European Union, around 191 TWh of energy from biogas was produced in 2020, of which approximately 32 TWh (equivalent to about 3.3 billion cubic metres) constituted biomethane—representing roughly 17% of total biogas output (Sulewski et al., 2023). In this context, biomethane accounted for less than 2% of total EU natural gas consumption in 2020.

Sustainable Biogas Production. DOI: https://doi.org/10.1016/B978-0-443-40557-0.00008-1

These figures indicate substantial room for expansion. The International Energy Agency estimates that, on a global scale, nearly 1 trillion cubic metres of biogases per year (in natural gas equivalent) could be produced sustainably—equivalent to approximately one quarter of current global natural gas demand (IEA, 2025). Similarly, Europe exhibits considerable long-term potential; numerous studies project biomethane production in excess of 1,000 TWh annually by 2050 (Sulewski et al., 2023).

In the aftermath of recent geopolitical disruptions, energy security has become an increasingly prominent policy priority. This shift is reflected in the REPowerEU plan, which outlines a target of achieving 35 billion cubic metres of biomethane production annually by 2030 and estimates the associated investment needs at approximately EUR 37 billion (European Commission, 2025).

This chapter presents an analysis of the economic viability of biogas projects and examines the structure of the biogas and biomethane markets, including current trends and development prospects. It further discusses available financial support mechanisms and policy instruments designed to stimulate growth within the biogas sector.

9.2 Cost–benefit analysis

The economic viability of investments in biogas plants depends on a comprehensive comparison of all life-cycle costs with the benefits achieved, understood as revenues and broader economic effects. The most important cost components include capital expenditures (CAPEX) associated with plant construction and operational expenditures (OPEX) related to day-to-day operation and maintenance (Budzianowski and Budzianowska, 2015; Gebrezgabher et al., 2010; Lombardi and Francini, 2020).

On the benefit side, the analysis primarily accounts for the value of the energy produced: electricity, heat, or upgraded fuel gas (biomethane). In addition, biogas projects may generate supplementary revenues or cost savings, such as waste treatment fees (so-called gate fees for accepting substrates for digestion), the value of avoided mineral fertilisers due to the use of digestate, and income from tradable rights or certificates (e.g., renewable energy certificates, emission reduction units, or similar instruments).

From a broader economic perspective, benefits may also include externalities that are more difficult to monetise through market prices, such as reductions in greenhouse gas emissions and other pollutants, improvements in waste management, and local development effects (e.g., job creation in rural areas). Although these impacts are often not fully captured by direct project cash flows, they frequently constitute a key rationale for public support schemes targeting the biogas sector (Kusz et al., 2024).

9.2.1 Capital expenditures

The construction of a biogas plant entails substantial capital investment. CAPEX typically includes the costs of installing digesters and storage tanks, CHP units (engine or turbine coupled with a generator) or, alternatively, biogas upgrading systems for biomethane production, as well as auxiliary infrastructure such as substrate and digestate storage facilities, grid connections, and related utilities. According to various sources, the indicative total investment cost of an agricultural biogas plant with an electrical capacity of 1 MW is approximately EUR 2.6–3.3 million (Klimek et al., 2021). For instance, Kusz et al. (2024), analysing a 1 MW biogas plant in Poland, report a cost range of EUR 2.6–3.3 million, while an investment model developed by the Ministry of Agriculture assumes EUR 3.3–3.5 million per 1 MW (Kusz et al., 2024).

Investment scale has a strong influence on unit costs due to economies of scale: larger installations (with higher capacity and throughput) tend to be relatively less expensive per unit of energy or biogas produced. In the Polish context, smaller projects are particularly capital-intensive relative to installed capacity, which extends payback periods and complicates access to financing. Capital barriers, high upfront expenditures and the associated investment risk are therefore frequently identified as one of the main constraints on biogas sector development in the absence of external support (Ertop et al., 2023; Kasinath et al., 2021; Kusz et al., 2024).

9.2.2 Operational expenditures

The operating costs of a biogas plant include expenditures related to substrate procurement and transport, staffing and labour, internal energy consumption such as digester heating and mixer operation, equipment maintenance and servicing, and the management or disposal of surplus digestate. Among these components, feedstock cost is frequently the decisive factor. If a plant relies on dedicated energy crops such as maize, the costs of cultivation or purchase must be included. In contrast, the use of waste and residue streams is often associated with low or even negative feedstock costs, since operators may receive payment for accepting waste in the form of a gate fee. In business models based on waste streams, including municipal biowaste or manure, substrates may be obtained at no cost or may generate additional revenue through gate fees, which can substantially improve the project's financial performance (Bhatt and Tao, 2020).

Economic assessments of agricultural biogas plants in Poland indicate that feedstock costs constitute one of the most influential determinants of profitability. The use of low cost waste based substrates such as manure improves key economic indicators, whereas models dependent on purchasing more expensive inputs such as maize silage significantly increase operating expenditures and reduce overall investment attractiveness (Klimek et al., 2021). Cost analyses also suggest that plants upgrading biogas to biomethane generally incur higher operating expenditures than plants without upgrading, due to the additional operational requirements associated with gas purification and separation processes (Sánchez-Martín et al., 2023).

9.2.3 Profitability assessment results and financial indicators

The economic performance of investments in biogas plants is commonly evaluated using indicators such as net present value (NPV) and internal rate of return (IRR). In a case study of an agricultural biogas plant in Poland, the level of economic effectiveness was shown to depend strongly on the financing structure. The authors indicate that without adequate public support the investment does not achieve satisfactory economic parameters. Improvements in NPV and IRR occur as the share of grant funding increases (Klimek et al., 2021; Kusz et al., 2024). In a case study of a 0.499 MW biogas plant, the project required subsidies in the range of 40%–60% in order to reach satisfactory economic performance (Kusz et al., 2024).

A model based assessment for a 1 MW biogas plant showed that, at a renewable energy certificate price of approximately 46 EUR per MWh, the IRR falls below 5%. This outcome suggests low investment profitability under minimal market based support (Klimek et al., 2021). Similarly, an assessment of an agricultural biogas plant in Poland based on NPV and IRR confirmed that project performance is highly sensitive to the financing structure and that, in the absence of sufficient public support, the investment fails to reach satisfactory economic results (Kusz et al., 2024). Overall, these findings indicate that the level of policy support and the conditions of financing significantly affect IRR and NPV values and therefore influence investment decisions regarding the implementation of biogas projects.

It is also important to note that the economic appraisal of biogas plants changes depending on which effects are included. Case study evidence shows that an assessment may incorporate environmental and social impacts, and that their inclusion affects the final evaluation of the project (Kusz et al., 2024). The literature further emphasises that the economic competitiveness of biogas and biomethane depends on energy policy conditions and on the specific support instruments applied (Chan Gutiérrez et al., 2018). It has also been noted that profitability for farmers remains uncertain, which may constrain sector development under market conditions without strong policy support (Hamman and Dziebowski, 2025). In addition, the biogas sector is described as structurally dependent on public subsidies and regulatory frameworks that shape business models and investment decisions (Hamman and Dziebowski, 2025).

Moreover, an analysis of operational results indicates that purchasing and operating a biogas plant can be financially unprofitable at the farm level, even though it generates environmental and social benefits (Belinska et al., 2024). A cost assessment reported that the cost of implementing a biogas plant in 2022 amounted to EUR 1,105,536 under the assumption that part of the work was carried out independently, whereas the cost of installing and commissioning a reference project in 2010 was EUR 3.6 million. The same publication reported an NPV over 15 years of operation of EUR 1.2 million. Based on operational data, only four years from 2014 to 2017 were profitable within the 2012–2020 period (Belinska et al., 2024). A summary of the key determinants of the economic viability of biogas projects is presented in Table 9.1.

In summary, biogas projects are characterised by high capital expenditures and substantial operating costs, which in many cases are difficult to cover solely through revenues from energy sales at market prices. Market analyses and international reports indicate that the cost of biomethane production remains higher than the cost of energy derived from fossil fuels, which limits its competitiveness under purely

Table 9.1 Key determinants of the economic viability of biogas projects.

Area	Main factors	Mechanism of impact on the project	Nature of impact
Capital expenditures (CAPEX)	Plant scale	Economies of scale	Larger scale reduces unit cost
	Technology choice	Additional technological modules	Increases initial investment outlays
	Access to financing	Cost of capital and financing structure	Lower cost of capital improves NPV and IRR
Operational expenditures (OPEX)	Feedstock price and type	Cost of fermentation input	Waste based substrates improve profitability
	Logistics and transport	Ongoing operating cost	Long transport distances worsen financial performance
	Maintenance and internal energy use	Fixed operating costs	High energy intensity reduces margins
Revenue streams	Sale of electricity	Market based revenue	Dependent on electricity prices and support schemes
	Sale of biomethane	Access to gas or transport markets	Greater flexibility of end use
	Waste acceptance fees, gate fee	Additional revenue stream	Stabilises the business model
	Certificates and environmental attributes	Monetisation of climate benefits	Often a key determinant of profitability
Regulatory environment	Stability of support policy	Regulatory risk	High stability encourages investment
	Renewable energy and climate targets	Demand creation	Increases demand for biomethane
Non-financial effects	Emission reduction	Environmental value	May justify public support
	Local development	Social acceptance	Influences feasibility of project implementation

market-based conditions (Lombardi and Francini, 2020). The literature emphasises that the profitability of biogas installations depends to a large extent on additional revenue streams, including waste management services, the valorisation of by-products, and regulatory support mechanisms. Without financial support instruments, a significant share of the technical biogas potential remains difficult to realise from the perspective of private investment returns. Under specific conditions, some projects may achieve cost competitiveness, particularly when installations operate at a large scale and benefit from favourable access to feedstocks. In practice, however, the development of the biogas sector remains closely linked to the regulatory environment, the level of public support, and the ability to generate revenues from environmental benefits and co-products. For this reason, further market analysis of biogas requires consideration of both economic and policy drivers that shape the pace and directions of sector development.

9.3 Market trends and opportunities in the biogas sector

9.3.1 Market size and growth dynamics

At present, biogas and biomethane play a relatively limited role in the global energy system. Approximately 40 billion cubic metres of natural gas equivalent biogas are produced annually, primarily in Europe, the United States, and China. Most of this volume is used for local heat and electricity generation and, in some cases, also for clean cooking. Biomethane is expanding rapidly at roughly 20% per year. Despite this growth, global biomethane production remains at only about 10 billion cubic metres of natural gas equivalent, which corresponds to approximately 0.2% of global natural gas demand. Biogases remain a small component of the global energy mix and account for around 3% of total modern bioenergy production. The observed 20% annual growth in biomethane use over the last five years has occurred from a low starting base (IEA, 2025).

Europe, and particularly the European Union, is the largest producer of biogas and biomethane and accounts for a substantial share of global output (IEA, 2025). At the EU level, review literature points to the rapid development of the biomethane segment. The first European biomethane installation was likely commissioned in the United Kingdom in 2009, and by 2020 nearly 900 installations in the EU produced 3.3 billion cubic metres of biomethane, equivalent to 32 TWh. In the same year, total energy production from biogas in the EU amounted to 191 TWh, of which 32 TWh was biomethane, corresponding to 17 percent. The remaining share reflected other biogas uses, including heat and electricity generation. The latest European Biogas Association data cited in the literature indicate that by August 2021 the number of biomethane installations in the EU had reached 992 (Sulewski et al., 2023).

Germany remains the largest producer of biogas and biomethane in Europe. In 2020, around 9,000 biogas plants operated in Germany, and biomethane production reached 11 TWh, equivalent to 1.13 billion cubic metres. In the same year, 242 biomethane upgrading plants were in operation (Sulewski et al., 2023; Kornatz et al., 2021). Denmark ranks second in terms of biomethane volume, with production exceeding 4 TWh in 2020 (Sulewski et al., 2023). Supported by strong policy incentives and sector expansion, biomethane accounted for more than 35% of total gas demand in Denmark in 2023 (IEA, 2025). France, Italy, and the Netherlands are also significant biomethane producers, with each country reaching production of approximately 2 TWh in 2020 (Sulewski et al., 2023). France is notable for having the highest number of biomethane installations in Europe. In 2020, 306 installations operated there with total production of 2 TWh. Combined, Germany and France accounted for nearly 55 percent of biomethane installations in the EU in 2020 (Sulewski et al., 2023).

Outside the EU, the United Kingdom is an important actor, where biomethane development has been supported, among other measures, by the Renewable Heat Incentive scheme (IEA, 2024). Nevertheless, in 2022 biomethane injected into the gas grid accounted for about 1 percent of gas consumption supplied through the grid in the United Kingdom (IEA, 2024). The United States has a large feedstock base, and the economics of biomethane projects, commonly referred to as renewable natural gas, have been strongly influenced by transport support mechanisms, including programs such as the Renewable Fuel Standard and the Low Carbon Fuel Standard. In 2023, the United States produced around 6 billion cubic metres of natural gas equivalent biogases, including about 3 billion cubic metres of biomethane. Biomethane production in 2023 represented a tripling relative to 2017 levels. Approximately half of total biogas production in the United States is derived from landfill gas (IEA, 2025).

In China, the biogas sector has developed since the 1980s through the large scale deployment of small household systems. By 2015, the number of household biogas installations reached 41.933 million, after which a decline occurred, with the figure falling to 30.077 million in 2020 (Wang et al., 2023). From the mid 2010s onwards, policy increasingly promoted larger and more centralised installations and biomethane projects. During 2015 to 2017, 65 pilot installations were launched, and under the Thirteenth Five Year Plan for 2016 to 2020, the construction of a total of 197 such facilities by 2020 was envisaged (Wang et al., 2023). According to the IEA, China produces around 10 billion cubic metres of natural gas equivalent biogases (IEA, 2025).

Beyond these leading regions, biogas development is also observed elsewhere. In India, the SATAT initiative was announced with the goal of commissioning 5,000 compressed biogas plants by 2025 (EPA, 2025). In the longer term, the literature indicates that in sector development scenarios Europe could exceed 1,000 TWh of biomethane production per year by 2050 (Sulewski et al., 2023).

9.3.2 Application pathways and emerging opportunities

Traditionally, biogas has been used locally for electricity and heat generation in combined heat and power systems. With the evolution of support policies in Europe, a gradual shift has been observed from electricity generation in CHP units towards biomethane production and its injection into the gas grid. This transition has often been facilitated by dedicated support schemes based on guaranteed tariffs or similar mechanisms for biomethane (Olczak, 2026). Different instruments have been applied at the national level. In the United Kingdom, a key support mechanism for biogas and biomethane development was the Non Domestic Renewable Heat Incentive introduced in 2011 (IEA, 2024). In Denmark, the integration of biomethane into the gas system has reached a particularly high level. In 2023, 0.7 billion cubic metres of natural gas equivalent biomethane accounted for more than 35% of total national gas demand (IEA, 2025). In Italy, the decree of 3 March 2018 introduced incentives for biomethane injected into the grid and for advanced biofuels in the transport sector (IEA, 2018).

The transition toward biomethane creates new market opportunities. Biomethane can substitute natural gas in final energy applications, and its use as a transport fuel is being developed within alternative fuel infrastructure, including compressed and liquefied natural gas solutions for heavy duty transport (Van Grinsven et al., 2021). In the transport sector, the revised Renewable Energy Directive establishes an important regulatory framework. By 2030, Member States must either reach a 29% share of renewable energy in transport or reduce the greenhouse-gas intensity of transport fuels by 14.5%; the revised framework also introduces a combined 5.5% sub-target for advanced biofuels and renewable fuels of non-biological origin, including a minimum RFNBO share (European Commission, 2018b).

In selected markets such as California, biomethane or renewable natural gas derived from waste streams, including animal manure management, can achieve highly favourable emission performance under Low Carbon Fuel Standard type methodologies. This outcome is partly attributable to the accounting of avoided methane emissions, which translates into high credit values within these systems (Nicholls, 2023). In the United States, the development of the renewable natural gas market is closely linked to transport support mechanisms, including the Renewable Fuel Standard and Low Carbon Fuel Standard programs. Projects monetise environmental attributes that are sold to obligated entities under these regulatory schemes (IEA, 2025; SEC, 2024).

At the same time, the role of biogas in waste management is increasing. Revised European Union regulations restrict landfill disposal, with the share of municipal waste sent to landfill to be reduced to 10 percent by 2035 (European Commission, 2018). This policy direction enhances the availability of feedstock streams for anaerobic digestion. Another emerging pathway involves integrating biogas and biomethane into the broader energy system through power to gas and power to methane solutions. In biological methanation processes, hydrogen produced via electrolysis is combined with carbon dioxide, for example from biogas, enabling the conversion of carbon dioxide into methane and increasing the methane content of the gas stream (IEA, 2018; Yörüklü et al., 2025). Demonstration projects combining electrolysis and biological methanation have been implemented in Germany, and larger scale power to gas applications based on biological methanation have been reported in Switzerland (Bioenergy International, 2023; IEA, 2018).

From a longer term perspective, the European Union has intensified cooperation with industry stakeholders in the biomethane field. In September 2022, the European

Commission and industry leaders launched the Biomethane Industrial Partnership to accelerate sector development (European Commission, 2022).

9.4 Support mechanisms

Biogas production is a sector that depends strongly on regulation and public policy. Given the profitability gaps discussed earlier, support mechanisms often determine whether an investment is initiated. Support can take multiple forms, including investment grants, feed in tariffs, price premiums, tax incentives, quota systems with tradable certificates, operating subsidies, concessional loans, and guarantees of origin.

9.4.1 European Union and Member States

In the European Union, support for biogas and biomethane is shaped both at the EU level and by Member States. The EU sets the legal framework, including targets and definitions, while the selection of support instruments largely remains a national responsibility (European Commission, 2023).

At the EU level, Directive (EU) 2018/2001, known as RED II, played a central role by establishing a binding target of at least 32% renewable energy in the EU energy mix by 2030 (European Commission, 2023). RED II also introduced a target of 14% renewable energy in final energy consumption in transport by 2030, together with an additional sub target for advanced biofuels derived from feedstocks listed in Part A of Annex IX. This sub target was set at 0.2% in 2022, 1% in 2025, and at least 3.5% in 2030 (Giuntoli, 2021). Biomethane can qualify as an advanced biofuel if it is produced from feedstocks included in Part A of Annex IX, such as selected fractions of waste and biological residues (European Commission, 2023; Giuntoli, 2021).

RED II also includes provisions on Guarantees of Origin for renewable energy. Their purpose is to enable the supplier to demonstrate to the final consumer the share or quantity of renewable energy in the supplier's energy mix. Guarantees of Origin provide the basis for a book and claim approach in which the certificate can be traded independently from the physical flow of energy (AIB, 2023; European Commission, 2023). The directive was subsequently revised in 2023 through RED III, formally Directive (EU) 2023/2413, which raised the binding EU target to at least 42.5% renewable energy by 2030, alongside an aspiration to reach 45% (European Commission, 2023; EUR Lex, 2023).

An additional political impetus for biomethane was provided by the REPowerEU plan in 2022, which set out a goal to increase sustainable biomethane production to 35 billion cubic metres by 2030 (European Commission, 2022). In this context, the European Commission and stakeholders launched the Biomethane Industrial Partnership on 28 September 2022 as a forum to identify barriers and support actions aimed at achieving the 35 billion cubic metres target (European Commission, 2022). The European Commission estimates that achieving the 35 bcm biomethane target by 2030 requires investment of approximately EUR 37 billion (European Commission, 2022).

9.4.2 Other countries

In the United States, support is strongly linked to market based instruments. A central pillar is the federal Renewable Fuel Standard, which creates demand for renewable fuels through compliance obligations settled with credits known as Renewable Identification Numbers (Lade et al., 2018). This framework is particularly relevant for biomethane, often referred to as renewable natural gas, when used in transport, as it enables the monetisation of the fuel's environmental attributes within the compliance system (Parker et al., 2017). A second key pillar is state level standards, with the most developed example being the California Low Carbon Fuel Standard. This crediting system is based on life cycle carbon intensity and supports low carbon transport fuels, including renewable natural gas (Yeh and Sperling, 2010). Additional measures include tax and investment provisions, such as incentives applicable to qualifying biogas facilities, and grant programs for rural areas that can support some agricultural projects.

In China, biogas policy is closely linked to environmental objectives and rural development. Historically, strong emphasis was placed on decentralised programs based on small household digesters. In recent years, more industrial approaches have also been promoted, often connected to waste management and efforts to introduce biomethane into the energy system (Zheng et al., 2020). At the same time, the literature highlights institutional and market barriers that hinder scaling, including quality standards, network access, and the availability of stable offtake mechanisms (Zheng et al., 2020).

In India, the development of biogas and compressed biomethane, commonly referred to as CBG, is supported through both social and environmental programs for small installations and market oriented initiatives designed to create predictable demand for CBG in transport (Singh et al., 2023). One of the most widely recognised instruments is SATAT, which is linked to the development of the CBG market and its supply chain, including production, distribution, and offtake by the fuel sector.

In Brazil, biogas and biomethane are increasingly linked to policies aimed at decarbonising transport fuels. In this context, systems of decarbonisation credits and the certification of climate performance play an important role. This approach is intended to reward fuels with a lower carbon footprint and to support investments in technologies that use waste from agriculture and the agri food industry (Grangeia et al., 2022).

In many developing countries, international mechanisms also play a role. Historically, methane reduction projects, including those related to biogas and landfill gas, could benefit from climate instruments under the Clean Development Mechanism (Lo and Cong, 2022). More recently, some projects have been financed through voluntary carbon markets. However, literature reviews point to heterogeneous quality and uncertainty regarding the climate impacts of many types of crediting projects (Probst et al., 2024).

9.5 Conclusions

This chapter demonstrates that the economic viability of biogas projects results from the interaction of three groups of factors. The first group concerns the CAPEX and

OPEX cost structure and the availability of suitable feedstocks. The second relates to the selected technological pathway, including the choice between combined heat and power and upgrading to biomethane. The third concerns the regulatory environment and the stability of support mechanisms. Biogas plants require substantial capital investment, while operating costs are largely driven by feedstock prices and feedstock logistics. For this reason, projects based on waste and residual biomass typically exhibit stronger economics than installations dependent on purchased energy crops. At the same time, facilities that upgrade biogas to biomethane may face higher operating costs, yet they can benefit from broader market access and greater flexibility in end use applications.

Under many market conditions, revenues from energy sales alone do not generate sufficient cash flows to justify investment. Additional value streams are therefore critical. These include waste management services, the utilisation of digestate, certificates and environmental attributes, and public support. This explains the sector's strong dependence on policy frameworks. From a global perspective, biogas and biomethane remain relatively small segments of the energy system, but they display clear growth momentum. Europe, including the European Union, remains the central hub of development. At the same time, market based instruments have gained importance in the United States, while strategies in China and India continue to evolve.

In conclusion, the sector has substantial feedstock and environmental potential. Realising this potential requires the simultaneous fulfilment of economic conditions, including scale, feedstock access, and logistics, technological conditions related to configuration choices, and regulatory conditions that provide coherent and predictable support and that enable the valuation of climate benefits.

AI disclosure

During the preparation of this work the author(s) used DEEPL, ChatGPT, Grammarly, Mendeley in order to Translation, language correction, bibliography management. After using this tool/service, the author(s) reviewed and edited the content as needed and take(s) full responsibility for the content of the publication.

References

AIB, 2023. Renewable energy guarantees of origin. AIB, Ireland. Available at https://www.aib-net.org/certification/certificates-supported/renewable-energy-guarantees-origin?utm_source=chatgpt.com.

Belinska, S., Bielik, P., Adamičková, I., Husárová, P., Onyshko, S., Belinska, Y., 2024. Assessment of environmental and economic-financial feasibility of biogas plants for agricultural waste treatment. Sustainability (Switzerland) 16 (7). Available at https://doi.org/10.3390/su16072740.

Bhatt, A.H., Tao, L., 2020. Economic perspectives of biogas production via anaerobic digestion. Bioengineering 7 (3), 1–19. Available at https://doi.org/10.3390/bioengineering7030074.

Bioenergy International, 2023. Swiss Power-to-Gas scoops Watt d'Or Award. Bioenergy International, Sweden. Available at https://bioenergyinternational.com/swiss-power-to-gas-project-scoops-watt-dor-award/?utm_source=chatgpt.com.

Budzianowski, W.M., Budzianowska, D.A., 2015. Economic analysis of biomethane and bioelectricity generation from biogas using different support schemes and plant configurations. Energy 88, 658–666. Available at https://doi.org/10.1016/j.energy.2015.05.104.

Chan Gutiérrez, E., Wall, D.M., O'Shea, R., Novelo, R.M., Gómez, M.M., Murphy, J.D., 2018. An economic and carbon analysis of biomethane production from food waste to be used as a transport fuel in Mexico. Journal of Cleaner Production 196, 852–862. Available at https://doi.org/10.1016/j.jclepro.2018.06.051.

EPA, 2025. Methane mitigation: lessons learned from India's compressed biogas (Renewable Natural Gas) projects. EPA, Washington, DC.

Ertop, H., Atilgan, A., Kocięcka, J., Krakowiak-Bal, A., Liberacki, D., Saltuk, B., et al., 2023. Calculation of the potential biogas and electricity values of animal wastes: Turkey and poland case. Energies 16 (22). Available at https://doi.org/10.3390/en16227578.

European Commission, 2018. Landfill waste - Environment - European Commission. European Commission. Available at https://environment.ec.europa.eu/topics/waste-and-recycling/landfill-waste_en?utm_source=chatgpt.com.

European Commission. (2023). Renewable Energy Directive. European Commission. Available at: https://energy.ec.europa.eu/topics/renewable-energy/renewable-energy-directive-targets-and-rules/renewable-energy-directive_en.

European Commission. (2025). Biomethane. European Commission. Available at https://energy.ec.europa.eu/topics/renewable-energy/bioenergy/biomethane_en.

European Commission, 2022. Communication from the commission to the European Parliament, the European Council, the Council, the European Economic and Social Committee and The Committee of the Regions REPowerEU. European Commission. Available at https://eur-lex.europa.eu/legal-content/EN/TXT/HTML/?uri=CELEX:52022DC0230&utm_source=chatgpt.com.

European Commission, 2023. ►B DIRECTIVE (EU) 2018/2001 OF THE EUROPEAN PARLIAMENT AND OF THE COUNCIL of 11 December 2018 on the promotion of the use of energy from renewable sources (recast) (Text with EEA relevance). 2018. European Commission.

EUR Lex. (2023). Directive (EU) 2023/2413 of the European Parliament and of the Council of 18 October 2023 amending Directive (EU) 2018/2001, Regulation (EU) 2018/1999 and Directive 98/70/EC as regards the promotion of energy from renewable sources. Official Journal of the European Union. Available at https://eur-lex.europa.eu/eli/dir/2023/2413/oj/

Gebrezgabher, S.A., Meuwissen, M.P.M., Prins, B.A.M., Lansink, A.G.J.M.O., 2010. Economic analysis of anaerobic digestion—a case of green power biogas plant in the Netherlands. NJAS - Wageningen Journal of Life Sciences 57 (2), 109–115. Available at https://doi.org/10.1016/j.njas.2009.07.006.

Giuntoli, J., 2021. Final Recast Renewable Energy Directive for 2021-2030 in the European Union. The International Council on Clean Transportation. Available at https://theicct.org/publication/final-recast-renewable-energy-directive-for-2021-2030-in-the-european-union/.

Grangeia, C., Santos, L., Lazaro, L.L.B., 2022. The Brazilian biofuel policy (RenovaBio) and its uncertainties: An assessment of technical, socioeconomic and institutional aspects. Energy Conversion and Management: X 13, 100156. Available at https://doi.org/10.1016/j.ecmx.2021.100156.

Hamman, P., Dziebowski, A., 2025. Building an agricultural biogas supply chain in Europe: organizational models and social challenges. Sustainability 17 (13), 5806. Available at https://doi.org/10.3390/su17135806.
IEA, 2018. National policy framework in Italy. IEA. Available at https://ec.europa.eu/energy/en/topics/renewable-energy/national-action-plans.
IEA., 2024. Implementation of bioenergy in the United Kingdom. Available at https://www.ieabioenergy.com/wp-content/uploads/2024/12/CountryReport2024_UK_final.pdf.
IEA, 2025. World energy outlook special report outlook for biogas and biomethane a global geospatial assessment. IEA. Available at https://www.iea.org.
Kasinath, A., Fudala-Ksiazek, S., Szopinska, M., Bylinski, H., Artichowicz, W., Remiszewska-Skwarek, A., et al., 2021. Biomass in biogas production: pretreatment and codigestion. Renewable and Sustainable Energy Reviews 150. Available at https://doi.org/10.1016/j.rser.2021.111509 111509.
Klimek, K., Kapłan, M., Syrotyuk, S., Bakach, N., Kapustin, N., Konieczny, R., et al., 2021. Investment model of agricultural biogas plants for individual farms in Poland. Energies 14 (21). Available at https://doi.org/10.3390/en14217375.
Kornatz, P., Gromke, J. D., & Rensberg, N. (2021). IEA Bioenergy Task 37: Country report Germany 2021. IEA Bioenergy Task 37. Available at https://task37.ieabioenergy.com/wp-content/uploads/sites/32/2022/10/Germany-Country-Report_05_2022.pdf.
Kusz, D., Kusz, B., Wicki, L., Nowakowski, T., Kata, R., Brejta, W., et al., 2024. The economic efficiencies of investment in biogas plants—a case study of a biogas plant using waste from a dairy farm in Poland. Energies 17 (15). Available at https://doi.org/10.3390/en17153760.
Keogh, N., Corr, D., Monaghan, R.F.D, 2022. Biogenic renewable gas injection into natural gas grids: A review of technical and economic modelling studies. Renewable and Sustainable Energy Reviews 168, 112818. Available at https://doi.org/10.1016/j.rser.2022.112818.
Lamolinara, Pérez-Martínez, A., Guardado-Yordi, E., Guillén Fiallos, C., Diéguez-Santana, K., Ruiz-Mercado, G.J., 2022. Anaerobic digestate management, environmental impacts, and techno-economic challenges. Waste Management 140, 14–30. Available at https://doi.org/10.1016/j.wasman.2021.12.035.
Lade, G.E., Lin Lawell, C.-Y.C., Smith, A.n., 2018. Policy Shocks and Market-Based Regulations: Evidence from the Renewable Fuel Standard. American Journal of Agricultural Economics 100 (3), 707–731. Available at https://doi.org/10.1093/ajae/aax097.
Lo, A.Y., Cong, R., 2022. Emission reduction targets and outcomes of the Clean Development Mechanism (2005–2020). PLOS Climate 1 (8), e0000046. Available at https://doi.org/10.1371/journal.pclm.0000046.
Lombardi, L., Francini, G., 2020. Techno-economic and environmental assessment of the main biogas upgrading technologies. Renewable Energy 156, 440–458. Available at https://doi.org/10.1016/j.renene.2020.04.083.
Nicholls, J., 2023. The Low Carbon Fuel Standard and Dairy Practices in the San Joaquin Valley Manure Biomethane, and Air Quality. Goldman School of Public Policy. Available at https://www.law.berkeley.edu/wp-content/uploads/archive/2023/06/LCFS-Dairy-Practices-Nicholls-6.13.23.pdf.
Olczak, M., 2026. Biomethane in Europe: Why Scaling Up is Harder Than it Looks. The Oxford Institute for Energy Studies. Available at https://www.oxfordenergy.org/publications/biomethane-in-europe-why-scaling-up-is-harder-than-it-looks/.

Patrizio, P., Leduc, S., Chinese, D., Dotzauer, E., Kraxner, F., 2015. Biomethane as transport fuel – a comparison with other biogas utilization pathways in northern Italy. Applied Energy 157, 25–34. Available at https://doi.org/10.1016/j.apenergy.2015.07.074.

Parker, N., Williams, R., Dominguez-Faus, R., Scheitrum, D., 2017. Renewable natural gas in California: An assessment of the technical and economic potential. Energy Policy 111, 235–245. Available at https://doi.org/10.1016/j.enpol.2017.09.034.

Probst, B.S., Toetzke, M., Kontoleon, A., Díaz Anadón, L., Minx, J.C., Haya, B.K., et al., 2024. Systematic assessment of the achieved emission reductions of carbon crediting projects. Nature Communications 15 (1). Available at https://doi.org/10.1038/s41467-024-53645-z.

Purkus, Gawel, Szarka, N., Lauer, M., Lenz, V., Ortwein, A., et al., 2018. Contributions of flexible power generation from biomass to a secure and cost-effective electricity supply—a review of potentials, incentives and obstacles in Germany. Energy, Sustainability and Society 8 (1). Available at https://doi.org/10.1186/s13705-018-0157-0.

SEC, 2024. United States Securities and Exchange Commission Washington, D.C. 20549. SEC. Available at https://www.sec.gov/Archives/edgar/data/1842279/000184227925000049/wdq-20241231.htm?utm_source=chatgpt.com.

Sánchez-Martín, L., Ortega Romero, M., Llamas, B., Suárez Rodríguez, M. del C., Mora, P., 2023. Cost Model for Biogas and Biomethane Production in Anaerobic Digestion and Upgrading. Case Study: Castile and Leon. Materials 16 (1), 359. Available at https://doi.org/10.3390/ma16010359.

Sulewski, P., Ignaciuk, W., Szymańska, M., Wąs, A., 2023. Development of the biomethane market in Europe. Energies 16 (4). Available at https://doi.org/10.3390/en16042001.

Singh, D., Tembhare, M., Kumar, S., 2023. Compressed biogas plants in India: Existing status, technological advances and challenges. Biomass and Bioenergy 178, 106964. Available at https://doi.org/10.1016/j.biombioe.2023.106964.

Van Grinsven, A., Otten, M., Van Den Toorn, E., Van Der Veen, R., Király, J., Van Den Berg, R., 2021. Research for TRAN Committee: alternative fuel infrastructures for heavy-duty vehicles. European Parliament, Policy Department for Structural and Cohesion Policies, Brussels. Available at https://www.europarl.europa.eu/RegData/etudes/STUD/2021/690901/IPOL_STU(2021)690901_EN.pdf.

Wang, Y., Zhi, B., Xiang, S., Ren, G.G., Feng, Y., Yang, G., et al., 2023. China's biogas industry's sustainable transition to a low-carbon plan—a socio-technical perspective. Sustainability 15 (6), 5299. Available at https://doi.org/10.3390/su15065299.

Weinand, J.M., McKenna, R., Karner, K., Braun, L., Herbes, C., 2019. Assessing the potential contribution of excess heat from biogas plants towards decarbonising residential heating. Journal of Cleaner Production 238. Available at https://doi.org/10.1016/j.jclepro.2019.117756 117756.

Yeh, S., Sperling, D., 2010. Low carbon fuel standards: Implementation scenarios and challenges. Energy Policy 38 (11), 6955–6965. Available at https://doi.org/10.1016/j.enpol.2010.07.012.

Yörüklü, H.C., Kamravamanesh, D., Köroğlu, E.O., Patel, G.H., Havukainen, J., Karjunen, H., et al., 2025. A comprehensive review on biological methanation processes: from gaseous feedstocks to biomethane. Energy Conversion and Management 341, 120075. Available at https://doi.org/10.1016/j.enconman.2025.120075.

Zheng, L., Cheng, S., Han, Y., Wang, M., Xiang, Y., Guo, J., et al., 2020. Bio-natural gas industry in China: Current status and development. Renewable and Sustainable Energy Reviews 128, 109925. Available at https://doi.org/10.1016/j.rser.2020.109925.

Conclusions and future perspectives

10

10.1 Future perspectives and conclusions

Biogas and biomethane can become material pillars of the global energy mix by 2030–50 as renewable, circular resources. Recent technical advances have improved production efficiency and broadened applications, supporting wider adoption globally (Satpathy, 2017). This chapter synthesises key insights on future cost trajectories, regulatory evolution and development opportunities for efficient biogas production. Currently, typical levelised biomethane costs are approximately €60–110 per MWh. Learning effects and economies of scale are projected to reduce these costs by roughly 20%–30% in coming decades, moving towards parity in markets where carbon is priced (Prussi et al., 2019). Supportive policies will be critical: many countries are shifting from feed-in tariffs to market-based incentives, implementing renewable gas mandates and pricing carbon to internalise environmental benefits (Gustafsson and Anderberg, 2022). Integrating biogas within a broader circular bioeconomy, via nutrient recycling from digestate and coupling with hydrogen (power-to-gas), can enhance sustainability and create new revenue streams. Challenges remain: methane leakage must be minimised, and small or emerging projects often face financing and scalability barriers without adequate support, underscoring the need for continued research and innovation. The outlook is optimistic: if current momentum in technology, policy and investment endures, biogas/biomethane can become a cost-effective pillar of renewable and circular energy systems, transforming organic waste into clean energy and fertilisers at scale while contributing meaningfully to climate and sustainability goals.

Evidence indicates that mature biogas and biomethane pathways, supported by climate policy, energy-security objectives and circular-economy priorities, can increase the conversion of organic residues into renewable energy and bio-fertiliser while delivering net greenhouse-gas reductions (Angelidaki et al., 2018; Hagman et al., 2018; Balagurusamy and Chandel, 2021; Kougias and Angelidaki, 2018). Realisation of this potential depends on bankable project structures, reliable feedstock logistics, contracted offtake, grid and registry interoperability, and verifiable methane-emissions management. Heterogeneity across sites implies that outcomes remain context-specific, yet the direction of travel is consistent with cost convergence as systems scale and integrate.

Priority actions are threefold: (1) establishment of bankable revenue stacks (e.g., CfD (contracts for difference) + LCFS (Low Carbon Fuel Standard)/RIN (Renewable Identification Number) + GoO (Guarantees of Origin)) to reduce WACC (weighted average cost of capital) and stabilise DSCR (debt service coverage ratio) at financing thresholds; (2) system integration, including standardised grid access and interoperable, cross-border GoO registries to unlock liquidity and lower transaction costs; and (3) measurable sustainability, via robust MRV (monitoring, reporting and

Sustainable Biogas Production. DOI: https://doi.org/10.1016/B978-0-443-40557-0.00009-3

verification) and LDAR (leak detection and repair) and digestate valorisation to secure environmental credibility and eligibility for performance-based incentives. Progress should be tracked using consistent monetary and emissions metrics (e.g., €/MWh; € (2023 real terms); tCO_2e; methane intensity), enabling transparent comparison across technologies and jurisdictions.

10.1.1 Cost trajectories and key drivers

Current and projected costs: Biogas and biomethane production costs are expected to decline over the coming decades. Today, typical biomethane costs in many markets (e.g., Europe) range from approximately €60–110/MWh, depending on feedstock and plant scale (Prussi et al., 2019; Lawson et al., 2021). In some markets these levels are already below retail natural gas prices, yet they remain several times higher than the cost of extracting fossil gas at the wellhead (Prussi et al., 2019; Balussou et al., 2018). Learning effects and economies of scale are projected to lower levelised costs by roughly 20%–30% by 2030–2050 (Skovsgaard and Jacobsen, 2017; Prussi et al., 2019). European scenario analyses foresee average biomethane production costs of around €70/MWh by 2050 (Lapiso and Roubík, 2025; Uranga-Valencia et al., 2025). Industry evidence already shows pronounced scale economies: *large installations (> 1 200 Nm³/h output)* report approximately €50–60/MWh, nearly half the unit cost of smaller plants at approximately €85–90/MWh (Lawson et al., 2021; Skovsgaard and Jacobsen, 2017; Stürmer, 2017; Pantaleo et al., 2013). These trends suggest that biogas could approach cost parity with conventional natural gas within the next two decades (Balussou et al., 2018; Prussi et al., 2019). Table 10.1 summarises cost benchmarks in different contexts and the associated drivers.

Global outlook and regional variations: The global sustainable biogas potential is on the order of 1 trillion m³ per year, approximately 25% of current natural gas demand, yet utilisation remains limited outside Europe (Lapiso and Roubík, 2025; Uranga-Valencia et al., 2025; Gustafsson and Anderberg, 2022). Roughly 90% of this potential could be produced at $10–30/GJ (~€36–108/MWh), indicating that with available feedstocks and technologies most biomethane can ultimately be supplied well below €100/MWh (Lapiso and Roubík, 2025; Uranga-Valencia et al., 2025). Large low-cost resource pools exist in specific regions; for example, agricultural and waste residues in Asia could yield tens of billions of cubic metres at production costs below $10/GJ, making biomethane competitive with local natural gas in those markets (Lapiso and Roubík, 2025; Uranga-Valencia et al., 2025). In contrast, small-scale digesters common in rural developing areas have much higher unit costs because of rudimentary designs and lack of scale; many household or community units in Africa and South Asia, while vital for cooking fuel and sanitation, are not economically competitive without subsidies or credit for co-benefits, with delivered energy costs often >€100/MWh (Lapiso and Roubík, 2025). Targeted support and innovation focused on low-cost digesters suited to decentralised use are therefore required (Uranga-Valencia et al., 2025). Between 2030 and 2040, as technology transfer proceeds and best practices diffuse, projections converge on a global biomethane cost range of roughly €50–80/MWh under favourable conditions, namely

Table 10.1 Biomethane cost benchmarks, key drivers, and references by context.

Context	Typical biomethane cost	Key drivers/notes	References
Large-scale AD (≥ 1000 Nm³/h) - EU (current and 2030–2050 trajectory)	~ €50–70/MWh today; ~ €40–60/MWh with further scale and integration	Pronounced economies of scale in CAPEX and OPEX; optimised feedstock logistics and co-digestion; technology learning and integration with upgrading reduce unit costs over time.	Skovsgaard and Jacobsen (2017), Pantaleo et al. (2013), Prussi et al. (2019), Balagurusamy and Chandel (2021)
Medium-scale AD (~ 500 Nm³/h) - EU (2020s)	~ €80–100/MWh	Costs sensitive to feedstock price and characteristics; choice of upgrading route (membranes, PSA, amines, water scrubbing) influences OPEX/CAPEX; efficiency comparable to industry averages.	Prussi et al. (2019), Upadhyay et al. (2023)
Small farm-scale AD (< 100 Nm³/h)	> €100/MWh (case-dependent); often not competitive without support	High specific CAPEX and operating intensity at small scales; limited heat and by-product valorisation; viability strongly dependent on incentives or niche revenue streams.	Scotto di Perta et al. (2019), Baena-Moreno et al. (2020b), Pantaleo et al. (2013)
AD + ex-situ biological methanation (power-to-gas) - pilot to early commercial	Indicative techno-economic assessment (TEA) for Denmark (€ (2023 real terms)): production cost ~ €47/MWh (base case) and ~ €31/MWh (large-scale); minimum selling price higher (~ €43–66/MWh).	Economics are dominated by the delivered H_2 cost, reactor space-time productivity, and gas-liquid mass-transfer performance; strong scale effects; integration with existing AD sites can lower unit costs where surplus renewable electricity is available.	Lawson et al. (2021), Tsapekos et al. (2021), Yörüklü et al. (2025), Feickert Fenske et al. (2023)

(*Continued*)

Table 10.1 (Continued)

Context	Typical biomethane cost	Key drivers/notes	References
Thermal gasification + methanation (advanced route; 2050 estimates)	~ €50–70/MWh (literature projections at maturity)	Second-generation route using lignocellulosic feedstocks; costs expected to decline with technology learning and supply-chain maturation; competitiveness contingent on scale and syngas cleaning efficiency.	Angelidaki et al. (2019), Prussi et al. (2019)
Large projects in emerging economies (landfill/ manure)	~ €35–55/MWh (low-end case studies)	Low-cost waste feedstocks and labour can reduce unit costs, though capital constraints and financing terms may offset advantages; outcomes vary by project design and policy context.	Uranga-Valencia et al. (2025), Marendra et al. (2019), Begum et al. (2023)

Notes: Costs are expressed as €/MWh of product energy (LHV) unless stated otherwise. 1 Nm^3 CH_4 = 9.97 kWh (LHV) for €/Nm^3 ↔ €/MWh conversions. *AD*, anaerobic digestion; *CfD*, contracts for difference; DSCR debt-service coverage ratio; ERGaR European Renewable Gas Registry; GoO guarantees of origin; *LCFS*, Low-Carbon Fuel Standard; *LDAR*, leak detection and repair; *LHV*, lower heating value; *MRV*, measurement, reporting and verification; *MSP*, minimum selling price; *P2G*, power-to-gas; *RIN (D3)*, renewable identification number credits for cellulosic biofuel; *WACC*, weighted average cost of capital.

waste-based feedstocks and efficient large plants, bringing biomethane broadly into cost parity, particularly where fossil fuel prices remain high or carbon costs are internalised (Uranga-Valencia et al., 2025; Prussi et al., 2019, 2021).

Key cost drivers: Across contexts, a small set of fundamentals largely determines biogas economics. Feedstock supply and logistics typically account for the largest share of the levelised cost, frequently in the range of 40%–60% or more (Stürmer, 2017; Prussi et al., 2019). Low- or negative-cost substrates, for example organic wastes that attract gate fees or manure requiring mandated management, markedly improve project viability, whereas reliance on dedicated energy crops introduces cultivation and opportunity costs that raise unit costs (Stürmer, 2017; Hagman et al., 2018). Deployment is therefore expected to continue shifting from crop-based digestion toward residues and manure on both cost and sustainability grounds (Hagman et al., 2018; Lapiso and Roubík, 2025).

Scale and plant design are the second major lever. Larger facilities benefit from economies of scale in digester construction, upgrading, and operations; empirical analyses indicate that enlarging plant capacity can reduce specific capital cost on the order of 30% and materially lower levelised costs (Skovsgaard and Jacobsen, 2017; Pantaleo et al., 2013). In practice, small farm plants at around 100 Nm³/h often struggle to reach cost competitiveness without support, while centralised digesters at roughly 1 000 Nm³/h can approach 50–60 €/MWh under favourable feedstock and integration conditions (Scotto di Perta et al., 2019; Prussi et al., 2019; Lawson et al., 2021). This contrast underlines scaling-up, or hub-and-spoke models, as one of the most effective routes to cost reduction, a trend already visible in countries such as Germany and Denmark (Gustafsson and Anderberg, 2022; Balussou et al., 2018).

Technology and process optimisation provide additional, incremental gains. While core anaerobic digestion is mature, targeted enhancements, such as physical, thermal or enzymatic pre-treatment, and co-digestion of complementary substrates, can raise methane yield and process stability, thereby lowering cost per unit of energy (Tabatabaei et al., 2020; Koniuszewska et al., 2020; Westerholm et al., 2018). Better heat integration and rising levels of automation are further reducing operating expenses and improving reliability, particularly as plants scale and adopt real-time monitoring and control (Offie et al., 2023; Ananda et al., 2025). Biomethane upgrading has also advanced: modern membrane, pressure-swing adsorption and amine systems exhibit higher efficiencies and declining specific costs, reducing their contribution to total production cost in well-optimised, larger plants; where biological methanation is integrated, hydrogen price and gas–liquid mass-transfer performance dominate the economics (Angelidaki et al., 2018; Upadhyay et al., 2023; Lawson et al., 2021; Yörüklü et al., 2025). Overall cost trajectories reflect gradual, system-wide improvements rather than step-change breakthroughs, with well-optimised projects converging in the 50–80 €/MWh range as scale, integration and policy frameworks mature (Skovsgaard and Jacobsen, 2017; Balussou et al., 2018; Prussi et al., 2019).

10.1.2 Regulatory evolution and market design

Policy momentum. Through the 2020s, policy support accelerated as biogas came to be recognised as a multi-benefit solution spanning energy security, waste management, and climate mitigation. Since 2020, more than 50 biogas-specific measures have been enacted worldwide (Lapiso and Roubík, 2025). In the European Union, ambition rose with energy-security and climate targets: REPowerEU (2022) set a 35 billion m³ biomethane target for 2030, nearly a tenfold increase over 2020, backed by the Biomethane Industrial Partnership and embedded in binding proposals. The revised Renewable Energy Directive, as amended by Directive (EU) 2023/2413, supports waste- and residue-based advanced biofuels in transport through the 2030 transport framework: Member States must either achieve a 29% renewable energy share in transport or a 14.5% reduction in transport fuel greenhouse-gas intensity, together with a combined 5.5% sub-target for advanced biofuels and renewable fuels of non-biological origin. Member States are translating these aims into national strategies (e.g., France targeting ~10% "green gas" in grids by 2030; Denmark reaching ~40% biomethane in 2022 under strong support schemes) (Gustafsson and Anderberg, 2022). Beyond Europe, India's SATAT programme targets 5000 biogas plants and a 5% biomethane blend in CNG by

2028, while Brazil's RenovaBio scheme awards credits for biomethane use in transport (Lapiso and Roubík, 2025; Taron et al., 2023). This policy push aligns with broader methane-mitigation efforts: more than 100 countries have signed the Global Methane Pledge (2021), which recognises the need for waste-sector methane reductions in which biogas can play a central role (Angelidaki et al., 2018; Hagman et al., 2018; Kougias and Angelidaki, 2018).

From feed-in tariffs to market-based support. Early deployment, especially in Europe, was catalysed by feed-in tariffs (FiTs) that guaranteed above-market prices. Germany's Renewable Energy Act in the 2000s offered at least €0.18 kWh^{-1} for biogas power, enabling thousands of farm digesters (Gowreesunker and Tassou, 2016). As the sector matured, countries including Germany, Italy, the United Kingdom, and France shifted from open-ended FiTs to competitive auctions and feed-in premiums that expose producers to market prices (Schröer and Latacz-Lohmann, 2024). Germany now tenders new capacity under auction-based support schemes, with support levels expressed in €/kWh rather than kW/h, while the UK replaced the Renewable Heat Incentive with the levy-funded Green Gas Support Scheme. Carbon Contracts for Difference are under discussion to bridge the gap between market prices and biomethane costs by remunerating avoided CO_2 emissions on a contract basis, analogous to proposed support schemes for green hydrogen and carbon capture (Baena-Moreno et al., 2020a). Grants and tax credits remain important: in the United States, the Inflation Reduction Act expanded federal tax incentives relevant to biogas projects, including investment tax credits for qualifying biogas property and electricity production credits for eligible renewable electricity projects; transport-fuel monetisation is mainly linked to fuel-credit mechanisms and, where applicable, clean-fuel production credits rather than a single universal renewable-natural-gas credit (Uranga-Valencia et al., 2025). Support is shifting from uniform subsidies toward hybrid incentives that reward environmental attributes and reduce market barriers via mandates, GoO, and infrastructure funding, an efficient path to mainstreaming biomethane while accounting for its full societal benefits (Salvador et al., 2022; Baena-Moreno et al., 2020a). A concise overview of key instruments appears in Table 10.2.

Carbon pricing and low-carbon fuel standards. As summarised in Table 10.2, carbon pricing and transport fuel-crediting are potent drivers of deployment. The EU Emissions Trading System historically excluded diffuse agricultural and waste emissions mitigated by biogas, but reforms are expanding coverage; by 2027 the EU plans to extend carbon pricing to road transport and building fuels, effectively imposing a carbon cost on fossil gas while biogenic CO_2 remains uncounted (Gustafsson and Anderberg, 2022). The proposed Gas Decarbonisation Package could introduce a binding renewable-gas quota for the grid by 2030, with certification and GoO providing the accounting backbone (Stephan, 2013). Several Member States are advancing demand-pull instruments, for example, France's obligations on gas suppliers to purchase biomethane GoO, and Denmark's rapid market penetration supported by schemes that also emphasise methane-emission control, further accelerating uptake (Gustafsson and Anderberg, 2022).

Credit markets in North America. LCFS and RFS programmes have reshaped project economics by issuing tradable credits for vehicle-fuel use of biomethane

Table 10.2 Key regulatory instruments for biomethane and market integration.

Instrument/ policy	Description and examples	References
Feed-in tariffs (legacy)	Guarantee above-market prices for a fixed term; catalysed early biogas deployment under national frameworks (e.g., EEG in Germany; analogous schemes in Great Britain). As markets mature, FiTs are phased out in favour of market-oriented designs.	Gowreesunker and Tassou (2016), Balussou et al. (2018), Schröer and Latacz-Lohmann (2024)
Competitive auctions and feed-in premiums	Allocate capacity via tenders that set clearing prices; feed-in premiums top up wholesale revenues. Designs expose producers to market signals while retaining revenue stability, shaping post-FiT choices (e.g., upgrading vs. flexibilisation).	Schröer and Latacz-Lohmann (2024), Balussou et al. (2018), D'Alpaos (2017)
Renewable-gas targets and mandates	Set quotas for biomethane in grids or transport fuels to create assured demand and guide infrastructure build-out; EU-level and national targets emphasise waste- and residue-based pathways with traceability requirements.	Gustafsson and Anderberg, (2022), Prussi et al. (2021), Núñez (2021)
Carbon pricing and environ-mental credits	Monetise lifecycle GHG benefits through carbon pricing and transport-fuel crediting to narrow the cost gap to fossil gas; effectiveness depends on valuation, accounting rules, and de-risking contracts (including CfD-type approaches).	Schröer and Latacz-Lohmann (2024), D'Alpaos (2017), Baena-Moreno et al. (2020b), Brandão and Santos (2023)
Guarantees of origin and certification	Issue tradable certificates that document renewable-gas attributes and enable cross-border transfers; underpin green-gas offers, compliance markets, and recognition of environmental value at the point of use.	Stephan (2013), Prussi et al. (2021), Prussi et al. (2019)
Grid-access rules and infrastructure support	Harmonise gas-quality and interconnection rules, streamline permitting, and share/subsidise connection costs to reduce barriers to pipeline injection (e.g., France, Denmark), enabling biomethane to reach distant users.	Graßmann (2008), Prussi et al. (2019), Gustafsson and Anderberg (2022)

(Continued)

Table 10.2 (Continued)

Instrument/ policy	**Description and examples**	**References**
Waste-sector regulation and organics diversion	Use landfill restrictions, separate biowaste collection, and obligations to capture landfill/wastewater gas to increase feedstock availability and strengthen AD project pipelines.	Duque-Acevedo et al. (2023), Salvador et al. (2022), Núñez (2021)
Methane-emission monitoring and LDAR requirements	Set performance standards for measuring and mitigating plant emissions (e.g., pressure-relief valves, handling systems); national programmes demonstrate measurable reductions.	Scheutz and Fredenslund (2019), Reinelt and Liebetrau (2020), Fredenslund et al. (2023), Hrad et al. (2015), Groth et al. (2015)
Public–private partnerships and finance de-risking	Deploy PPP models, concessional finance, and risk-sharing mechanisms to improve bankability, especially in emerging markets; appraisal frameworks align incentives with market exposure.	Taron et al. (2023), D'Alpaos (2017), Brandão and Santos (2023)
Transport-sector instruments	Apply advanced biofuel quotas and sectoral certificate schemes to steer biomethane to vehicle-fuel markets; complement with fleet-oriented support (e.g., bio-CNG/bio-liquefied natural gas for buses and waste-collection vehicles).	Prussi et al. (2021, 2019), Gustafsson and Anderberg (2022)

Notes: Renewable identification number (cellulosic); bio-CNG/LNG: compressed/liquefied biomethane; EN 16723: European biomethane quality standard. Instruments often co-exist and may be combined in revenue stacks (e.g., CfD + LCFS/RIN + GoO).

based on lifecycle carbon intensity. Dairy-manure biomethane used as CNG in freight can generate high credit revenues because its lifecycle CI is strongly negative, reflecting avoided methane from storage (Scheutz and Fredenslund, 2019). Under LCFS and RFS programmes, credit values may exceed the energy commodity value of biomethane, especially for manure-derived pathways with strongly negative carbon intensities. However, actual revenues depend on current credit prices, pathway-specific carbon intensity values and regulatory eligibility. Landfill-gas- and wastewater-derived biomethane, while less carbon-negative, also capture significant premia, explaining the orientation of new projects toward vehicle-fuel markets.

European market instruments and trade policy. Beyond EU-wide GoO, Italy's *Certificati di Immissione in Consumo* create tradable certificates for advanced transport biofuels that biomethane producers can monetise, and voluntary "green gas" certificate markets are emerging as corporates decarbonise gas portfolios, often with premia for manure-based biomethane due to higher climate benefit

(Stephan, 2013; Gustafsson and Anderberg, 2022). Trade policy may add demand: measures such as carbon border adjustments can raise the value of low-carbon inputs in energy-intensive value chains, indirectly strengthening the case for renewable gas in industry (Brandão and Santos, 2023). Valuing biomethane's climate benefits, via carbon pricing, fuel credits, and certificates, has become central to bankability, helping to close the residual cost gap to fossil gas while aligning incentives with methane-mitigation and waste-management benefits (Baena-Moreno et al., 2020b; Gustafsson and Anderberg, 2022; Scheutz and Fredenslund, 2019).

Grid access and market integration. Ensuring non-discriminatory access to gas infrastructure remains a core regulatory task. Early facilities often relied on on-site combined heat and power because pipeline injection was technically and administratively cumbersome. Harmonised gas-quality standards for biomethane, EN 16723 for grid injection and vehicle fuel, together with streamlined connection/interconnection procedures have shifted practice; France simplified permitting/applications and subsidised connection costs, while Denmark partially socialises new biomethane grid hook-up costs across all gas consumers (Graßmann, 2008; Gustafsson and Anderberg, 2022; Prussi et al., 2021). France's MAPPED project demonstrated territory-scale digital coordination of biogas production and grid-injection management across multiple sites (Acarion et al., 2017). By 2022, Europe counted >1200 biomethane injection sites, more than twice the number only a few years earlier, enabling biomethane to be transported through and traded via the gas grid to reach distant users (Lapiso and Roubík, 2025).

Infrastructure constraints and cross-border architecture. Where pipeline networks are sparse (large parts of Asia, Africa, and Latin America), local utilisation for power or transport dominates until infrastructure expands; policies increasingly focus on practical interconnection standards and bankable offtake suited to decentralised contexts (Uranga-Valencia et al., 2025; Lapiso and Roubík, 2025; Taron et al., 2023).

Toward international trade. Liquefied biomethane (bio-liquefied natural gas (LNG)) and other logistics solutions can support flows between waste-rich regions and demand centres as standards and certification mature (Angelidaki et al., 2019; Prussi et al., 2019). A trajectory toward mainstream commodity status, transparent pricing, tradable certificates, and routine grid access, implies increasing exposure to normal market dynamics for producers, while regulators must ensure non-discriminatory interconnection and robust certification to scale biomethane from local niche to internationally integrated fuel (Stephan, 2013; Prussi et al., 2021).

Regulators are also building the market architecture for cross-border trade. The European Renewable Gas Registry is interlinking national registries to allow transfer of GoO across countries, advancing toward a pan-European renewable-gas market (Stephan, 2013; Prussi et al., 2021). Long-term offtake contracts are emerging, for example, utilities in the Netherlands and Denmark offering customers 100% renewable-gas supply by sourcing biomethane certificates. Over the longer term, international trade may develop via bio-LNG shipments or dedicated pipelines, mirroring LNG logistics where certification and tracking are robust. These developments signal maturation from local niche to mainstream commodity, with transparent pricing and wider distribution. For producers, this entails exposure to normal market dynamics, including price volatility and competition. For policymakers, it underscores the importance of guaranteed grid access and interoperable certification to scale biomethane efficiently (Stephan, 2013; Prussi et al., 2021; Schröer and Latacz-Lohmann, 2024).

10.1.3 Development opportunities in a circular bioeconomy

Digestate valorisation. Digestate, the nutrient-rich residue from anaerobic digestion, functions as a bio-fertiliser that can displace energy-intensive mineral inputs and close regional nutrient loops (Lubańska and Kazak, 2023). Under waste- and residue-focused scenarios, maximising biogas output and returning processed digestate to soils could meet a material share of agricultural nutrient demand, estimated at ~15% in the EU by 2050 and ~10% in India when deployed systemically (Lapiso and Roubík, 2025).

Pathways to products. Solid–liquid separation concentrates phosphorus and organic matter in the solid fraction, enabling composting, drying, or pelletisation into soil-amendment products, while the liquid fraction, rich in ammoniacal nitrogen and potassium, lends itself to ammonia stripping, membrane concentration, or evaporation to yield marketable fertiliser solutions or recover ammonium salts. Phosphorus can be precipitated from digestate as struvite, producing storable fertiliser granules; several biogas-based biorefinery concepts already integrate such nutrient-recovery steps alongside energy generation (Valentino et al., 2021; Salvador et al., 2022).

Process intensification. Tailored pretreatments and co-digestion strategies that stabilise operation and raise methane yields indirectly improve fertiliser economics by increasing overall plant throughput and enabling more consistent digestate quality (Tabatabaei et al., 2020; Koniuszewska et al., 2020; Hagman et al., 2018). Bio-augmented ensiling of sweet sorghum has been evaluated as a pretreatment, with measured effects on methane production during anaerobic digestion (Dong et al., 2025).

Policy alignment. Policies increasingly recognise digestate as a tradable fertilising product rather than waste, reducing market frictions and facilitating cross-border use within regional bioeconomy strategies (Salvador et al., 2022; Lubańska and Kazak, 2023). Positioning digestate as a co-product creates additional revenue streams for plant operators, strengthens links with local farming systems, and improves public acceptance by coupling renewable energy with soil health and nutrient-recycling outcomes (Valentino et al., 2021; Lubańska and Kazak, 2023; Duque-Acevedo et al., 2023).

Integration with other systems. Modern biogas plants are increasingly embedded in broader biorefinery and waste-management value chains, evolving from single-output energy facilities into multi-product resource-recovery hubs (Valentino et al., 2021). A prominent synergy couples anaerobic digestion with algal cultivation: CO_2-rich off-gas from upgrading and nutrient-laden liquid digestate are routed to ponds or photobioreactors, where microalgae convert CO_2 and nutrients into biomass for animal feed, biofuels, or other high-value chemicals. This approach captures biogenic carbon, polishes the liquid fraction, and yields protein-rich coproducts that may substitute soy or fishmeal; pilot projects have demonstrated technical feasibility in integrated biorefinery settings (Kougias and Angelidaki, 2018; Angelidaki et al., 2018). Biogas upgrading can also yield a purified biogenic CO_2 by-product. Rather than venting or flaring the ~40% CO_2 in raw biogas, operators can capture it during upgrading for food-grade uses (beverages, food processing, cooling), Power-to-X routes with green H_2, and mineralisation uses such as concrete curing (Angelidaki et al., 2018, 2019; Yörüklü et al., 2025). Where biological methanation is integrated, the captured CO_2 is directly converted to CH_4 on-site using renewable H_2; economics are driven by hydrogen cost and gas-liquid mass-transfer performance (Tsapekos et al., 2021; Feickert Fenske et al., 2023; Lawson et al., 2021). Techno-economic analysis also shows that integrating biogas upgrading with co-production of biomethane and fertiliser-grade urea can be feasible under suitable conditions (Baena-Moreno et al., 2020b).

Utilities and wastewater. Integrating biogas systems with municipal wastewater treatment is a mature opportunity: many plants already anaerobically digest sewage sludge; co-digesting food waste and other organics can turn facilities into net energy producers while delivering stabilised biosolids and improved nutrient management (Begum et al., 2023; Saikat et al., 2023). Utilities are deploying advanced monitoring, digital optimisation, and model-based control to raise energy circularity and reduce emissions, supporting the transition to energy-positive water resource recovery facilities (Farzin and Sadri Moghaddam, 2025).

Decentralised scale. As smaller-scale solutions in the Global South, household and community digesters convert kitchen scraps and manure into clean cooking gas and nutrient-rich slurry for fields, reducing reliance on fuelwood, improving sanitation, and enhancing soil fertility; these local co-benefits are central to their value proposition (Lapiso and Roubík, 2025).

Policy and products. Policy frameworks are beginning to recognise the multi-functionality of biogas systems. Beyond energy support, instruments that value waste-treatment services and nutrient recycling, such as gate-fee mechanisms, materials-market access, and public-private partnerships, improve bankability and align incentives across sectors (Salvador et al., 2022; Taron et al., 2023; Duque-Acevedo et al., 2023). Continued innovation is expanding the product slate: lines of work include inducing production of bioplastics precursors (e.g., polyhydroxyalkanoates) within biorefinery platforms and coupling digestion with pyrolysis of digestate to produce biochar for carbon sequestration and soil enhancement while recovering energy (Valentino et al., 2021; Nkuna et al., 2024; Duque-Acevedo et al., 2023). Taken together, well-integrated biogas systems contribute simultaneously to clean energy, climate action, sustainable cities, soil restoration, and public health, making them attractive for cross-sector finance and regional circular-economy strategies (Brandão and Santos, 2023; Salvador et al., 2022; Lubańska and Kazak, 2023).

10.1.4 Power-to-gas (bio-power-to-gas) integration

Concept and routes. Integrating biogas with variable renewables links surplus electricity to methane synthesis. In a Bio-P2G configuration, electrolysis produces hydrogen that is combined with the CO_2 fraction of raw or upgraded biogas to form additional CH_4 via the Sabatier reaction, increasing methane output without additional primary feedstock and providing long-duration energy storage through the gas grid (Angelidaki et al., 2018, 2019; Yörüklü et al., 2025). Two implementation routes dominate: catalytic methanation in fixed-bed or similar reactors at elevated temperature with purified gas streams, and biological methanation using hydrogenotrophic consortia either in dedicated ex-situ reactors (e.g., trickle-bed systems) or by in-situ H_2 dosing to the digester headspace (Angelidaki et al., 2019; Yörüklü et al., 2025; Tsapekos et al., 2021; Shrestha et al., 2025). European pilots have demonstrated efficient CO_2-to-CH_4 conversion and stable operation in trickle-bed configurations, while reactor hydrodynamics and gas-liquid mass transfer remain key scale-up levers (Tsapekos et al., 2021; Feickert Fenske et al., 2023).

Economics and flexibility. Where low-cost surplus power is available, biological methanation can approach levelised costs of ~€60–80/MWh by 2030, converging toward ~€50–60/MWh as systems scale and integrate by mid-century; outcomes are driven chiefly by H_2 price, reactor productivity, and integration with existing AD infrastructure (Lawson et al., 2021; Yörüklü et al., 2025; Prussi et al., 2019). The attraction is threefold: higher total methane yield by converting biogenic CO_2, seasonal storage via existing gas networks, and a closed-carbon loop aligned with deep-renewables power systems (Angelidaki et al., 2018; Yörüklü et al., 2025). A complementary pathway reforms biogas or biomethane to renewable hydrogen with CO_2 capture, leveraging established gas-to-H_2 process trains to supply low-carbon H_2 for fuel cells or industry; integration choices hinge on local H_2 demand, policy incentives, and heat/CO_2 valorisation options (Angelidaki et al., 2019; Yörüklü et al., 2025). As operational complexity rises, advanced monitoring and control can support flexible, multi-output hubs that switch between biomethane injection, P2G operation, and H_2 production in response to market signals and grid conditions (Offie et al., 2023; Kunatsa, 2024; Westerholm et al., 2018; Ananda et al., 2025).

10.1.5 Bio-liquefied natural gas and transport fuels

Upgrading biogas to vehicle-grade biomethane enables use as compressed bio-CNG for urban buses and trucks or as liquefied bio-LNG for long-haul road freight and maritime applications, targeting segments that are difficult to electrify (Prussi et al., 2021; Angelidaki et al., 2018; Prussi et al., 2019). When produced from wastes and residues, lifecycle greenhouse-gas intensities are markedly lower than for fossil fuels, and the near-absence of sulphur and aromatics delivers air-quality co-benefits; these attributes underpin growing interest from fleet operators and ports (Angelidaki et al., 2018, 2018; Prussi et al., 2021; Kougias and Angelidaki, 2018). Infrastructure and certification are advancing in tandem. Europe's expanding CNG/LNG refuelling network and emerging bunkering capability provide physical access for biomethane, while GoO and related certification schemes allow suppliers and end-users to claim renewable content at the point of use or along supply chains (Prussi et al., 2021; Stephan, 2013). Policy instruments, advanced biofuel quotas, transport-sector targets, and market-based incentives, are steering biomethane into heavy-duty road and, increasingly, maritime niches, with national frameworks aligning deployment with energy-security and climate objectives (Gustafsson and Anderberg, 2022; Prussi et al., 2019, 2021). Cost competitiveness depends on feedstock, upgrading route, and the additional steps required to liquefy methane for bio-LNG. Learning and scale effects in biomethane upgrading, together with supportive market design, narrow the cost gap to fossil fuels; crediting mechanisms and targeted support further improve bankability where needed (Upadhyay et al., 2023; Balussou et al., 2018; Schröer and Latacz-Lohmann, 2024; Baena-Moreno et al., 2020b). Taken together, the combination of maturing infrastructure, credible certification, and focused sectoral policy creates a practical pathway for biomethane to decarbonise hard-to-electrify transport segments at scale (Prussi et al., 2021; Gustafsson and Anderberg, 2022).

10.1.6 Methane emissions: compliance and mitigation

Controlling methane leakage is pivotal to the climate credibility of biogas systems. Methane is a potent greenhouse gas, so relatively small loss rates can erode or negate benefits derived from avoided emissions at source and displacement of fossil fuels. Earlier investigations reported non-trivial emissions at some facilities, particularly older or poorly maintained plants (Groth et al., 2015; Reinelt and Liebetrau, 2020). A survey of 23 biogas plants by Scheutz and Fredenslund, (2019) reported facility-level loss rates historically ranged from <1% (best practice) to ~15% (worst cases), with an average ~4%–5% in earlier surveys; large agricultural plants clustered near ~2%–3%, wastewater plants often higher (~7%–8%). Modern designs and LDAR bring routine losses <1%–2% under well-managed conditions (Scheutz and Fredenslund, 2019; Reinelt and Liebetrau, 2020; Fredenslund et al., 2023). Routine emissions have fallen at newer installations owing to improved designs, for example enclosed digestate storage and double-membrane covers, yet fugitive losses can still arise at seals, flanges, and pipework, pressure-relief valves, digestate pits, engine exhausts (unburnt CH_4), flares and during digestate handling (Reinelt and Liebetrau, 2020; Groth et al., 2015; Hrad et al., 2015).

Regulatory and industry responses increasingly treat methane control as a compliance requirement. National programmes in countries such as Denmark and Germany have systematised measurement and mitigation at plant level (Fredenslund et al., 2023). Denmark's voluntary programme used infrared cameras and tracer techniques to pinpoint leaks and prioritise repairs, achieving material emission reductions (Fredenslund et al., 2023). The European Union is finalising a Methane Regulation that will mandate leak detection and repair at biogas facilities, improve monitoring practice and set performance standards for components including pressure relief valves (Prussi et al., 2021). These measures make low methane loss a condition for operation and reinforce the sector's climate contribution.

Operational mitigation focuses on high-impact sources. Evidence indicates that a limited number of components or practices, for example open digestate lagoons or malfunctioning valves, often dominate site-wide emissions (Reinelt and Liebetrau, 2020; Scheutz and Fredenslund, 2019). Targeted interventions, covering and, where feasible, post-digesting digestate storages, installing secondary containment for pressure relief valves, ensuring engines are fitted with effective oxidation catalysts, maintaining seal integrity and improving flare systems for emergency venting, substantially reduce losses (Reinelt and Liebetrau, 2020). Continuous or high-frequency monitoring, including perimetre sensors and drone-based surveys, is being deployed to identify and correct anomalies rapidly (Hrad et al., 2015). As these practices diffuse, overall losses can be driven below 1% of production under well-managed conditions (Reinelt and Liebetrau, 2020).

Compliance with stringent emissions standards is set to become part of best practice and a prerequisite for market access. When losses are minimised, for example below 1%–2%, biogas systems deliver very high net greenhouse-gas savings, often 70%–200% relative to fossil comparators once avoided methane from wastes is accounted for (Scheutz and Fredenslund, 2019). The combination of tighter regulation, better monitoring technologies and proactive industry programmes is expected to push methane losses to minimal levels by 2030, safeguarding climate integrity while capturing more product value for operators (Reinelt and Liebetrau, 2020; Fredenslund et al., 2023).

10.1.7 Risks, barriers, and financing

Despite a positive outlook, several risks and barriers could hinder biogas and biomethane deployment if not addressed. Policy and market uncertainty is a primary concern. Projects are capital-intensive with lifetimes of 15–20 years or longer, so abrupt changes in tariffs, subsidies or renewable-gas targets can undermine investment viability (D'Alpaos, 2017; Balussou et al., 2018). Boom-bust cycles observed in countries such as Italy and the United Kingdom were driven by policy shifts that eroded investor confidence. Stable, long-term signals of the type remain essential for financing.

Economic scalability and feedstock supply represent a second class of risk, particularly for smaller projects. Many agricultural plants have struggled when feedstock availability or support payments changed (Stürmer, 2017). Investing in a digester materially alters a farm's risk profile, diversifying in some cases while adding operating and market risks in others (Kellner et al., 2012). High feedstock costs or scarcity can quickly make anaerobic digestion uneconomic (Pantaleo et al., 2013). The absence of economies of scale at small size leads to higher specific costs, and without cooperative models or consolidation these operators find it difficult to compete (Skovsgaard and Jacobsen, 2017). Analyses from southern Spain highlight that small biomethane plants were not profitable without subsidies, with government incentives decisive for viability in that context (Baena-Moreno et al., 2020a). Financing such plants is challenging unless innovative ownership and aggregation approaches are used, for example community models, agri-cooperatives or bundling multiple sites for investment (Csiba and Szigeti, 2023).

Administrative and technical barriers also constrain deployment. Permitting and compliance can be lengthy in jurisdictions without streamlined procedures; environmental impact assessments, grid-connection approvals and stringent waste-handling rules can delay projects and deter capital (Taron et al., 2023). In some developing contexts, the absence of clear energy regulations for biogas adds uncertainty. Operational expertise is another constraint: successful plants require microbiological, engineering and process-control capabilities, and skills shortages have contributed to suboptimal performance or early failures in certain programmes (Uranga-Valencia et al., 2025). Feedstock logistics pose additional challenges, as collecting and transporting bulky wastes can erode scale benefits when catchment areas become too large, creating a sizing trade-off for centralised plants (Skovsgaard and Jacobsen, 2017). Building reliable supply networks requires coordination across many farms, an economic and organisational hurdle beyond technical feasibility (Guenther-Lübbers and Theuvsen, 2015). Public acceptance can further limit projects where odour or traffic are concerns; transparent engagement and benefit-sharing, for example offering local heat or bio-fertiliser, help secure local support (Gustafsson and Anderberg, 2022).

Financing models and solutions: Innovative financing and business structures are emerging to underpin expansion. Aggregation and portfolio approaches bundle multiple small projects to reach investor-relevant scale; cooperative models in countries such as Denmark and Germany pool feedstock from many farms into larger digesters, sharing costs and revenues and capturing economies of scale (Gowreesunker and Tassou, 2016; Gustafsson and Anderberg, 2022). Public–private partnerships are gaining traction for municipal solid waste and wastewater applications, where cities contribute guaranteed waste streams or offtake for power or vehicle fuel and private partners provide capital and operational expertise; concessional loans and results-based finance from development institutions support such structures in the Global South (Taron et al., 2023; Lapiso and Roubík, 2025).

The Energy Service Company model lowers entry barriers for feedstock owners by shifting capex and performance risk to a specialist operator who sells energy or waste-treatment services under contract. This aligns incentives for professional operation and continuous optimisation, including digital monitoring and control where appropriate (Csiba and Szigeti, 2023; Kunatsa, 2024).

Green finance instruments broaden the investor base. Green bonds and sustainability-linked loans are increasingly used for circular-bioeconomy infrastructure portfolios, while carbon-credit revenues from compliance or voluntary markets can enhance bankability and strengthen cash-flow cover where policy frameworks recognise avoided emissions and environmental attributes (Brandão and Santos, 2023; Taron et al., 2023; Baena-Moreno et al., 2020a). Robust feasibility assessment and risk allocation remain central: long-term feedstock agreements, contracted offtake, and clear exposure to policy and market prices materially affect project value (D'Alpaos, 2017; Kellner et al., 2012; Pantaleo et al., 2013). Financing assumptions should be reported explicitly as WACC (pre-tax, nominal) and target DSCR, with key sensitivities indicated.

Soft-cost reductions complement financial engineering. Streamlined, one-stop permitting and standardised templates for feedstock supply and energy or certificate sales can cut development timelines and transaction costs, while cooperative and hub-and-spoke logistics help reconcile scale with local feedstock realities (Csiba and Szigeti, 2023; Salvador et al., 2022; Gustafsson and Anderberg, 2022; Stürmer, 2017). Aligning projects with broader waste-management modernisation, rural development and climate strategies opens additional funding channels and stakeholder support, helping to move first-of-a-kind and small projects through the development bottleneck and into stable operation (Salvador et al., 2022; Taron et al., 2023; Lapiso and Roubík, 2025).

10.1.8 Research and innovation agenda (2030–2050)

Sustaining progress in efficient biogas production requires a coordinated research and innovation agenda that spans microbiology, pretreatment, reactor and biomethane upgrading, digital optimisation, emissions monitoring and circular biorefinery integration.

Microbial and feedstock innovations: The microbiome remains a primary lever for yield and stability. Work on engineering or tailoring consortia, together with refined process ecology, is expected to improve resilience to inhibitors and broaden the usable feedstock spectrum (Tabatabaei et al., 2020; Agustini and Gutterres, 2017). Bioaugmentation is advancing through both materials science and microbial selection. Studies using natural carrier materials to immobilise beneficial communities show promise for methane enhancement and operational robustness (Dadic et al., 2024), while metagenomics-guided enrichment of lignin-degrading consortia targets recalcitrant biomass streams (Ozsefil et al., 2024). By 2030, pilot-scale validation of specifically cultured or modified microbial consortia that tolerate inhibitors or accelerate lignocellulose breakdown is plausible. Feedstock pretreatments, including thermal, enzymatic and physical methods, require gains in energy efficiency and cost performance through better heat integration and targeted enzyme cocktails (Koniuszewska et al., 2020; Kemka et al., 2024). Progress in low-cost routes for agricultural residues could materially expand the feedstock base, consistent with circular-waste strategies (Uranga-Valencia et al., 2025). Temperature-adaptation management of communities offers another optimisation axis for stability and throughput (Westerholm et al., 2018). As an illustration of the potential step change, mild thermo-mechanical processes that double methane yield from straw at reasonable cost would transform farm-scale digestion economics.

Process and reactor intensification, upgrading and power-to-methane: Biological methanation and tighter integration of upgrading with digestion are central innovation areas. Pilot and demonstration studies report efficient CO_2-to-CH_4 conversion in ex-situ reactors such as trickle-bed systems, highlighting the importance of hydrodynamics and gas–liquid mass transfer for scale-up (Tsapekos et al., 2021; Feickert Fenske et al., 2023). Techno-economic assessments of biological upgrading indicate competitive trajectories where low-cost renewable electricity and system integration are available, with performance and cost governed by hydrogen price, reactor productivity, and site integration (Lawson et al., 2021; Yörüklü et al., 2025; Shrestha et al., 2025). Continuous improvement in conventional upgrading, membranes, PSA and amines, remains essential for cost and efficiency gains, including hybrid and biocatalytic options (Angelidaki et al., 2018, 2019; Upadhyay et al., 2023).

Digitalisation, automation and AI: Data-driven operation can reduce OPEX, raise uptime and support flexible, grid-responsive running. AI-enabled frameworks for micro- and full-scale plants demonstrate potential for predictive control, anomaly detection and feedstock scheduling, and can optimise energy circularity in water resource recovery facilities (Offie et al., 2023; Farzin and Sadri Moghaddam, 2025; Ananda et al., 2025). Toolkits that connect supervisory control with process models and market signals are an emerging focus for plant-to-system integration, including bio-P2G dispatch (Jain et al., 2025; Kunatsa, 2024).

Emissions monitoring and mitigation: Method development for accurate, cost-effective methane monitoring remains a priority. Facility-scale campaigns using laser absorption spectroscopy and tracer dispersion methods quantify losses and guide mitigation planning (Groth et al., 2015; Gao et al., 2009; Hrad et al., 2015). Evidence on dominant emission points, including pressure-relief valves and storage handling, underpins targeted LDAR and component standards (Reinelt and Liebetrau, 2020; Scheutz and Fredenslund, 2019). National programmes show that systematic detection and repair can deliver measurable reductions, reinforcing climate integrity as plants scale (Fredenslund et al., 2023).

Circular biorefinery products and nutrient recovery: Valorisation pathways that turn digestate into marketable fertilisers and materials strengthen economics and regional nutrient cycles. Platforms that integrate solid–liquid separation, ammonia recovery and phosphorus precipitation, for example struvite, align with urban waste-to-product concepts and enable new revenue streams (Valentino et al., 2021; Salvador et al., 2022). Broader biorefinery concepts, including materials and carbon products such as biochar from coupled thermochemical steps, can complement energy outputs where systems are designed for multi-product optimisation (Duque-Acevedo et al., 2023; Nkuna et al., 2024; Lubańska and Kazak, 2023).

Socio-technical integration and finance: Research on business models, risk allocation and policy design is needed to accelerate bankability while safeguarding performance. Public-private partnership structures, portfolio financing and certification-linked revenue models interact with social acceptance and regional development goals in ways that merit rigorous evaluation (Taron et al., 2023; Brandão and Santos, 2023; D'Alpaos, 2017). Priority areas include standardising contracts for feedstock and offtake, integrating environmental attributes into finance and ensuring MRV frameworks that support trust in low-carbon gas markets.

A coordinated agenda across these fronts can deliver higher yields, lower costs, robust emissions performance and diversified product slates, positioning biogas systems as adaptable infrastructure within a circular bioeconomy through 2030–2050.

Digitalisation and process optimisation: Digital transformation can lift performance beyond what basic control loops deliver. Advanced sensing, automation and AI-driven control offer predictive insights and earlier interventions, enabling operators to stabilise biology and optimise yields (Kunatsa, 2024; Ananda et al., 2025). Real-time tools, online gas analysers, volatile fatty acid monitoring and, potentially, on-site genomic assays for microbial health, combined with machine-learning models can flag incipient instability and recommend feed or mixing adjustments. In a micro-scale demonstration, an AI-based optimisation framework increased biogas generation by 43% through operational tuning, illustrating the headroom available from data-driven control (Offie et al., 2023). Scaling such approaches to commercial plants is expected to deliver incremental gains, higher output and availability, lower specific OPEX, without major capital additions (Farzin and Sadri Moghaddam, 2025; Jain et al., 2025; Ananda et al., 2025). Priorities include user-centred decision-support systems integrated with SCADA, remote O&M based on digital twins for fleet operators, and flexible dispatch concepts that align plant operation with power and gas markets, as explored under Germany's "flexibilisation" pathway (Schröer and Latacz-Lohmann, 2024; Offie et al., 2023; Kunatsa, 2024). As connectivity rises, cyber-physical security and robust fail-safe design will be integral to best practice.

Power-to-X and biochemical pathways: Power-to-gas integration links surplus renewable electricity to methane synthesis by supplying electrolytic H_2 to convert biogenic CO_2 into additional CH_4, expanding output without more primary feedstock and enabling long-duration storage via gas infrastructure (Angelidaki et al., 2018, 2019; Yörüklü et al., 2025). Two routes dominate: catalytic methanation in high-temperature reactors with purified gas streams, and biological methanation using hydrogenotrophic consortia either ex-situ, for example trickle-bed reactors, or in-situ via H_2 dosing (Angelidaki et al., 2019; Tsapekos et al., 2021; Yörüklü et al., 2025). European pilots demonstrate efficient CO_2-to-CH_4 conversion, while reactor hydrodynamics and gas-liquid mass transfer are key scale-up levers (Tsapekos et al., 2021; Feickert Fenske et al., 2023). Techno-economic assessments point to competitive ranges where low-cost renewable electricity and tight integration are available, with costs governed chiefly by H_2 price, reactor productivity and site configuration (Lawson et al., 2021; Yörüklü et al., 2025). Parallel work on catalytic systems targets impurity tolerance and dynamic operation to follow variable power (Angelidaki et al., 2019). Additional power-to-X options use biogas as a feedstock for other products, including reforming to renewable hydrogen with CO_2 capture, which leverages established gas-to-hydrogen process trains while substituting biogenic carbon flows (Angelidaki et al., 2019; Yörüklü et al., 2025). For example, case studies in Spain have explored coupling waste-derived biomethane production with hydrogen storage to enhance renewable fuel supply flexibility (Pous de la Flor et al., 2025). R&D priorities include modular reformers compatible with biogas compositions, improved sorbents and membranes for CO_2 capture, and control strategies that allow plants to switch between biomethane injection, P2G operation and H_2 production in response to market signals (Kunatsa, 2024; Offie et al., 2023; Ananda et al., 2025).

Materials and environmental research: Cost and reliability gains are achievable through materials innovation in core plant components. Work on more durable, lower-cost digester construction, such as concrete additives or liners that resist biogenic corrosion, can cut maintenance and extend overhaul intervals. Upgrading trains remain a focal point: improved membrane materials and solvent systems that deliver higher CH_4 purity in a single pass or lower regeneration energy would reduce specific upgrading costs (Upadhyay et al., 2023). Environmental priorities include longitudinal studies on repeated digestate applications across soil types to verify agronomic benefits and identify risks where feedstocks carry contaminants, for example microplastics or heavy metals. Facility design also merits attention. Compact layouts for urban settings, high-integrity covers for digestate storage, and odour-minimising conveyance can shrink emissions footprints and improve social acceptance. Social-science work is integral to deployment: research on integrating biogas into farming systems, structuring incentives for feedstock cooperation and communicating co-benefits to communities can inform practical programmes and permitting strategies (Salvador et al., 2022; Gustafsson and Anderberg, 2022).

Explicit circular-economy frameworks: A recognised gap is the lack of metrics that fully value biogas systems' multi-output services. Assessment tools that combine life-cycle analysis with economic modelling are needed by 2030 to quantify energy, waste-treatment, fertiliser, and greenhouse-gas benefits in a unified way, aligning support instruments with delivered system value (Lapiso and Roubík, 2025; Salvador et al., 2022). Business-model research should translate that accounting into bankable revenue streams, for example markets for carbon-negative biofertilisers derived from digestate and payments for ecosystem services such as avoided methane from improved manure management (Csiba and Szigeti, 2023). Framing biogas within explicit circular metrics can guide procurement, certification, and finance, ensuring policies reward the complete set of societal benefits rather than kilowatt-hours alone.

By 2040–2050, the innovation agenda encompasses high-impact, sometimes speculative, directions. Synthetic biology could enable bespoke consortia or engineered enzymes able to depolymerise recalcitrant biomass streams end-to-end. Automation and robotics may reduce labour and variability across preprocessing and feeding, for example autonomous collection of farm residues and automated dosing into digesters. Virtual power plant concepts are likely to mature, with biogas engines, batteries and other flexible assets dispatched in concert so that biogas units ramp strategically to balance variable wind and solar (Skovsgaard and Jacobsen, 2017). Achieving net-zero lifecycle emissions by mid-century will be a central theme, encompassing near-zero methane loss (Fredenslund et al., 2023) and reductions in ancillary CO_2 from transport and on-site electricity use. Research on electrifying plant operations with renewable power and on low-carbon drying or dewatering of digestate can contribute to this goal. Through the 2030 s and 2040 s, biogas is expected to embed within smart, circular regional systems; work undertaken in the present decade lays the foundation for this integration.

10.1.9 Synthesis and actionable recommendations

Synthesis: The outlook for efficient biogas production is strong, contingent on consolidating recent gains and addressing persistent hurdles. Technologically, the sector is shifting from single-purpose energy production to multi-output circular platforms that deliver renewable energy, nutrient recycling and emissions reduction. Costs are trending downward; learning and scale effects are expected to position typical biomethane at roughly €50–70/MWh in the 2030 s, particularly in leading regions (Prussi et al., 2019; Skovsgaard and Jacobsen, 2017). Policy momentum is robust, with ambitious targets and evolving market frameworks, blending mandates, carbon pricing and certification, creating favourable conditions, while policy stability and coherence remain essential for investor confidence (Gustafsson and Anderberg, 2022; Schröer and Latacz-Lohmann, 2024). Integration into energy markets and infrastructure is accelerating as biomethane injection and cross-border trading expand, setting the stage for mainstream commodity status. Coupling with waste management and agriculture amplifies sustainability outcomes by closing nutrient cycles and reducing local pollution (Lubańska and Kazak, 2023; Duque-Acevedo et al., 2023). To realise this trajectory through 2030–2050, coordinated effort across technology, policy, finance and systems integration is required.

Actionable recommendations: Based on the analysis in this chapter, the following recommendations are put forward for policymakers, industry practitioners and researchers:

Scale up and optimise projects. Promote larger, efficient installations and cooperative models; aggregate feedstocks (co-ops, hub-and-spoke) to capture economies of scale; fund audits to raise yields (mixing, heat integration). Support optimisation at existing plants through training and technical audits that identify efficiency gains, including improved mixing and heat integration.

Ensure stable, value-oriented policy support: Maintain consistent long-term policies while shifting incentives to reward environmental value. Provide clear targets and stable support, for example contracts or guarantees of at least 10 years, to mitigate investment risk (D'Alpaos, 2017; Balussou et al., 2018). Reorient subsidies toward mechanisms such as carbon payments, renewable-gas quotas and credits that internalise climate and waste-management benefits (Scheutz and Fredenslund, 2019; Baena-Moreno et al., 2020a).

Facilitate market integration and infrastructure: Invest in the infrastructure and regulatory frameworks required for biomethane market growth. Priorities include funding injection points in underserved areas, harmonising gas-quality standards internationally and expanding trading platforms for GoO (Stephan, 2013). Treat renewable gas on fair terms in network access and create conditions for cross-border trading of biomethane volumes.

Expand the circular-economy scope: Maximise co-product use. Develop markets and standards for digestate-based fertilisers and soil amendments, including recognition under fertiliser regulations (Duque-Acevedo et al., 2023; Valentino et al., 2021). Promote pilots that integrate biogas with other valorisation routes, for example algae cultivation or CO_2 utilisation, and disseminate successful models. Provide incentives or recognition for plants that deliver multiple services, namely energy, waste treatment, and nutrient recycling.

Tighten methane-emission controls: Implement rigorous monitoring and mitigation. Enact leak-detection and repair requirements for all biogas and biomethane facilities, and adopt best practice for gas containment (Reinelt and Liebetrau, 2020; Fredenslund et al., 2023). Prioritise R&D on low-cost continuous monitoring, for example sensor networks and laser detection, to enable near real-time management. Aim for an industry standard of less than 1% methane loss by 2030.

Address financing and investment barriers: Innovate financing to support small and emerging projects. Expand green credit lines or guarantee schemes through development banks or green funds, support public–private partnerships in waste-to-energy and facilitate community investment via cooperatives or municipal bonds (Taron et al., 2023; Guenther-Lübbers and Theuvsen, 2015). Integrate biogas into agricultural and rural-development funding programmes given its cross-sector role.

Advance research and training: Invest in R&D and human capital for next-generation systems. Align public and private funding with the innovation agenda outlined in this chapter, including microbial enhancements, power-to-gas and AI-enabled optimisation (Tabatabaei et al., 2020; Offie et al., 2023). Expand training and knowledge-transfer programmes to build a skilled workforce, for example certification of plant operators and technical assistance for new projects, especially in the Global South (Lapiso and Roubík, 2025; Uranga-Valencia et al., 2025). Encourage international cooperation, for example, through IEA Bioenergy and the World Biogas Association, to accelerate diffusion of best practices (Uranga-Valencia et al., 2025; Brandão and Santos, 2023).

Holistic planning and integration: Embed biogas within energy-system and circular-economy strategies. Integrate biogas into electricity and gas network planning for balancing, storage and firm supply roles, and into climate strategies as a source of negative emissions through methane mitigation. Align waste-management policy to prioritise anaerobic digestion as a primary route for organics rather than landfilling or incineration (Salvador et al., 2022). Treat biogas as a multi-faceted asset so that investments and policies capture its full range of benefits.

Efficient biogas production sits at the intersection of renewable energy, waste reduction and sustainable agriculture, and is a core element of the emerging circular bioeconomy. The period 2030–2050 will determine whether the sector scales from a modest contributor to a material component of decarbonisation and resource efficiency. Earlier national analyses anticipated major biogas expansions, for example, a 2011 study projected substantial biogas-based power generation in Austria by 2020, reflecting early expectations for the sector's growth (Stürmer and Schmid, 2011). Evidence assembled in this chapter indicates steady progress: costs trend down, supportive policies take effect, and innovation widens viable pathways. Delivery now rests with industry, policymakers, researchers and communities to scale deployment responsibly and in local context. Momentum will depend on coordinated action to address persistent barriers.

Implementation priorities are clear. Scale up prudently through aggregation and high-efficiency plants; align policy and markets to reward verified environmental value; valorise co-products such as digestate; minimise methane emissions through rigorous monitoring and leak abatement; expand fit-for-purpose financing; and invest in research, digitalisation and workforce training. If these measures are delivered, biogas and biomethane can by 2050 be widely deployed, converting millions of tonnes of organic residues into clean energy and valuable products, with measurable contributions to climate targets, energy access and sustainable development. This trajectory is consistent with the regenerative, circular economy examined in this book.

AI disclosure

During the preparation of this work the author(s) used DEEPL, ChatGPT, Grammarly, Mendeley in order to Translation, language correction, bibliography management. After using this tool/service, the author(s) reviewed and edited the content as needed and take(s) full responsibility for the content of the publication.

References

Acarion G., Beline F., Charnier C., Dabert P., Richard C.M., Cadre E.L.E., et al., 2017, Mapped project: Digital innovative solutions for intelligent biogas production and injection over a territory. International Gas Research Conference Proceedings, 1, 345-346.

Ananda, A., Sujeeth, R.K., Archana, S., 2025. Integrating automation in biomass transformation: pportunities, challenges, and future directions. Bioenergy Research 18 (1). Available at https://doi.org/10.1007/s12155-025-10864-6 Article 63.

Agustini, C.B., Gutterres, M., 2017. Chapter 4. Acting microbiota on solid wastes in biogas production. Biogas: Production, Applications and Global Developments. Nova Science Publishers, Inc, Hauppauge, NY, pp. 103–122.

Angelidaki, I., Treu, L., Tsapekos, P., Luo, G., Campanaro, S., Wenzel, H., et al., 2018. Biogas upgrading and utilization: current status and perspectives. Biotechnology Advances 36 (2), 452–466. Available at https://doi.org/10.1016/j.biotechadv.2018.01.011.

Angelidaki, I., Xie, L., Luo, G., Zhang, Y., Oechsner, H., Lemmer, A., et al., 2019. Chapter 33 - Biogas upgrading: current and emerging technologies. In: Pandey, A., Larroche, C., Dussap, C.-G., Gnansounou, E., Khanal, S.K., Ricke, S. (Eds.), Biomass, Biofuels, Biochemicals: Biofuels: Alternative Feedstocks and Conversion Processes for the Production of Liquid and Gaseous Biofuels. Academic Press, pp. 817–843. Available at https://doi.org/10.1016/B978-0-12-816856-1.00033-6.

Baena-Moreno, F.M., Malico, I., Rodríguez-Galán, M., Serrano, A., Fermoso, F.G., Navarrete, B., 2020. The importance of governmental incentives for small biomethane plants in South Spain. Energy 206. Available at https://doi.org/10.1016/j.energy.2020.118158 Article 118158.

Baena-Moreno, F.M., Sebastia-Saez, D., Wang, Q., Reina, T.R., 2020. Is the production of biofuels and bio-chemicals always profitable? Co-production of biomethane and urea from biogas as a case study. Energy Conversion and Management 220. Available at https://doi.org/10.1016/j.enconman.2020.113058 Article 113058.

Balagurusamy, N., Chandel, A.K., 2021. Biogas Production: From Anaerobic Digestion to a Sustainable Bioenergy Industry. Springer, Cham, pp. 1–469. Available at https://doi.org/10.1007/978-3-030-58827-4.

Balussou, D., McKenna, R., Möst, D., Fichtner, W., 2018. A model-based analysis of the future capacity expansion for German biogas plants under different legal frameworks. Renewable and Sustainable Energy Reviews 96, 119–131. Available at https://doi.org/10.1016/j.rser.2018.07.041.

Brandão, A.S., Santos, J.M.R.C.A., 2023. On the societal impact of publicly funded circular bioeconomy research in Europe. Research Evaluation 32 (2), 441–457. Available at https://doi.org/10.1093/reseval/rvad002.

Begum, S., Juntupally, S., Arelli, V., Anupoju, G.R., 2023. Waste to biomethane: advent of circular economy. In Biogas to biomethane: Engineering, production, sustainability 225–242. Available at https://doi.org/10.1016/B978-0-443-18479-6.00007-7.

Csiba, P., Szigeti, C., 2023. Business models for spanish biogas production. In: Arsenyeva, O., Romanova, T., Sukhonos, M., Biletskyi, I., Tsegelnyk, Y. (Eds.), Smart Technologies in Urban Engineering. STUE 2023. Lecture Notes in Networks and Systems, vol 808. Springer, Cham, pp. 363–370. Available at https://doi.org/10.1007/978-3-031-46877-3_32.

Dadic, B., Ivankovic, T., Spelic, K., Hrenovic, J., Jurisic, V., 2024. Natural materials as carriers of microbial consortium for bioaugmentation of anaerobic digesters. Applied Sciences (Switzerland) 14 (16). Available at https://doi.org/10.3390/app14166883.

D'Alpaos, C., 2017. Methodological approaches to the valuation of investments in biogas production plants: incentives vs. market prices in Italy [Approcci metodologici alla valutazione degli investimenti in produzione di biogas: incentivi vs prezzi di mercato in Italia]. Valori e Valutazioni 19, 53–64.

Dong, B., Yin, X., Ren, H., Tian, H., Wang, Y., Li, J., 2025. Effects of bio-augmented ensiling pretreatment on the methane production of sweet sorghum by anaerobic digestion. Nongye Gongcheng Xuebao/Transactions of the Chinese Society of Agricultural Engineering 41 (13), 272–280. Available at https://doi.org/10.11975/j.issn.1002-6819.202412080.

Duque-Acevedo, M., Ulloa-Murillo, L.M., Belmonte-Ureña, L.J., Camacho-Ferre, F., Mercl, F., Tlustoš, P., 2023. Sustainable and circular agro-environmental practices: a review of the management of agricultural waste biomass in Spain and the Czech Republic. Waste Management and Research 41 (5), 955–969. Available at https://doi.org/10.1177/0734242X221139122.

Farzin, F., Sadri Moghaddam, S., 2025. Enhancing energy circularity and achieving carbon neutrality in water resource recovery facilities through an AI-driven optimization framework for biogas production. Journal of Environmental Management 393, 127076. Available at https://doi.org/10.1016/j.jenvman.2025.127076.

Feickert Fenske, C., Md, Y., Strübing, D., Koch, K., 2023. Preliminary gas flow experiments identify improved gas flow conditions in a pilot-scale trickle bed reactor for H_2 and CO_2 biological methanation. Bioresource Technology 371, 128648. Available at https://doi.org/10.1016/j.biortech.2023.128648.

Fredenslund, A.M., Gudmundsson, E., Maria Falk, J., Scheutz, C., 2023. The Danish national effort to minimise methane emissions from biogas plants. Waste Management 157, 321–329. Available at https://doi.org/10.1016/j.wasman.2022.12.035.

Gao, Z., Mauder, M., Desjardins, R.L., Flesch, T.K., van Haarlem, R.P., 2009. Assessment of the backward Lagrangian Stochastic dispersion technique for continuous measurements of CH_4 emissions. Agricultural and Forest Meteorology 149 (9), 1516–1523. Available at https://doi.org/10.1016/j.agrformet.2009.04.004.

Gowreesunker, B.L., Tassou, S.A., 2016. The impact of renewable energy policies on the adoption of anaerobic digesters with farm-fed wastes in Great Britain. Energies 9 (12). Available at https://doi.org/10.3390/en9121038.

Graßmann, N., 2008. New regulations concerning the feed-in of biogas into the natural gas network in EEG, GasNZV and GasNEV [Neue regelungen für die einspeisung von biogas in das erdgasnetz in EEG, GasNZV und GasNEV]. GWF, Gas - Erdgas 149 (4), 222–226.

Groth, A., Maurer, C., Reiser, M., Kranert, M., 2015. Determination of methane emission rates on a biogas plant using data from laser absorption spectrometry. Bioresource Technology 178, 359–361. Available at https://doi.org/10.1016/j.biortech.2014.09.112.

Guenther-Lübbers, W., Theuvsen, L., 2015. Regional economic effects of biogas production: an analysis based on the example of Lower Saxony [regionalwirtschaftliche Effekte der biogasproduktion: Eine analyse am beispiel niedersachsens]. Berichte uber Landwirtschaft 93 (2).

Gustafsson, M., Anderberg, S., 2022. Biogas policies and production development in Europe: a comparative analysis of eight countries. Biofuels 13 (8), 931–944. Available at https://doi.org/10.1080/17597269.2022.2034380.

Hagman, L., Blumenthal, A., Eklund, M., Svensson, N., 2018. The role of biogas solutions in sustainable biorefineries. Journal of Cleaner Production 172, 3982–3989. Available at https://doi.org/10.1016/j.jclepro.2017.03.180.

Hrad, M., Piringer, M., Huber-Humer, M., 2015. Determining methane emissions from biogas plants: operational and meteorological aspects. Bioresource Technology 191, 234–243. Available at https://doi.org/10.1016/j.biortech.2015.05.016.

Jain, N., Rohatgi, A., Suransh, J., Kumar, D., 2025. Artificial Intelligence in Biofuel Applications. Optimizing Biofuel Production with Artificial Intelligence. wiley, India, pp. 1–32. Available at https://doi.org/10.1002/9781394301287.ch1.

Kellner, U., Musshoff, O., Wostbrock, D., 2012. Effects of investing in a biogas plant on overall risk for agricultural enterprises [Auswirkung einer Investition in eine Biogasanlage auf das Gesamtrisiko eines landwirtschaftlichen Betriebes]. Berichte uber Landwirtschaft 90 (1), 70–83.

Kemka, U.N., Ogbulie, T.E., Oguzie, K., Akalezi, C.O., Oguzie, E.E., 2024. Impact of substrate heterogeneity on anaerobic co-digestion process: a review. EQA-International Journal of Environmental Quality 62, 18–34. Available at https://doi.org/10.6092/issn.2281-4485/18498.

Koniuszewska, I., Korzeniewska, E., Harnisz, M., Czatzkowska, M., 2020. Intensification of biogas production using various technologies: a review. International Journal of Energy Research 44 (8), 6240–6258. Available at https://doi.org/10.1002/er.5338.

Kougias, P.G., Angelidaki, I., 2018. Biogas and its opportunities: a review. Frontiers of Environmental Science and Engineering 12 (3), 14. Available at https://doi.org/10.1007/s11783-018-1037-8.

Kunatsa, T., 2024. 13 - The role of artificial intelligence in greening biogas operations. In: Matambo, T.S., Christian, R. (Eds.), Innovations in the Global Biogas Industry: Applications of Green Principles. Woodhead Publishing, pp. 361–397. Available at https://doi.org/10.1016/B978-0-443-22372-3.00014-5.

Lapiso, T.T., Roubík, H., 2025. A systematic scoping review of biogas as an instrumental technology for a circular economy and a way forward to ensure sustainable development in the context of global south nations. Sustainable Development 33 (6), 9073–9105. Available at https://doi.org/10.1002/sd.70080.

Lawson, N., Alvarado-Morales, M., Tsapekos, P., Angelidaki, I., 2021. Techno-economic assessment of biological biogas upgrading based on danish biogas plants. Energies 14 (24), 8252. Available at https://doi.org/10.3390/en14248252.

Lubańska, A., Kazak, J.K., 2023. The role of biogas production in circular economy approach from the perspective of locality. Energies 16 (9). Available at https://doi.org/10.3390/en16093801.

Marendra, F., Prasetya, A., Cahyono, R.B., Ariyanto, T., 2019. A Sustainability assessment of biogas plant based on fruit waste in Indonesia: case study of Biogas Plant Gamping, Yogyakarta. IOP Conference Series: Materials Science and Engineering 543 (1), 012059. Available at https://doi.org/10.1088/1757-899x/543/1/012059.

Nkuna, R., Mohlomi, N., Mukhuba, M., Thathana, M.G., Puri, A.K., 2024. 11 - Green aspects of integrating different processes with anaerobic digestion for biogas production. In: Matambo, T.S., Christian, R. (Eds.), Innovations in the Global Biogas Industry: Applications of Green Principles. Woodhead Publishing, pp. 299–340. Available at https://doi.org/10.1016/B978-0-443-22372-3.00012-1.

Núñez, M.P., 2021. Spain launches its first Circular Economy Action Plan, 2021-2030: How will we decouple economic growth from resource consumption in the next decade? [ESPAÑA ESTRENA SU PRIMER PLAN DE ACCIÓN DE ECONOMÍA CIRCULAR 2021-2030 ¿CÓMO DESACOPLAREMOS EL CRECIMIENTO ECONÓMICO DEL CONSUMO DE RECURSOS EN LA PRÓXIMA DÉCADA?]. Actualidad Juridica Ambiental 2021 (114), 99–113.

Offie, I., Piadeh, F., Behzadian, K., Campos, L.C., Yaman, R., 2023. Development of an artificial intelligence-based framework for biogas generation from a micro anaerobic digestion plant. Waste Management 158, 66–75. Available at https://doi.org/10.1016/j.wasman.2022.12.034.

Ozsefil, I.C., Miraloglu, I.H., Ozbayram, E.G., Ince, B., Ince, O., 2024. Bioaugmentation of anaerobic digesters with the enriched lignin-degrading microbial consortia through a metagenomic approach. Chemosphere 355. Available at https://doi.org/10.1016/j.chemosphere.2024.141831.

Pantaleo, A., De Gennaro, B., Shah, N., 2013. Assessment of optimal size of anaerobic co-digestion plants: an application to cattle farms in the province of Bari (Italy). Renewable and Sustainable Energy Reviews 20, 57–70. Available at https://doi.org/10.1016/j.rser.2012.11.068.

Pous de la Flor, J., Martínez-Hernando, M.P., Paredes, R., Garcia-Franco, E., Cabello, J.P., Ortega, M.F., 2025. Waste as a source of fuel and developments in hydrogen storage: applied cases in spain and their future potential. Applied Sciences (Switzerland) 15 (13), 7514. Available at https://doi.org/10.3390/app15137514.

Prussi, M., Julea, A., Lonza, L., Thiel, C., 2021. Biomethane as alternative fuel for the EU road sector: analysis of existing and planned infrastructure. Energy Strategy Reviews 33, 100612. Available at https://doi.org/10.1016/j.esr.2020.100612.

Prussi, M., Padella, M., Conton, M., Postma, E.D., Lonza, L., 2019. Review of technologies for biomethane production and assessment of Eu transport share in 2030. Journal of Cleaner Production 222, 565–572. Available at https://doi.org/10.1016/j.jclepro.2019.02.271.

Reinelt, T., Liebetrau, J., 2020. Monitoring and mitigation of methane emissions from pressure relief valves of a biogas plant. Chemical Engineering and Technology 43 (1), 7–18. Available at https://doi.org/10.1002/ceat.201900180.

Saikat, B., Sivamani, S., Prasad, B.S.N., 2023. Biogas as a value generation in industrial wastewater, A review. In. Biorefinery for water and wastewater treatment 49–69. Available at https://doi.org/10.1007/978-3-031-20822-5_3.

Stürmer, B., Schmid, E., 2011. Assessment of electric power production potentials from biogas in Austria by 2020 [Abschätzung des österreichischen Biogasproduktionspotentials zur Stromerzeugung in 2020]. Journal of the Austrian Society of Agricultural Economics 20 (2), 149–158.

Salvador, R., Barros, M.V., Donner, M., Brito, P., Halog, A., De Francisco, A.C., 2022. How to advance regional circular bioeconomy systems? Identifying barriers, challenges, drivers, and opportunities. Sustainable Production and Consumption 32, 248–269. Available at https://doi.org/10.1016/j.spc.2022.04.025.

Satpathy, P., 2017. Recent Developments in Biogas Technology Biogas: Production, Applications and Global Developments. Nova Science Publishers, Inc, Germany, pp. 209–239.

Scheutz, C., Fredenslund, A.M., 2019. Total methane emission rates and losses from 23 biogas plants. Waste Management 97, 38–46. Available at https://doi.org/10.1016/j.wasman.2019.07.029.

Schröer, D., Latacz-Lohmann, U., 2024. Flexibilization or biomethane upgrading? Investment preference of German biogas plant operators for the follow-up of guaranteed feed-in tariffs. GCB Bioenergy 16 (2). Available at https://doi.org/10.1111/gcbb.13111.

Scotto di Perta, E., Cervelli, E., Pironti di Campagna, M., Pindozzi, S., 2019. From biogas to biomethane: techno-economic analysis of an anaerobic digestion power plant in a cattle/buffalo farm in central Italy. Journal of Agricultural Engineering 50 (3), 127–133. Available at https://doi.org/10.4081/jae.2019.939.

Shrestha, S., Aryal, N., Ghimire, A., Lohani, S.P., 2025. Evaluation of biomethanation process with intermittent hydrogen purging. Next Energy 8. Available at https://doi.org/10.1016/j.nxener.2025.100325.

Skovsgaard, L., Jacobsen, H.K., 2017. Economies of scale in biogas production and the significance of flexible regulation. Energy Policy 101, 77–89. Available at https://doi.org/10.1016/j.enpol.2016.11.021.

Stephan, B., 2013. 19 - Market development and certification schemes for biomethane. In: Wellinger, A., Murphy, J., Baxter, D. (Eds.), The Biogas Handbook: Science, Production and Applications. Woodhead Publishing, pp. 444–462. Available at https://doi.org/10.1533/9780857097415.3.444.

Stürmer, B., 2017. Feedstock change at biogas plants – impact on production costs. Biomass and Bioenergy 98, 228–235. Available at https://doi.org/10.1016/j.biombioe.2017.01.032.

Tabatabaei, M., Aghbashlo, M., Valijanian, E., Kazemi Shariat Panahi, H., Nizami, A.S., Ghanavati, H., et al., 2020. A comprehensive review on recent biological innovations to improve biogas production, part 1: Upstream strategies. Renewable Energy 146, 1204–1220. Available at https://doi.org/10.1016/j.renene.2019.07.037.

Taron, A., Majumder, A., Bodach, S., Agbefu, D., 2023. Public-private partnerships for the circular bio-economy in the global south: lessons learned. Resource Recovery and Reuse 2023 (22), 1–50. Available at https://doi.org/10.5337/2023.205.

Tsapekos, P., Treu, L., Campanaro, S., Centurion, V.B., Zhu, X., Peprah, M., et al., 2021. Pilot-scale biomethanation in a trickle bed reactor: process performance and microbiome functional reconstruction. Energy Conversion and Management 244. Available at https://doi.org/10.1016/j.enconman.2021.114491.

Upadhyay, A., Kovalev, A.A., Zhuravleva, E.A., Kovalev, D.A., Litti, Y.V., Masakapalli, S.K., et al., 2023. Recent development in physical, chemical, biological and hybrid biogas upgradation techniques. Sustainability (Switzerland) 15 (1). Available at https://doi.org/10.3390/su15010476.

Uranga-Valencia, L.P., Pérez-Álvarez, S., Gabriel-Parra, R., Chávez-Medina, J.A., Magallanes-Tapia, M.A., Sánchez-Chávez, E., et al., 2025. Biogas production from organic waste in the forestry and agricultural context: challenges and solutions for a sustainable future. Energies 18 (12), 3174. Available at https://doi.org/10.3390/en18123174.

Valentino, F., Lorini, L., Pavan, P., Majone, M., 2021. Development of a biorefinery platform for urban waste valorisation into biogas and added-value products. Chemical Engineering Transactions 86, 13–18. Available at https://doi.org/10.3303/CET2186003.

Westerholm, M., Isaksson, S., Karlsson Lindsjö, O., Schnürer, A., 2018. Microbial community adaptability to altered temperature conditions determines the potential for process optimisation in biogas production. Applied Energy 226, 838–848. Available at https://doi.org/10.1016/j.apenergy.2018.06.045.

Yörüklü, H.C., Kamravamanesh, D., Köroğlu, E.O., Patel, G.H., Havukainen, J., Karjunen, H., et al., 2025. A comprehensive review on biological methanation processes: from gaseous feedstocks to biomethane. Energy Conversion and Management 341. Available at https://doi.org/10.1016/j.enconman.2025.120075.

Index

Note: Page numbers followed by *f* indicate figures, *t* indicate tables.

C

D

E

9780443405570